# Carbon Nanostructures

**Series Editor**

Paulo Araujo, The University of Albama, Tuscaloosa, AL, USA

**Editorial Board**

Antonio Gomes Sousa Filho, Universidade Federal do Ceara—UFC, Fortaleza, Brazil

Stephen K. Doorn, Los Alamos National Laboratory—LANL, Los Alamos, USA

Aaron D. Franklin, Electrical and Computer Engineering, Hudson Hall 129, Duke University, Durham, USA

Achim Hartschuh, Ludwig-Maximilians-Universität, München, Germany

AF818318

Carbon is intimately connected to almost everything we deal with in a daily basis. Due to its outstanding properties, such as high stability at environmental conditions, different hybridizations, strong covalent bond formation and easy of compounds formation, carbon has been a topic of scientific interest in several areas. Indeed, starting in the 19th century, chemists have devoted a whole field to study carbon-based compounds, which is, nowadays, known as Organic Chemistry. Remarkably, the last 30 years have been witnessing an exponential advance in the science involving carbon and carbon structures. Since the discovery of Fullerenes in 1985, which was awarded the Nobel Prize in Chemistry in 1996, carbon nanostructures have been attracting a great deal of attention from the research community. This public interest dramatically increased after the publications by the Iijima and Bethune groups on single-wall carbon nanotubes in 1993 and found a "new research era" with the isolation of a monolayer of carbon atoms, also called graphene, which conducted groundbreaking experiments demonstrating outstanding phenomena such as the Klein-Tunneling and the fractional quantum hall effect. No wonder, graphene was the object of the 2010 Nobel Prize in Physics.

The "Carbon Nanostructures" book series covers the state-of-art in the research of nanocarbons and their applications. Topics related to carbon allotropes such as diamond, graphite, graphene, fullerenes, metallofullerenes, solid C60, bucky onions, foams, nanotubes and nanocones, including history, theory, synthesis, chemistry & physics, Biophysics & engineering, characterization methods, properties and applications are welcome. Within the "Carbon Nanostructures" book series, the reader will find valuable, up-to-date account of both the newer and traditional forms of carbon. This book series supports the rapid expansion of this field and is a valuable resource for students and professionals in several different areas.

Springer welcomes new book ideas from potential authors. If you are interested in publishing your book in this series, please contact us via mayra.castro@springer.com

**** Indexed by Scopus (2018) ****

Carlos Pérez Bergmann ·
Fernando Machado Machado
Editors

# Carbon Nanomaterials as Adsorbents for Environmental and Biological Applications

Second Edition

Springer

*Editors*
Carlos Pérez Bergmann
Escola de Engenharia
Universidade Federal do Rio Grande do Sul (UFRGS)
Porto Alegre, Rio Grande do Sul, Brazil

Fernando Machado Machado
Technology Development Center
Federal University of Pelotas (UFPEL)
Pelotas, Brazil

ISSN 2191-3005 ISSN 2191-3013 (electronic)
Carbon Nanostructures
ISBN 978-3-032-08668-6 ISBN 978-3-032-08669-3 (eBook)
https://doi.org/10.1007/978-3-032-08669-3

1st editions: © Springer International Publishing Switzerland, 2015
2nd edition: © The Editor(s) (if applicable) and The Author(s), under exclusive license to Springer Nature Switzerland AG 2026

This work is subject to copyright. All rights are solely and exclusively licensed by the Publisher, whether the whole or part of the material is concerned, specifically the rights of translation, reprinting, reuse of illustrations, recitation, broadcasting, reproduction on microfilms or in any other physical way, and transmission or information storage and retrieval, electronic adaptation, computer software, or by similar or dissimilar methodology now known or hereafter developed.
The use of general descriptive names, registered names, trademarks, service marks, etc. in this publication does not imply, even in the absence of a specific statement, that such names are exempt from the relevant protective laws and regulations and therefore free for general use.
The publisher, the authors and the editors are safe to assume that the advice and information in this book are believed to be true and accurate at the date of publication. Neither the publisher nor the authors or the editors give a warranty, expressed or implied, with respect to the material contained herein or for any errors or omissions that may have been made. The publisher remains neutral with regard to jurisdictional claims in published maps and institutional affiliations.

This Springer imprint is published by the registered company Springer Nature Switzerland AG
The registered company address is: Gewerbestrasse 11, 6330 Cham, Switzerland

If disposing of this product, please recycle the paper.

# Preface

Carbon-based nanomaterials and their innovative applications continue to attract significant attention from both academic and industrial sectors, particularly for their use as nanoadsorbents. The increasing demand for efficient and sustainable technologies for water purification and environmental monitoring has driven research into novel carbon nanostructures with enhanced adsorption capacity, high selectivity, and superior performance.

In this context, the Second Edition of *Carbon Nanomaterials as Adsorbents for Environmental and Biological Applications* presents an updated and expanded synthesis of current knowledge on the subject. This edition highlights recent advances in the field, including the removal of contaminants of emerging concern from wastewater and the development of new carbon nanoadsorbents such as carbon quantum dots and three-dimensional graphene family materials.

The book offers a comprehensive review of the properties, synthesis, and applications of carbon nanomaterials as advanced adsorbents, combining theoretical fundamentals with practical insights. In addition to revisiting and updating core concepts from the first edition, this volume introduces novel content that integrates environmental remediation with state-of-the-art methodologies in materials science and data analysis.

Throughout its chapters, this edition explores various classes of carbon nanomaterials—such as carbon nanotubes, graphene-based structures, carbon quantum dots, and three-dimensional architectures—highlighting their properties and capabilities for removing organic and inorganic pollutants from water systems. It also presents detailed discussions on theoretical adsorption models, experimental techniques, and the peculiarities of adsorption processes, alongside innovative approaches for biological applications. Furthermore, this edition introduces a new chapter focusing on the use of mathematical modeling and artificial intelligence to optimize adsorptive processes, reflecting the evolving landscape of materials science and computational methods.

As with the first edition, the book is written in clear and accessible language with a practical, application-oriented focus. It serves as a valuable resource for students, researchers, and professionals interested in exploring the real-world applications of carbon nanomaterials in environmental and biological systems, fostering collaboration between academia and industry to address global challenges in water treatment and environmental sustainability.

Porto Alegre, Brazil
Pelotas, Brazil
July 2025

Carlos Pérez Bergmann
Fernando Machado Machado

# Contents

# Carbon-Based Nanomaterials: Advanced Materials for Water Treatment and Environmental Monitoring

**Fernando Machado Machado, Eder Claudio Lima, and Carlos Pérez Bergmann**

**Abstract** Ensuring the provision of clean and affordable water to fulfill human sustenance is recognized as one of the paramount challenges of the twenty-first century. The global water supply is increasingly incapable of meeting the rapidly escalating demand, exacerbated by factors such as population growth, climate change, and deterioration in water quality. The advancement of technological innovations in integrated water management is of paramount importance. Nanotechnology emerges as a highly promising approach to augment water and wastewater treatment processes, thereby enhancing treatment efficacy and facilitating the monitoring of water quality sources. This chapter aims to elucidate recent advancements in nanotechnology, with a specific emphasis on carbon-based nanomaterials (CBNs) for the monitoring and treatment of water and wastewater. In this introductory chapter, we transition from foundational concepts of adsorption to an examination of promising applications of CBNs. The chapter culminates with a discussion regarding the initial limitations, ongoing challenges, and potential opportunities associated with the application of CBNs for environmental remediation concerning water quality.

**Keywords** Adsorption · Monitoring · Carbon-based nanomaterials · Nanoadsorbents

F. Machado Machado (✉)
Materials Research Laboratory, Technology Development Center, Federal University of Pelotas (UFPel), Pelotas, RS, Brazil
e-mail: fernando.machado@ufpel.edu.br

E. C. Lima
Institute of Chemistry, Federal University of Rio Grande do Sul (UFRGS), Porto Alegre, RS, Brazil
e-mail: profederlima@gmail.com

C. P. Bergmann
Programa de Pós-Graduação Em Engenharia de Minas, Metalúrgica e de Materiais, Laboratório de Cerâmica (LACER), Universidade Federal do Rio Grande do Sul (UFRGS), Porto Alegre, Brazil
e-mail: bergmann@ufrgs.br

© The Author(s), under exclusive license to Springer Nature Switzerland AG 2026

C. P. Bergmann and F. Machado Machado (eds.), *Carbon Nanomaterials as Adsorbents for Environmental and Biological Applications*, Carbon Nanostructures,
https://doi.org/10.1007/978-3-032-08669-3_1

# 1 Understanding Adsorption

Adsorption encompasses processes occurring at various interfaces [1]. It holds considerable practical importance across diverse fields, ranging from daily life activities to specialized industrial applications, environmental protection, and biological systems. This process, where molecules adhere to a surface, underpins a vast array of technologies and natural phenomena.

In our daily lives, adsorption manifests in surprisingly common yet essential ways. Classic examples include household water filters that utilize activated carbon to effectively remove impurities, chlorine, and undesirable odors, thereby enhancing the water's palatability. Similarly, silica gel sachets and activated carbon, frequently placed in refrigerators or closets, function as effective deodorizers and dehumidifiers by adsorbing odor-causing molecules and excess moisture.

Within the industrial sector, adsorption plays a fundamental role due to its exceptional efficiency and capacity for large-scale purification and separation. It is crucial in water and wastewater treatment, where it removes pollutants with high efficacy, often utilizing activated carbon. In the petrochemical industry, molecular sieves play a crucial role in purifying gases and liquids, ensuring product quality, and protecting equipment from damage. Furthermore, in food and beverage production, adsorption plays a vital role in ensuring product purity, as exemplified in the refining of sugar. Across various other sectors, advanced technologies such as Pressure Swing Adsorption (PSA) are essential for gas separation and purification [2], including the production of industrial gases, as well as for drying gas streams where moisture removal is critical. In essence, adsorption stands as a cornerstone technology, significantly contributing to environmental protection, operational safety, and overall efficiency across countless industrial processes.

Numerous chemical, physical, and biological processes manifest at the boundary between two distinct phases. Fundamentally, adsorption is defined as the increase in concentration of a given substance at an interface compared to its concentration in the adjacent phases [1]. This process can be observed in various systems based on the types of phases in contact: liquid–gas, liquid–liquid, solid–solid, solid–liquid, and solid–gas. For the scope of this discussion, we will primarily focus on the solid–liquid system [3].

Adsorption is intrinsically a surface phenomenon involving two primary components [1]:

- **The adsorbate**: This refers to the species present in the fluidic phase (gas or liquid) that becomes attached to the condensate phase (solid or liquid).
- **The adsorbent**: This is the condensate phase (solid or liquid) upon which the adsorption process occurs.

Common adsorbents include biochar, activated carbon, alumina, silica, and polymeric resins [1]. When the adsorbent consists of inactive biomass (e.g., fungi, dead algae, agricultural residues like fruit peels or shells), it is termed a biosorbent [4].

Conversely, if the biomass is active (living), the retention process is known as bioaccumulation, which should not be confused with biosorption.

It is crucial to distinguish adsorption from absorption, where adsorbate molecules penetrate into the bulk of the solid phase. When both processes occur simultaneously or are difficult to differentiate, the more encompassing terms sorption, sorbent, sorbate, and sorptive are often employed [4]. This relationship can be expressed as Sorption = Adsorption + Absorption, indicating that these phenomena can occur concurrently [4].

Fundamentally, adsorption can be categorized into two types [4]:

- **Chemical Adsorption (Chemisorption)**: This type involves the formation of a chemical bond between the sorbate molecule and the surface of the adsorbent. The enthalpy of adsorption for chemisorption is significantly higher, ranging from 200 to 800 kJ $mol^{-1}$, comparable to the energy of a covalent bond. Typically, the adsorption of inorganic compounds onto CBNs is seen as chemisorption.
- **Physical Adsorption (Physisorption)**: This type involves relatively weak intermolecular forces, with an enthalpy of adsorption typically ranging from 0.4 to 80 kJ $mol^{-1}$. Adsorption of organic molecules onto CBNs is generally classified as physisorption.

Table 1 presents the fundamental aspects of physisorption and chemisorption.

Table 1 clearly shows that chemical and physical adsorption processes exhibit distinct characteristics, except for reversibility. It is a common misconception that chemisorption necessarily implies irreversible adsorption or that physisorption always implies reversible adsorption—these notions are, in fact, misconceptions frequently encountered in the adsorption literature (a more in-depth discussion of the characteristics of adsorption processes will be presented in chapter "Theoretical Fundamentals for the Batch Contact Adsorption Method").

**Table 1** Key Characteristics of physisorption and chemisorption

| Features | Chemisorption | Physisorption |
|---|---|---|
| Enthalpy of adsorption | 200 to 800 kJ $mol^{-1}$ | 0.4 to 80 kJ $mol^{-1}$ |
| Origin of interactions | Chemical bonds (common electronic orbitals) | Van der Waals bonds (no common electronic orbitals) |
| Heat of adsorption | High heat of adsorption (>2- or 3-times the latent heat of evaporation) | Low heat of adsorption (<2- or 3-times the latent heat of evaporation) |
| Specificity | Specific | Nonspecific |
| Number of adsorbed layers | Monolayer | Monolayer or multilayer |
| Reversibility | Irreversible or reversible | Irreversible or reversible |
| Electrons transfer | Electron transfer leading to bond formation between sorbate and surface | No electrons transfer, although polarization of sorbate may occur |

Adapted from [5]

The successful application of adsorption in wastewater treatment largely depends on the proper selection of the adsorbent material. Among the most critical factors influencing adsorption performance are the adsorbent's textural properties—such as specific surface area, pore size distribution, and total pore volume—and its surface chemistry, including the presence and nature of functional groups. As adsorption is inherently a surface-driven phenomenon, these properties directly affect the efficiency with which pollutants are removed from aqueous environments.

In this context, carbon-based nanomaterials (CBNs) are particularly noteworthy [9]. These materials are highly attractive for adsorption applications due to their exceptionally high specific surface areas, well-developed porous structures, and the versatility of surface modification. These combined attributes render them promising adsorbents for practical wastewater monitoring and purification applications [5–8].

## 2 Nanotechnology and Carbon Nanoadsorbents for Environmental Remediation

Nanotechnology stands as one of the twenty-first century's most attractive and promising fields of technological development, offering unparalleled opportunities for research and innovation derived from the unusual properties of matter organized at nanometer dimensions. According to ISO/TC 229 Nanotechnologies [10], this field encompasses:

- *"1. Understanding and control of matter and processes at the nanoscale, typically, but not exclusively, below 100 nm in one or more dimensions, where the onset of size-dependent phenomena usually enables novel applications,*
- *2. Utilizing the properties of nanoscale materials that differ from the properties of individual atoms, molecules, and bulk matter, to create improved materials, devices, and systems that exploit these new properties."*

Nanotechnology has gained significant interest due to its potential to improve our quality of life and its vital role in environmental preservation, particularly in water monitoring [11] and treatment-purification [6–8]. In recent years, various nanotechnology-derived materials have been evaluated as promising nanoadsorbents for environmental [6, 8] and biological applications [12].

Among the different types of nanomaterials, CBNs stand out due to carbon's unparalleled versatility [6–8]. Carbon, recognized as the most versatile element in the periodic table, forms a vast array of structures due to its unique bonding characteristics, including various hybridizations and allotropes [13]. Its capacity to form single, double, or triple bonds, alongside catenation, facilitates the creation of diverse architectures encompassing zero-dimensional (0D), one-dimensional (1D), two-dimensional (2D), and three-dimensional (3D) structures [13]. These nanoscale allotropes, including Carbon Quantum Dots (CQDs), Carbon Nanotubes (CNTs), Graphene-Family Materials (GFM), and 3D graphene materials (see Fig. 1), are

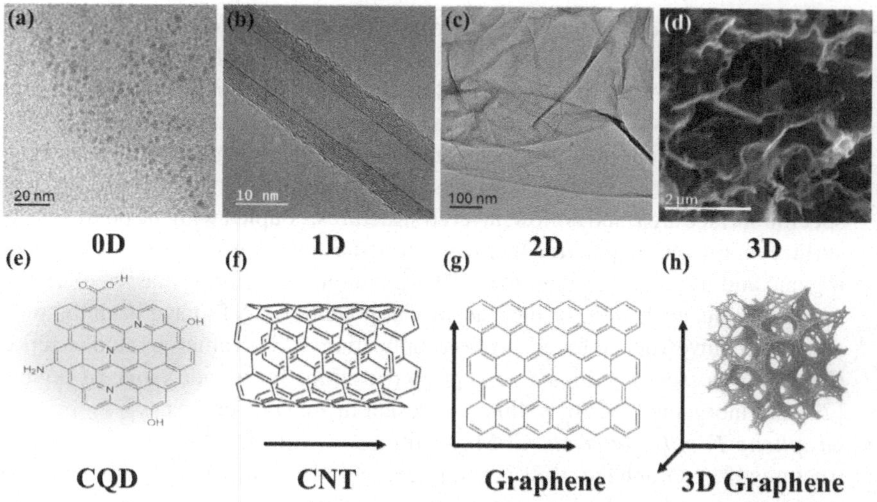


**Fig. 1** Microscopic images and schematic representations of 0D, 1D, 2D, and 3D CBNs structures with potential environmental applications: **a**, **e** CQDs, **b**, **f** carbon nanotubes, **c**, **g** graphene, and **d**, **h** 3D graphene

highly pertinent as advanced adsorbents in environmental protection and various industrial applications [6–8].

From a properties and characteristics standpoint, each CBN offers distinct advantages for applications in wastewater monitoring and purification:

- ***Carbon Quantum Dots***: These novel carbon nanoparticles, typically characterized as spherical or semi-spherical structures with dimensions less than 10 nm, possess a significant surface area and unique photoluminescent properties [14]. Since their discovery in 2004 [15], CQDs have garnered significant attention for their dual role in water treatment: functioning as efficient adsorbents and as highly sensitive sensors [16, 17]. They exhibit a range of advantageous properties, including excellent hydrophilicity, biocompatibility, high electrical conductivity, low toxicity, and tunable fluorescence [18, 19]. These combined characteristics render them exceptionally promising for both detecting and removing hazardous contaminants from aqueous environments [18, 19]. CQDs are particularly well-suited for sensing applications due to their tunable photoluminescent (PL) properties [11, 20]. Their ability to emit high-intensity light across various colors, contingent upon their size, makes them ideal for fluorescence-based detection of diverse contaminants such as heavy metals, pesticides, pharmaceuticals, and other organic and inorganic pollutants [20, 21]. This capability enables sensitive detection even at low concentrations, facilitating accurate and real-time monitoring of water quality [14, 20–22]. Beyond their sensing capabilities, CQDs actively participate in adsorption processes [14, 17]. Their ample surface area and intrinsic chemical reactivity contribute to their high efficiency as adsorbents [23].

- ***Carbon Nanotubes***: Cylindrical allotropes of carbon molecules, elucidated in 1991 [24], CNTs are essentially envisioned as graphitic carbon sheets rolled into hollow cylinders. They are categorized as single-walled (SWCNTs) [25, 26] and multi-walled (MWCNTs), distinguished by the number of concentric tubes [24]. CNTs exhibit exceptional properties, including high strength, unique electrical, mechanical, and thermal characteristics [27]. Their superior hydrophobicity, high specific surface area, and hollow, layered structures, coupled with the possibility of surface modification, position them as promising adsorbents for various synthetic organic and inorganic contaminants [6, 8]. Among the various classes of carbon nanomaterials evaluated for their adsorbent potential, CNTs have been the most extensively investigated for adsorption applications [28]. Numerous studies report that CNTs possess significant adsorption capacities for a wide range of organic [29] and inorganic [29, 30] pollutants present in aqueous phases.
- ***Graphene-Family Materials***: GFM constitutes a significant class of carbon nanostructures [31]. Graphene itself is a single atomic layer of $sp^2$-hybridized carbon atoms arranged in a 2D hexagonal lattice, serving as the fundamental building block for other graphitic materials [32]. Graphene boasts an exceptionally large specific surface area and exhibits unique properties, such as high intrinsic mobility and high optical transmittance [32]. These attributes lend graphene substantial potential for numerous applications, including environmental remediation [29, 33]. Beyond the pristine structure, the graphene family encompasses various promising materials for adsorption processes, ranging from few-layer graphene (FLG) to graphene oxide (GO) and reduced graphene oxide (rGO) [31]. Although GFM members have not yet been studied as extensively as CNTs, several authors have demonstrated their capacity to successfully remove various contaminants [29, 33], including antibiotics, heavy metals, and synthetic dyes, from aqueous solutions [29, 33].
- **3D Graphene materials**: These represent a notable advancement, effectively overcoming the restacking challenge inherent to graphene sheets [34, 35] that often limits their practical applicability in environmental settings [36, 37]. These materials are formed by interconnecting graphene-family material sheets into a three-dimensional network [34–37]. This distinctive 3D architecture confers novel collective physicochemical properties that differ significantly from those of individual building blocks, including remarkably low density, high porosity, and a large surface area [38]. They can be further classified into graphene foams and sponges, with sponges broadly categorized as organogels, hydrogels, aerogels, and xerogels [38]. Numerous research groups are actively focusing on the monitoring, treatment, and purification of wastewater using CBNs as nanoadsorbents [7, 37].

Given these characteristics, numerous research groups worldwide have focused on the application and improvement of CBNs [39, 40]. The primary objective of these efforts is to demonstrate the superior efficiency of CBNs compared to traditional adsorbents, such as activated carbon, silica gel, activated alumina, zeolites, and synthetic polymer resins, as well as providing a clearer understanding of the interaction mechanisms involved [39, 40]. This pursuit is significantly motivated by the urgent global imperative to achieve the Sustainable Development Goals (SDGs) of the United Nations 2030 Agenda, particularly those related to clean water and sanitation (SDG 6), industry, innovation, and infrastructure (SDG 9), sustainable cities and communities (SDG 11), and responsible consumption and production (SDG 12). A comprehensive review of the literature published on this topic highlights the growing importance attributed to CBNs in this field. Analysis of Fig. 2, which presents the results of this survey, reveals a significant annual increase in the number of publications. Notably, from 2015 (the year of the first edition of this book) to the current year (2025), 123,475 publications on this topic have already been recorded, representing an increase of ~963% compared to the data presented in the first edition (11,616 until 2014).

In terms of monitoring and biological applications, substantial research has been conducted regarding novel methods for detecting and treating diseases using CBNs for drug delivery and biosensor monitoring, especially propelled by the Coronavirus pandemic and its consequences [41]. Considerable effort is currently focused on identifying ideal materials to ensure the optimal characteristics of nanostructured sensors and drug delivery systems. CBNs hold great potential for such applications, as evidenced by the growing number of publications in this area. Figures 3 and 4 illustrate, respectively, a historical survey of published literature concerning the use of CBNs as sensors and in drug delivery systems for the period between 2015 and 2025. The analysis of these figures also highlights the exponential growth of research interest in these fields, with a substantial increase in publications on CBN-based sensors (totaling 71,781 results) and drug delivery systems (totaling 39,001)

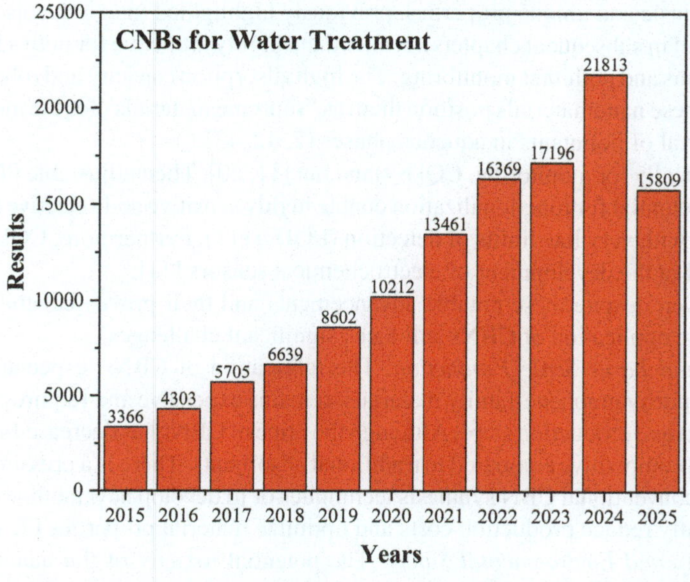


**Fig. 2** Articles published from 2015 to 2025 regarding the application of CBNs as nanoadsorbents for water treatment (www.sciencedirect.com, last accessed 15th June 2025)

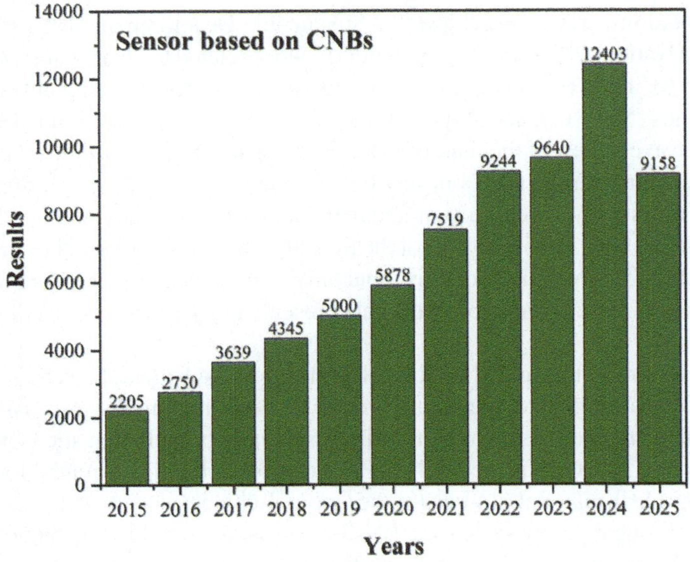


**Fig. 3** Articles published between 2015 and 2025 regarding the development of CBNs-based sensors (www.sciencedirect.com, last accessed 15th June 2025)

between 2015 and 2025. These data underscore the promising potential of CBNs in these applied areas.

CBNs emerge as a promising frontier for addressing environmental pollution. Their intrinsic and unique properties, previously highlighted (and to be discussed in more detail in subsequent chapters), render them highly attractive for both adsorption applications and pollutant monitoring. The high adsorption capacity and robust selectivity of these nanomaterials position them as "supreme materials" for the monitoring and removal of pollutants in aqueous phases [7, 42, 43].

Specifically for monitoring, CQDs stand out [11, 20]. Their adjustable PL properties and potential for functionalization enable highly sensitive and selective pollutant detection with very low limits of detection (LODs) [11]. Furthermore, CQDs can be employed in the development of electrochemical sensors [11].

However, despite these notable advancements and their proven capabilities, the large-scale application of CBNs still faces significant challenges.

*Cost and Large-Scale Production*: The fabrication of CBNs, especially CNTs and high-purity graphene-family materials, remains expensive and requires complex equipment and production steps. Although the price of CBNs has decreased sharply, it is still not competitive compared to traditional adsorbents. There is a pressing need to improve conventional CBN synthesis techniques or to develop new methods that can significantly reduce production costs and optimize material properties [7, 42, 43].

*Toxicity and Environmental Safety*: The potential toxicity of the nanomaterials themselves is a fundamental concern. Comprehensive risk assessments for human

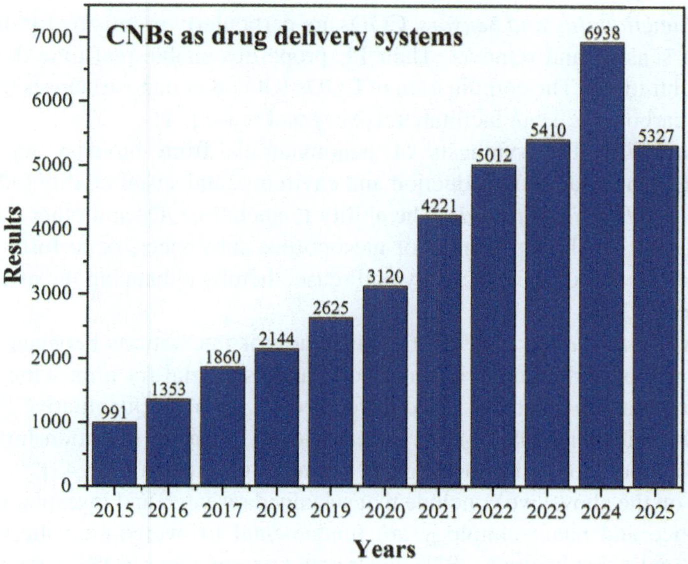


**Fig. 4** Articles published between 2015 and 2025 regarding the application of CBNs in drug delivery systems (www.sciencedirect.com, last accessed 15th June 2025)

health and ecosystems are necessary, including studies of cellular toxicity and life cycle analyses [11, 42].

*Complexity of Real Systems and Modeling*: Most current research focuses on laboratory conditions, whereas real water is highly heterogeneous and contains complex combinations of pollutants, natural organic matter, and variable pH and ionic strength. This heterogeneity complicates the development of universal monitoring and adsorption models [7, 42, 43].

*Characterization and Data*: There is a need for more detailed characterization of CBNs after pre-treatment and functionalization, as information from manufacturers of pristine CBNs can lead to erroneous interpretations of adsorption results. Furthermore, improved reporting of adsorption data is essential to facilitate data mining and the development of advanced models in the "Big Data" era [43] and for Artificial Intelligence models [44].

Future opportunities and perspectives are promising and focus on key areas.

*Superior Properties*: CBNs possess exceptional textural properties, chemical and mechanical stability, and the ability to be functionalized, making them promising candidates for use as sensors and/or adsorbents. They demonstrate superior adsorption capacity compared to activated carbon in many situations [7, 42, 43, 45].

*Application Versatility*: CBNs can be utilized to detect and remove a wide range of pollutants, including heavy metals, dyes, and emerging contaminants of concern. Additionally, the possibility of surface modification of these nanostructures, such as oxidation and electrochemical assistance, can significantly improve the rate and capacities of adsorption and selectivity [7, 42, 43, 45].

*Dual Functionality and Sensors*: CQDs are particularly promising for dual applications in sensing and removal. Their PL properties enable real-time detection at low concentrations. The combination of CQDs with other nanostructures (polymers, activated carbon, etc.) can facilitate recovery and reuse [11].

*Sustainability*: The synthesis of nanomaterials from biomass represents a promising path toward cost reduction and environmental sustainability [42, 45].

*Innovative Hybrid Materials*: The ability to anchor CQDs and other CBNs onto solid materials, such as polymers or mesoporous substances, or to form magnetic composites facilitates their recovery and reuse, thereby enhancing the effectiveness of treatments [11].

*Process Optimization*: The continuous study of interactions between the functional groups of carbon materials and pollutants is crucial for improving and optimizing preparation processes. Predictive models, such as Quantitative Structure–Activity Relationship (QSAR) and Linear Solvation Energy Relationship (LSER), are valuable tools for understanding and optimizing adsorption [7, 43].

Based on the above, we conclude that sustained research and investment in materials science and nanotechnology are fundamental to overcoming the remaining obstacles and consolidating CBNs as strategic components in the development of more sustainable and efficient environmental technologies.

**Acknowledgements** The authors gratefully acknowledge financial support from the Brazilian funding agencies: Conselho Nacional de Desenvolvimento Científico e Tecnológico (CNPq), Coordenação de Aperfeiçoamento de Pessoal de Nível Superior (CAPES), and Fundação de Amparo à Pesquisa do Estado do Rio Grande do Sul (FAPERGS). Prof. Fernando M. Machado thanks CNPq for research grant support (Processes 308203/2022-4, 408413/2022-0, 403522/2025-0, and 443031/2023-1) and FAPERGS (Process 23/2551-0001863-2). Prof. Edér C. Lima acknowledges support from CNPq (Process 303612/2021-5), CAPES-PROEX (Processes 88881.844968/2023-01 and 001), and FAPERGS (Process 24/2551-0001291-5).

## Abbreviations

| | |
|---|---|
| 0D | Zero-Dimensional |
| 1D | One-Dimensional |
| 2D | Two-Dimensional |
| 3D | Three-Dimensional |
| CBNs | Carbon-Based Nanomaterials |
| CNTs | Carbon Nanotubes |
| CQDs | Carbon Quantum Dots |
| FLG | Few-Layer Graphene |
| GFM | Graphene-Family Materials |
| GO | Graphene Oxide |
| LODs | Limits of Detection |
| LSER | Linear Solvation Energy Relationship |
| MWCNTs | Multi-Walled Carbon Nanotubes |

| | |
|---|---|
| NOM | Natural Organic Matter |
| PFAS | Per- and Polyfluoroalkyl Substances |
| PL | Photoluminescent |
| PSA | Pressure Swing Adsorption |
| QSAR | Quantitative Structure–Activity Relationship |
| rGO | Reduced Graphene Oxide |
| SDGs | Sustainable Development Goals |
| SWCNTs | Single-Walled Carbon Nanotubes |

## References


1. Helfferich, F.G.: Principles of adsorption & adsorption processes, by D. M. Ruthven, John Wiley & Sons, 1984, xxiv + 433 pp. AIChE J. **31**, 523–524 (1985). https://doi.org/10.1002/aic.690310335
2. Elshaday, M., M'Arimi, M.M., Ramkat, R.C.: A review of recent developments in application of low cost natural materials in purification and upgrade of biogas. Renew. Sustain. Energy Rev. **145**, 111081 (2021). https://doi.org/10.1016/j.rser.2021.111081
3. Abe, I., Hayashi, K., Kitagawa, M., et al.: Adsorptive mechanism on activated carbon in the liquid phase. I. Free energy change for adsorption of organic compounds from aqueous solution on activated carbon. Bull. Chem. Soc. Jpn. **52**, 1899–1904 (1979). https://doi.org/10.1246/bcsj.52.1899
4. Lima, E.C., Dehghani, M.H., Guleria, A., et al.: Adsorption: fundamental aspects and applications of adsorption for effluent treatment. In: Dehghani, M.H., Karri, R., Lima, E.C. (eds.) Green Technologies for the Defluoridation of Water, pp. 41–88. Elsevier (2021). https://doi.org/10.1016/B978-0-323-85768-0.00004-X
5. Bergmann, C.P., Machado, F.M., Lima, É.C.: Introduction. In: Bergmann, C., Machado, F. (eds.) Carbon Nanomaterials as Adsorbents for Environmental and Biological Applications. Carbon Nanostructures. Springer, Cham (2015). https://doi.org/10.1007/978-3-319-18875-1_1
6. Bergmann, C.P., Machado, F.M., Lima, É.C.: Carbon Nanomaterials as Adsorbents for Environmental and Biological Applications. Carbon Nanostructures. Springer, Cham (2015). https://doi.org/10.1007/978-3-319-18875-1
7. Sabzehmeidani, M.M., Mahnaee, S., Ghaedi, M., et al.: Carbon based materials: a review of adsorbents for inorganic and organic compounds. Mater. Adv. **2**, 598–627 (2021). https://doi.org/10.1039/D0MA00087F
8. Machado, F.M., Lima, É.C.: Carbon nanoadsorbents for removal of organic contaminants from water. In: Bhardwaj Mishra, S., Mishra, A. (eds.) Bio- and Nanosorbents from Natural Resources. Springer Series on Polymer and Composite Materials. Springer, Cham (2018). https://doi.org/10.1007/978-3-319-68708-7_2
9. Duncan, J., Savage, N., Street, A., et al.: Nanotechnology Applications for Clean Water: Solutions for Improving Water Quality, 2nd edn. Micro & Nano Technologies, William Andrew Inc., Norwich, NY (2014)
10. ISO/TC 229 Nanotechnologies. https://www.iso.org/committee/381983.html. Accessed 12 June 2025
11. Silva, E.C., de Souza, N.F., Machado, F.M., et al.: Contribution of colloidal materials to air, water and soil environmental sustainability from basic concepts to advanced applications. In: Riegel-Vidotti, I.C., Zimmermann, L.M., de Barros, H.R. (eds.) Royal Society of Chemistry, ch. 10, pp. 219–243 (2025). https://doi.org/10.1039/9781837676149-00219

12. Eivazzadeh-Keihan, R., Noruzi, E.B., Chidar, E., et al.: Applications of carbon-based conductive nanomaterials in biosensors. Chem. Eng. J. **442**, 136183 (2022). https://doi.org/10.1016/j.cej.2022.136183
13. Terrones, M., Botello-Méndez, A.R., Campos-Delgado, J., et al.: Graphene and graphite nanoribbons: morphology, properties, synthesis, defects and applications. Nano Today **5**, 351–372 (2010). https://doi.org/10.1016/j.nantod.2010.06.010
14. Wei, T., Ni, H., Ren, X., et al.: Fabrication of nitrogen-doped carbon dots biomass composite hydrogel for adsorption of Cu (II) in wastewater or soil and DFT simulation for adsorption mechanism. Chemosphere **361**, 142432 (2024). https://doi.org/10.1016/j.chemosphere.2024.142432
15. Xu, X., Ray, R., Gu, Y., et al.: Electrophoretic analysis and purification of fluorescent single-walled carbon nanotube fragments. J. Am. Chem. Soc. **126**, 12736–12737 (2004). https://doi.org/10.1021/ja040082h
16. Biranje, A., Azmi, N., Tiwari, A., et al.: Quantum dots based fluorescent probe for the selective detection of heavy metal ions. J. Fluoresc. **31**, 1241–1250 (2021). https://doi.org/10.1007/s10895-021-02755-8
17. Deng, Y., Ok, Y.S., Mohan, D., et al.: Carbamazepine removal from water by carbon dot-modified magnetic carbon nanotubes. Environ. Res. **169**, 434–444 (2019). https://doi.org/10.1016/j.envres.2018.11.035
18. Bimberg, D., Ledentsov, N.N., Grundmann, M., et al.: Luminescence properties of semiconductor quantum dots. J. Lumin. **72–74**, 34–37 (1997). https://doi.org/10.1016/S0022-2313(97)00084-7
19. Ahirwar, R.C., Mehra, S., Reddy, S.M., et al.: Progression of quantum dots confined polymeric systems for sensorics. Polymers (Basel) **15**, 405 (2023). https://doi.org/10.3390/polym15020405
20. Praseetha, P.K., Litany, R.I.J., Alharbi, H.M., et al.: Green synthesis of highly fluorescent carbon quantum dots from almond resin for advanced theranostics in biomedical applications. Sci. Rep. **14**, 24435 (2024). https://doi.org/10.1038/s41598-024-75333-0
21. Xu, Y., Lan, J., Wang, B., et al.: Simple fabrication of carbon quantum dots and activated carbon from waste wolfberry stems for detection and adsorption of copper ion. RSC Adv. **13** (2023). https://doi.org/10.1039/d3ra04026g
22. Esmail, L.A., Jabbar, H.S.: Violuric acid carbon dots as a highly fluorescence probe for ultrasensitive determination of Zn (II) in tomato paste. Food Chem. **413**, 135638 (2023). https://doi.org/10.1016/j.foodchem.2023.135638
23. Zhao, J., Zhu, J., Li, Y., et al.: Quantum dot reinforced electrospun carbon nano-fiber fabrics with high surface area for ultrahigh rate supercapacitors. ACS Appl. Mater. Interfaces **12**, 11669–11678 (2020). https://doi.org/10.1021/acsami.9b22408
24. Iijima, S.: Helical microtubules of graphitic carbon. Nature **354**, 56–58 (1991). https://doi.org/10.1038/354056a0
25. Iijima, S., Ichihashi, T.: Single-shell carbon nanotubes of 1-nm diameter. Nature **363**, 603–605 (1993). https://doi.org/10.1038/363603a0
26. Bethune, D.S., Klang, C.H., De Vries, M.S., et al.: Cobalt-catalysed growth of carbon nanotubes with single-atomic-layer walls. Nature **363**, 605–607 (2003). https://doi.org/10.1038/363605a0
27. O'Connell, M.J.: Carbon Nanotubes Properties and Applications. CRC Press, New York (2006). https://doi.org/10.1201/9781315222127
28. Sajid, M., Asif, M., Baig, N., et al.: Carbon nanotubes-based adsorbents: properties, functionalization, interaction mechanisms, and applications in water purification. J. Water Process. Eng. **47**, 102815 (2022). https://doi.org/10.1016/j.jwpe.2022.102815
29. Sophia, A.C., Lima, E.C.: Removal of emerging contaminants from the environment by adsorption. Ecotoxicol. Environ. Saf. **150**, 1–17 (2018). https://doi.org/10.1016/j.ecoenv.2017.12.026
30. Zhang, J., Zhai, J., Zheng, H., et al.: Adsorption, desorption and coadsorption behaviors of sulfamerazine, Pb(II) and benzoic acid on carbon nanotubes and nanosilica. Sci. Total. Environ. **738**, 139685 (2020). https://doi.org/10.1016/j.scitotenv.2020.139685

31. Bianco, A., Cheng, H.M., Enoki, T., et al.: All in the graphene family—a recommended nomenclature for two-dimensional carbon materials. Carbon **65**, 1–6 (2013). https://doi.org/10.1016/j.carbon.2013.08.038
32. Novoselov, K.S., Geim, A.K., Morozov, S.V., et al.: Electric field effect in atomically thin carbon films. Science **306**, 666–669 (2004). https://doi.org/10.1126/science.1102896
33. Chowdhury, S., Balasubramanian, R.: Recent advances in the use of graphene-family nanoadsorbents for removal of toxic pollutants from wastewater. Adv. Colloid Interface Sci. **204**, 35–56 (2014). https://doi.org/10.1016/j.cis.2013.12.005
34. Chen, Z., Ren, W., Gao, L., et al.: Three-dimensional flexible and conductive interconnected graphene networks grown by chemical vapour deposition. Nat. Mater. **10**, 424–428 (2011). https://doi.org/10.1038/nmat3001
35. Qin, S.Y., Liu, X.J., Zhuo, R.X., et al.: Microstructure-controllable graphene oxide hydrogel film based on a pH-responsive graphene oxide hydrogel. Macromol. Chem. Phys. **213**, 2044–2051 (2012). https://doi.org/10.1002/macp.201200281
36. Shen, Y., Fang, Q., Chen, B.: Environmental applications of three-dimensional graphene-based macrostructures: adsorption, transformation, and detection. Environ. Sci. Technol. **49**, 67–84 (2015). https://doi.org/10.1021/es504421y
37. Leão, M.B., Rosa, P.C.C., Corte, C.L.D., et al.: Eco-friendly, non-toxic and super adsorbent hydrogels based on graphene. Mater. Chem. Phys. **288**, 126408 (2022). https://doi.org/10.1016/j.matchemphys.2022.126408
38. Sun, Z., Fang, S., Hu, Y.H.: 3D graphene materials: from understanding to design and synthesis control. Chem. Rev. **120**, 10336–10453 (2020)
39. Ren, X., Li, J., Tan, X., et al.: Comparative study of graphene oxide, activated carbon and carbon nanotubes as adsorbents for copper decontamination. Dalton Trans. **42**, 5266–5274 (2013). https://doi.org/10.1039/C3DT32969K
40. Krishna, R.H., Chandraprabha, M.N., Samrat, K., et al.: Carbon nanotubes and graphene-based materials for adsorptive removal of metal ions—a review on surface functionalization and related adsorption mechanism. Appl. Surf. Sci. Adv. **16**, 100431 (2023). https://doi.org/10.1016/j.apsadv.2023.100431
41. Rodrigues, D.L.C., Fuhr, A.C.F.P., Guido, J.A., et al.: Olive biomass-derived magnetic activated biochar for ciprofloxacin removal: integrated kinetic, isotherm, thermodynamic, and spectroscopic analysis. Sep. Purif. Technol. **360**, 131014 (2025). https://doi.org/10.1016/j.seppur.2024.131014
42. Nasrollahzadeh, M., Sajjadi, M., Iravani, S., et al.: Carbon-based sustainable nanomaterials for water treatment: state-of-art and future perspectives. Chemosphere **263**, 128005 (2021). https://doi.org/10.1016/j.chemosphere.2020.128005
43. Apul, O.G., Karanfil, T.: Adsorption of synthetic organic contaminants by carbon nanotubes: a critical review. Water Res. **68**, 34–55 (2015). https://doi.org/10.1016/j.watres.2014.09.032
44. Mijwel, A.A.S., Ahmed, A.N., Afan, H.A., et al.: Artificial intelligence models for methylene blue removal using functionalized carbon nanotubes. Sci. Rep. **13**, 18260 (2023). https://doi.org/10.1038/s41598-023-45032-3
45. Sheoran, K., Kaur, H., Siwal, S.S., et al.: Recent advances of carbon-based nanomaterials (CBNMs) for wastewater treatment: synthesis and application. Chemosphere **299**, 134364 (2022). https://doi.org/10.1016/j.chemosphere.2022.134364

# Carbon Quantum Dots: Advanced Sensors and Adsorbents for the Removal of Toxic Chemicals

**Nicholas F. de Souza, Emilly C. Silva, Leandro A. Pocrifka, Fernando Machado Machado, André R. Fajardo, and Artur J. M. Valente**

**Abstract** Since their discovery in 2004, carbon quantum dots (CQDs) have attracted significant attention due to their unique properties, including excellent hydrophilicity, biocompatibility, high electrical conductivity, low toxicity, and tunable fluorescence. These characteristics make them highly promising for detecting and removing hazardous contaminants. This chapter explores the innovative applications of CQDs in environmental sensing and pollutant removal. It presents a reader-friendly overview of the various synthesis methods for CQDs, ranging from top-down to bottom-up approaches, emphasizing sustainable and cost-effective strategies. The discussion then focuses on their use as electrochemical sensors for detecting toxic compounds and their role in the adsorption of organic and inorganic pollutants from water. Finally, the chapter addresses the current challenges and perspectives

N. F. de Souza · F. Machado Machado
Materials Research Laboratory (LPM), Technology Development Center, Federal University of Pelotas (UFPel), Pelotas, Brazil
e-mail: nicholasfs97@gmail.com

F. Machado Machado
e-mail: fernando.machado@ufpel.edu.br

E. C. Silva
Laboratory of Materials for Adsorption and Sensing Studies, Center for Energy Research (Nupenerg), Federal University of Roraima (UFRR), Boa Vista, Brazil
e-mail: emilly.silva@ufrr.br

L. A. Pocrifka
Laboratory of Electrochemistry and Energy (LEEN), Graduate Program in Chemistry, Federal University of Amazonas (UFAM), Manaus, Brazil

A. R. Fajardo (✉)
Laboratory of Technology and Development of Composites and Polymeric Materials (LaCoPol), Federal University of Pelotas (UFPel), Pelotas, Brazil
e-mail: andre.fajardo@ufpel.edu.br

A. J. M. Valente (✉)
Department of Chemistry, CQC-IMS, University of Coimbra, Coimbra, Portugal
e-mail: avalente@ci.uc.pt

© The Author(s), under exclusive license to Springer Nature Switzerland AG 2026

C. P. Bergmann and F. Machado Machado (eds.), *Carbon Nanomaterials as Adsorbents for Environmental and Biological Applications*, Carbon Nanostructures,
https://doi.org/10.1007/978-3-032-08669-3_2

regarding the environmental applications of CQDs and highlights future directions for advancing their use in environmental remediation.

**Keywords** Detection · Adsorption · Synthesis · Biomass · Photoluminescent

## Abbreviations

| | |
|---|---|
| CEC | Contaminants of emerging concern |
| CNTs | Carbon nanotubes |
| CQDs | Carbon quantum dots |
| CPE | Carbon paste electrode |
| GCE | Glassy carbon electrode |
| GQDs | Graphene quantum dots |
| LSV | Linear sweep voltammetry |
| PL | Photoluminescent |
| QY | Quantum yield |
| SPEEK | Sulfonated poly (ether ether ketone) |
| TEM | Transmission electron microscopy |
| XRD | X-ray diffraction |

## 1 Introduction

Water is undeniably essential for all known life forms on this planet, playing a vital role in bodily functions, ecosystems, and a wide array of human and non-human activities, including agriculture, food production, and hygiene. Ensuring water availability and quality is fundamental to providing access and potability for the many organisms that depend on this resource, either directly or indirectly. However, pollution, over-exploitation, and climate change have significantly threatened water resources worldwide. Adding to these challenges, increasing resource consumption and the development of new products have led to the discharge of various chemicals (both inorganic and organic) into water bodies and wastewater systems, presenting new and complex challenges for monitoring and treatment.

Monitoring water quality is critical for ensuring potability, safeguarding public health and the environment, and informing policies and strategies for preservation and treatment. Water and wastewater treatment face significant obstacles due to the presence of organic substances classified as contaminants of emerging concern (CECs) [1, 2], as well as inorganic species such as metals, salts, and radionuclides [3, 4]. The scientific consensus indicates that traditional treatment processes often fall short in efficiently removing specific contaminants [5], underscoring the urgent need to improve existing methods or develop new, more effective approaches. Similar concerns extend to the methods and processes used to monitor and quantify these contaminants in various water systems.

Over the past decades, substantial research efforts have been devoted to developing various chemical, physical, and biological processes to remove toxic chemicals from water. Among these, (nano)filtration, coagulation, ion exchange, photo- or electro-assisted oxidation, and adsorption stand out as the most prominent techniques [6, 7]. Adsorption, in particular, has garnered significant attention due to its operational simplicity, cost-effectiveness compared to other methods, broad applicability, and potential for reusing the adsorbent material [8, 9]. The efficiency of adsorption is closely tied to the compatibility between the adsorbent material and the target contaminant, which can be an atom, ion, or molecule [10, 11]. This compatibility largely depends on the physicochemical properties of the adsorbent material, including high surface area, porosity, and chemical and mechanical stability [12, 13]. Furthermore, ideal adsorbents should be non-toxic, regenerable, and reusable [12].

Given these requirements, significant research has focused on designing and developing advanced adsorbent materials. Numerous classes of compounds have been explored, including biomass, polymers, mineral clays, silicates, metal–organic frameworks, and carbon derivatives [14–16]. Among carbon-based materials, activated carbon, bio- and hydrochars, graphene, fullerene, and carbon nanotubes have been widely used, either independently or in the production of devices for treating [17, 18] and monitoring water quality [19, 20]. More recently, CQDs, a novel class of carbon nanoparticles, have gained significant attention for their dual role in water and wastewater treatment due to their ability to act as efficient adsorbents for various contaminants [21, 22] and their unique photoluminescent properties, which enable sensing applications [22].

CQDs are carbon derivatives typically synthesized through top-down or bottom-up processes, yielding spherical or semi-spherical nanoparticles with a diameter of less than 10 nm (see Fig. 1) and a high surface area. Depending on the synthesis method, CQDs can be doped with heteroatoms (e.g., N, B, S), enhancing their reactivity and interaction with other species, which is particularly beneficial for adsorption applications [23]. Moreover, CQDs exhibit exceptional stability against photobleaching, chemical inertness, high water solubility, biocompatibility, and low toxicity [24]. Thanks to these properties, CQDs have found applications across diverse fields, including optoelectronics, catalysis, bioimaging, biosensing, drug delivery, food safety, and agriculture [25–27].

CQDs can interact with contaminants through physical, chemical, or electrostatic mechanisms, making them promising adsorbents for various chemical pollutants in water. Additionally, they can be anchored onto solid materials such as polymers [27], MOFs [28], or mesoporous substances [29], facilitating their recovery and reuse. Beyond their adsorption capabilities, CQDs photoluminescent (PL) properties enable their use in sensing applications, allowing real-time detection and quantification of contaminants in aqueous media [30]. These characteristics position CQDs as versatile, bivalent materials capable of monitoring and removing toxic contaminants from water.

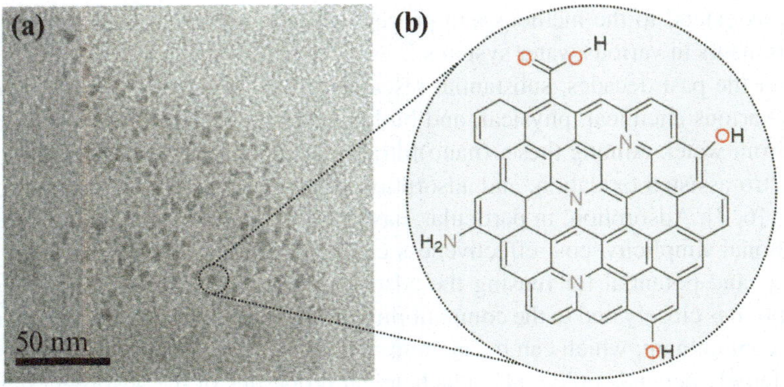


**Fig. 1** **a** Microscopic image and **b** schematic representation of CQDs structure

This chapter offers an overview of this emerging research field to provide a comprehensive perspective on the use of CQDs for sensing and adsorbing toxic chemicals in water. It compiles key advancements in applying CQDs and CQD-based nanocomposites, discusses fundamental methods for synthesizing CQDs and anchoring them onto other materials, and examines the main challenges and prospects for their application in water quality management.

## 2 Synthesis of Carbon Quantum Dots (CQDs)

In 2004, while using the arc discharge method to synthesize carbon nanotubes (CNTs), Xu et al. accidentally discovered carbon nanoparticles with fluorescent properties during the separation and purification process via electrophoresis [31]. Following this discovery, researchers have significantly expanded the variety of methods used to prepare CQDs. The synthesis routes of CQDs are essential to determine their structure and, consequently, their final properties. This is because CQDs are composed mainly of carbon, hydrogen, and oxygen, and these are influenced by the synthesis process, which generally determines the proportion of these elements in these nanomaterials [32]. In other words, variations in size, morphology and type or amount of doping will generate changes in the final properties of these nanomaterials and, therefore, the synthesis method plays a significant role in the entire process to ensure homogeneity in size and structure [33].

CQDs can be synthesized via top-down or bottom-up approaches [34, 35]. Top-down methodologies involve breaking larger carbonaceous materials, such as graphite [35], CNTs [36], activated carbon [37], or graphene [38], using techniques like laser ablation [39], arc discharge [40], acidic oxidation [41], or electrochemical oxidation [42]. These methods face challenges, including low quantum yield (QY), limited spectral efficiency, harsh conditions, and significant variation in size and morphology [43].

Bottom-up methods synthesize CQDs from organic precursors with $sp^2$ structures, such as citric acid [44], glucose [45], L-glutamic acid [46], and sucrose [47], by heating them above their melting points to trigger chemical reactions [48]. Biomass materials like plant leaves [49] and seeds [50] are also used, where a more in-depth discussion on the use of biomass as a feedstock for CQD production will be covered later. Common techniques include microwave synthesis [51, 52], hydro-thermal/solvothermal methods [53, 54], and pyrolysis [55, 56]. These approaches provide control over the composition and PL properties of CQDs, as the structure and size of these nanomaterials are influenced by several factors, such as the molecular structure of the carbon precursor, the choice of solvent, and the reaction conditions, including time, temperature, and pressure. These crucial parameters significantly affect the reactants and the nucleation process of CQDs [56]. Additionally, bottom-up methods tend to be simpler than top-down approaches, which enables the use of a broader variety of raw materials [44, 57]. These methods also offer high QY and facilitate efficient doping with single or multiple heteroatoms derived from organic precursors during the synthesis process [57]. Table 1 compares the advantages and disadvantages of top-down and bottom-up procedures as reported in the literature.

## 2.1 *Biomass-Derived CQDs*

Biomass, such as food scraps and agricultural residues, presents a promising option for converting low-value materials into high-value nanomaterials [58]. With 40–50% carbon content, along with nitrogen and oxygen, these residues are ideal precursors for CQDs synthesis, offering the benefits of being cheap, plentiful, easy to meet synthesis conditions, and the possibility of scaling up CQDs production [43]. Biomass waste products previously used to generate CQDs successfully include lemon and orange peels [59, 60], green tea [61], eggshell [62], coffee grounds [63], sugarcane bagasse [59], manure [60], among others [64].

The formation of CQDs from biomass involves complex chemical processes, including dehydration, decomposition, polymerization, condensation, aromatization, and nucleation [65, 66]. Limited understanding arises from challenges in characterizing intermediates [67]. Variations in precursor structures and factors like carbon atom count, chain length, and surface groups influence the structure and properties of biomass-derived CQDs [68]. These properties can be tailored through size/shape control, surface passivation, heteroatom doping, and blending with composites [48].

## 2.2 *Size and Shape Control*

The PL emission of CQDs is strongly influenced by size due to quantum confinement effects, as shown by Jiang et al. [68]. Size and shape adjustments are crucial, as they affect the sp2 structure distribution and electron energy levels [69]. Smaller,

**Table 1** Overview of reported methods for synthesizing CQDs from various carbon-based precursors via top-down and bottom-up approaches: procedures, advantages, and limitations

| Method (top-down/ bottom-up) | Procedure | Advantages | Limitations |
|---|---|---|---|
| Arc discharge (top-down) | Two graphite rods, serving as the anode and cathode, are placed in an open vessel, with an electric arc generated between them | The arc discharge method can produce CQDs with good PL properties highly soluble in water | This results in uneven particle sizes, require high temperatures, and produces CQDs with low QY and impurities that are difficult to remove |
| Laser ablation (top-down) | Laser irradiation fragments larger carbon source particles, suspended or immersed in a liquid medium, into smaller ones, forming nanoparticles and CQDs | This method is simple, does not involve chemical agents, and efficiently produces CQDs with good water solubility quickly | This method is not environmentally friendly, provides limited control over particle sizes, is expensive, and results in low QY for CQDs |
| Acidic oxidation (top-down) | Larger carbon precursors are broken down into smaller fragments through oxidation using strong oxidizing agents such as $H_2O_2$, $H_2SO_4$, and $HNO_3$ | The generated CQDs exhibit tunable PL performance, high QY, high purity, and low cost | It requires harsh conditions with strong acids and complex processes, leading to poor control over CQDs sizes |
| Microwave method (bottom-up) | Microwave methods use electromagnetic waves to heat the carbon source quickly, speeding up carbonization and pyrolysis and forming CQDs | A rapid, simple, and environmentally friendly synthesis method | Appropriate for small-scale CQDs production, with limited industrial applications |

(continued)

**Table 1** (continued)

| Method (top-down/ bottom-up) | Procedure | Advantages | Limitations |
|---|---|---|---|
| Hydrothermal/ solvothermal synthesis (bottom-up) | Carbon sources are dissolved in water or an organic solvent and placed in a Teflon-lined autoclave, where high pressure and temperature synthesize CQDs | The QY of CQDs can reach 80% using a simple, eco-friendly, one-step process with various carbon sources | Demands extended processing times, delaying large-scale production, and necessitates further purification steps post-synthesis |
| Pyrolysis method (bottom-up) | In pyrolysis, organic compounds decompose at high temperatures in an oxygen-free environment, sometimes with catalysts | It is a fast, low-cost, and eco-friendly method that uses various carbon sources, enabling large-scale CQDs production without post-treatment modification | Requires high temperatures, making scalability challenging and resulting in CQDs with a broad size distribution |

uniformly distributed CQDs result from rapid nucleation, while growth depends on surface properties and the solubility of the precursors [65]. Controlling the carbon source and solvent is essential for determining the size of biomass-derived CQDs [70, 71]. Kalita et al. synthesized size-controlled CQDs from rice grains by varying heating times (3–10 min), achieving sizes from 2 to 6.5 nm, with QY rising from 16 to 24% as sizes decreased [72].

## 2.3 *Surface Passivation*

Fluorescent CQDs can be synthesized from biomass precursors, but surface defects in bare CQDs lead to low fluorescence stability, intensity, and low QY, necessitating surface passivation [65]. Passivation fine-tunes material properties by modifying surface defects while preserving the core composition [48]. It also enhances electron–hole recombination efficiency by blocking traps, thereby increasing QY. Common passivating agents include organic molecules and polymers [73, 74]. In 2017, Pan et al. developed a fluorescence sensor by enhancing CQD fluorescence through surface passivation with 1-ethyl-3-(3-dimethylaminopropyl)-carbodiimide (EDC), increasing QY from 18.83% to 41.10% as EDC concentration rose from 2.5 to 25 mmol/L [73]. While effective, surface passivation is complex and often relies on cytotoxic and environmentally harmful organic reagents or catalysts [57].

## 2.4 Heteroatom Doping

Heteroatom doping is a simple and efficient method to enhance the fluorescent properties of biomass-derived CQDs, such as QY and emission color [75, 76]. Unlike passivation, which often involves toxic reagents, non-metallic doping preserves the CQD's electronic structure due to the size similarity with carbon atoms, enabling uniform doping via simpler synthesis [57]. Doping with elements like N, P, S, and B modifies carbon frameworks, introduces active sites, and adjusts the band gap, creating new energy levels and enhancing fluorescence and other CQDs properties [58, 77]. Liu et al. synthesized biomass-derived CQDs with nitrogen doping via a microwave-assisted hydrothermal process, increasing QY from 2.95% to 35.39% [74]. Similarly, Hu et al. [78] produced nitrogen and sulfur co-doped CQDs from green tea leaf residue using high-temperature pyrolysis and $H_2SO_4$ oxidation, achieving a QY of 14.8%, over three times higher than undoped CQDs [78].

## 2.5 Composite Material Blending

Recently, extensive research has focused on developing innovative composite materials combining CQDs with inorganic nanoparticle cores, such as $TiO_2$ [75], $Cu_2O$ [76], $Ag_3PO_4$ [77], among others. Composite materials incorporating CQDs are also applicable in adsorption processes for the removal of environmental pollutants [79], such as when combined with cellulose hydrogels and aerogels for the adsorption of heavy metals in water [80, 81]. The resulting nanohybrids combine the optical benefits and functional groups of CQDs with these materials, offering great promise as high-performance adsorbents and photocatalysts [82]. In 2018, Tadesse et al. developed N-doped carbon quantum dots/magnetite nanocomposites ($Fe_3O_4$@NCQDs NCs) through a simple synthesis method, utilizing non-toxic and affordable lemon juice as a precursor for removing pollutants from organic dyes. The experiment demonstrated rapid removal of methyl blue dye from an aqueous solution within 20 m, achieving an adsorption efficiency exceeding 90.84% [83]. Wang et al. utilized microalgae as precursors to produce CQDs, which were then employed as a green modifier to mediate the formation of nano-MnS/FeS, enhancing $Cd^{2+}$ removal. By adding only 1 wt % CQDs, the $Cd^{2+}$ adsorption capacity of CQDs-MnS reached 481 mg/g at 25 °C, exceeding that of most previously reported metal sulfides and providing a new perspective on enhancing heavy metal removal using CQDs [84].

Table 2 shows a selection of the main synthetic approaches involving biomass or other natural starting materials, outlining reaction conditions, particle sizes, QY values, and reported applications. This comparison emphasizes the efficiency and practical utility of each synthesis technique.

**Table 2** Biomass-derived CQDs: synthesis approaches, properties, and applications

| Biomass | Method | Reaction conditions | Size (nm) | QY (%) | Applications | References |
|---|---|---|---|---|---|---|
| Pine wood | Hydrothermal | 180 °C, 3 h | 2–6# | 4.69 | Detection of $Fe^{3+}$ | [80] |
| Banana peel | Hydrothermal | 200 °C, 24 h | 4–6# | 20 | Bioimaging | [81] |
| Discarded cigarette butts | Hydrothermal | 200 °C, 6 h | 3.8 | – | Detection of $Fe^{3+}$ | [85] |
| Jackfruit seeds | Microwave | 600 W, 1 min | 3–7# | 17.91 | Detection of $Au^{3+}$ and multicolor cell imaging | [86] |
| Waste cotton linter | Microwave | 150 °C, 4 min | 10.1 | – | Bioimaging | [87] |
| Carica papaya waste | Pyrolysis | 200 °C, 15 min | 7.0 | 23.7 | Chromium detection in water | [88] |
| Mango peel | Pyrolysis | 300 °C, 6 h | 2–6# | 8.5 | Detection of $Fe^{2+}$ | [89] |
| Muskmelon | Acid Oxidation | 100 °C, 1 h | 3.5 | 7.07 | Detection of $Hg^{2+}$ | [90] |

# For CQDs with size range

# 3 Electrochemical Sensors Based on CQDs

Electrochemical sensors based on CQDs have gained prominence in recent decades due to their ability to provide sensitive, selective, and rapid detection of various substances. CQDs are carbon nanoparticles with unique fluorescent properties, making them ideal for sensor applications [91]. Their high surface-to-volume ratio and the possibility of functionalization with different chemical groups further broaden the spectrum of possible detections, from metal ions to complex organic compounds.

This class of nanomaterials is widely used as detection and sensing platforms. CQDs have excellent optical and electrical properties, making them promising materials for such applications. Depending on their size, they have high QY, narrow emission spectra, and variable fluorescence. Furthermore, due to their chemically active surface, they can be bound to other species, ensuring sensor specificity [92]. Among the methods described in the literature, CQDs have already been used in fluorescent, electrochemical, and infrared absorption spectroscopy strategies to sense emerging chemical pollutants [93, 94].

The main advantage of CQDs sensors lies in their simplicity of fabrication and reduced cost. These sensors can be synthesized from low-cost materials and relatively simple methods. Moreover, their chemical stability and robust PL ensure reliable performance under adverse conditions [92]. The versatility of CQDs allows for the fabrication of sensors in different forms, such as thin films, membranes, and even portable devices, enhancing their potential use in various areas, including medicine, environmental control, and food safety [95].

In the medical field, CQDs sensors have been explored to detect specific biomarkers, aiding in the early diagnosis of diseases. For example, glucose detection in blood using CQDs sensors has shown promise, offering a more accessible and less invasive alternative compared to traditional methods [96]. In environmental control, these sensors can monitor the presence of heavy metals in water bodies, providing real-time data crucial for environmental preservation. In food safety, rapidly detecting contaminants such as pesticides and antibiotics ensures consumer safety and prevents potential disease outbreaks [97].

The combination of CQDs with other nanomaterials, such as gold nanoparticles or metal oxides, has led to the development of sensors with even more enhanced capabilities [98, 99]. This hybrid approach allows for fine-tuning of sensory properties and the possibility of multiplexed detection, where different analytes can be detected simultaneously. Furthermore, incorporating CQDs into portable device platforms and smartphones paves the way for developing personal use sensors, democratizing access to advanced monitoring technologies.

Despite significant advances, challenges persist in the area of electrochemical sensors based on CQDs. The reproducibility and standardization of synthesis methods are critical issues that must be addressed for large-scale commercialization. Additionally, a deep understanding of the interacting mechanisms between CQDs and analytes is essential to optimize sensor performance. Continuous investments in research and development are fundamental to overcoming these obstacles and consolidating the application of CQD sensors in various fields [92].

Electrochemical sensors are based on measuring the electrical response generated when the analyte interacts with the working electrode to which potential is applied. There are various electroanalytical methods, such as voltammetry, potentiometry, and amperometry. All of them offer high sensitivity and selectivity, rapid response, and ease of handling compared to other analytical methods [100]. In recent years, these electrochemical sensors have experienced significant advancements and improvements in their performance by integrating nanomaterials. The properties of nanomaterials can be ideal for a specific analyte, and their high surface area-to-volume ratio enhances the sensor's sensitivity.

Numerous articles on voltametric analysis with electrodes modified by the quantum dots of pharmaceuticals are published annually. Publications on the modification of quantum dots can be exemplified as follows. Tang et al. developed an electrochemical sensor using a glassy carbon electrode (GCE) modified with graphene quantum dots (GQDs) for the determination of hydroquinone and catechol [95]. This sensor was fabricated using the electrodeposition method and characterized by electrochemical impedance spectra. The proposed GQD sensor demonstrated excellent sensitivity, reproducibility, and reliability in electrochemical measurement, achieving a detection limit of up to 0.08 μmol/L in the range of 0.5 to 100 μmol/L. The simultaneous detection of HQ and CC with the GQD/GC electrode was carried out in river water samples, and there was a good recovery.

Another highly sensitive and selective quantum dot modified carbon paste electrode (CPE) was developed by Pasandideh-Nadamani et al. [101]. The synthesis of CdS-QDs, which were quite stable, was characterized by X-ray diffraction (XRD) techniques and transmission electron microscopy (TEM). The CdS-QDs were obtained by an in-situ technique using a thiosulfate precursor [101].

Algarra et al. [102] developed CQDs in a modified GCE to electrochemically determine dopamine and uric acid. The CQDs were obtained from graphite using the modified Hummers method. The electrochemical determination of both compounds revealed a significant increase in peak current in the CQDs-GCE compared to the unmodified glassy carbon electrode. Using linear sweep voltammetry (LSV), the proposed sensor exhibited high sensitivity. The detection limits were found to be 1.3 μmol/L for uric acid and 2.7 μmol/L for dopamine [102].

Nanomaterials are ideal for creating sensors. In quantum dots, size is a controllable parameter and, when combined with the "quantum confinement" effect, results in exceptional optical and electrical properties. As the size of quantum dots varies due to the quantum confinement effect, the color of their luminescence also changes. Quantum dots can be used as fluorescent probes for medical diagnosis and imaging. However, heavy metals like CdSe, CdTe, and CdS tend to degrade under physiological conditions, and the release of ions is toxic [103]. The disadvantages and advantages of quantum dots in an electrochemical sensor are summarized in Fig. 2.

In summary, electrochemical sensors based on CQDs represent a promising frontier in detecting highly sensitive and specific substances. Their unique properties and versatile applications make them ideal candidates to meet the growing medical diagnostics, environmental monitoring, and food safety demands. With continuous advances in material science and nanobiotechnology, the expectation is that these

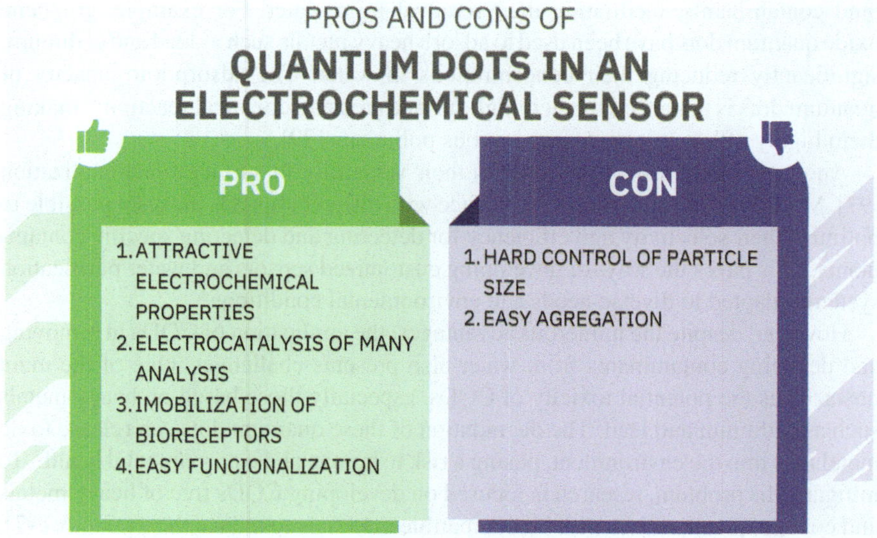


**Fig. 2** Disadvantages and advantages of CQDs in an electrochemical sensor

sensors will play an increasingly crucial role in our society, ensuring a safer and more sustainable future.

## 4 Detection and Removal of Hazardous Contaminants from Water

As noticed in the literature, CQDs have gained attention as a promising technology for both the detection and removal of water contaminants, owing to their unique physicochemical properties described earlier in this chapter. These semiconductor nanomaterials exhibit remarkable optical and electronic characteristics resulting from the quantum confinement effect [104], which allows precise size tuning and enables a wide range of applications, particularly in interfacial chemistry (sensing, energy transfer, and surface modification). The focus of the research is on water purification and monitoring water quality.

One of the main advantages of CQDs is their ability to emit high-intensity light in various colors, depending on their size [105]. This characteristic makes them ideal candidates for fluorescence-based detection of contaminants. When combined with specific materials, quantum dots can be used as fluorescent sensors, detecting the presence of contaminants such as heavy metals, pesticides, pharmaceuticals, and other organic and inorganic pollutants. The sensitivity of quantum dots allows the detection of low concentrations of contaminants, providing more accurate and effective monitoring of water quality.

In addition to detection, CQDs are also used in contaminant removal [94]. Studies have shown that functionalized CQDs with specific chemical groups can selectively bind contaminants, facilitating their removal from water. For example, graphene oxide quantum dots have been used to adsorb heavy metals such as lead and cadmium, significantly reducing their concentrations in water. The adsorption capacity of quantum dots is enhanced by their high surface area and chemical reactivity, making them highly efficient in removing various pollutants [79].

Another critical aspect of CQDs is their versatility in surface functionalization [97]. Modifying the quantum dots' surface with different ligands makes it possible to optimize their selectivity and efficiency for detecting and detecting specific contaminants. This paves the way for developing customized sensors and water purification systems adapted to diverse needs and environmental conditions.

However, despite the numerous advantages, the application of CQDs in removing and detecting contaminants from water also presents challenges. One of the main obstacles is the potential toxicity of CQDs, especially those based on heavy metals such as cadmium and lead. The degradation of these quantum dots can release toxic metal ions into the environment, posing a risk to human and environmental health. To mitigate this problem, research is focused on developing CQDs free of heavy metals and coating quantum dots with biocompatible materials to reduce their toxicity [97].

There are clear challenges in detecting organic and inorganic compounds. Organic contaminants present structural complexity, often requiring advanced separation techniques and highly sensitive detection methods to distinguish between structurally similar compounds and trace concentrations. The latter, in turn, presents the challenge of interference from other ions present in the sample matrix, which may require more elaborate sample preparation techniques. Issues such as binding specificity. CQDs are often modified with ligands that have a high affinity for specific metal ions, allowing highly selective detection and robustness since, in aqueous solutions, the detection of inorganic compounds can be more robust due to the lower interference from other complex organic molecules. Turning to organic contaminants, it is essential to consider that their detection can be more sensitive since CQDs can detect slight variations in the concentration of organic molecules due to their optical properties [97].

## *4.1 Inorganic Contaminants*

Among the waste discharged by industries, rich in organic and inorganic pollutants, potentially toxic metals stand out, as they present harmful properties to the entire trophic chain and biomes. The species of most significant concern include Pb, Hg, Cu, Cd, Cr, Ni, and Mg [106]. These species can originate from cars, dyes, fertilizers, pesticides, bleaching agents, and some metal alloys. Legislation in some countries has become stricter in limiting these products in water, but this is not a global reality, especially in emerging countries.

Currently, there are several processes for detecting and monitoring heavy metals, allowing the prevention of environmental damage. Due to their properties, CQDs appear to be the best fluorescent platform for performing chemical sensing processes. CQDs offer an abundance of raw materials, as well as a variety of synthetic routes available, so they can offer moldable functional groups for sensing distinct species, resulting in a material that, in addition to adsorption capabilities, allows visual detection of metallic species by modifying their luminescent properties [107]. Regarding metals, the assembly of CQDs in other materials has been explored in the literature, mainly for sensing applications. The demand for surface area and porosity to aggregate the highest possible concentration of metals requires a matrix that houses and preserves the fluorescence of these CQDs. Functionalized CQDs can form host–guest supramolecular adducts [108]. This feature is also advantageous for the adsorption of metals, as described in some works in the literature [109].

A relevant example is the study conducted by Liu et al. [110] research. Their study focused on using CQDs as a low-cost photo-responsive semiconductor material to detect heavy metal ions, specifically hexavalent chromium [Cr(VI)] in water. The detection system effectively identified Cr(VI) with a detection limit of 3.86 μmol/L and a recovery rate of 95.4%–109.5%. The fluorescence-quenching mechanism of the Cr (VI) reaction with CQDs was studied using ultraviolet, infrared, and fluorescence spectra. The method has proven to be convenient and efficient for detecting Cr(VI) in water environments [110].

Liu et al. in 2022 described the development of a fluorescence sensor array based on CQDs to detect multiple heavy metal ions, including Cr(VI), Fe(III), Fe(II), and Hg(II) in environmental samples. The sensor array, assisted by stepwise prediction and machine learning, detected metal ions within a concentration range of 1–50 μmol/L with an accuracy of 95%. The method demonstrated high efficiency and accuracy in detecting single and mixed heavy metal ions in deionized water and real samples [111].

Concluding with examples of advances in the detection of inorganic composts, Sgreccia et al. recently explored the combination of ion exchange membranes with CQDs for detecting and removing heavy metal ions, such as Pb(II) and Cd(II), from water. The composite membranes formed by sulfonated poly(ether ether ketone) (SPEEK) with embedded CQDs showed a pronounced turn-off effect for lead and achieved 100% removal efficiency for Pb and Cd when the concentration was below half of the ion exchange capacity of SPEEK. The study highlighted the potential of CQDs in water treatment applications [112].

## *4.2 Organic Contaminants*

The increasing contamination of water sources by organic pollutants, including personal care products, pharmaceuticals, and dyes, is a major environmental concern. Even at low levels, the presence of these substances in aquatic ecosystems poses significant risks to both the environment and human health [113]. As discussed, CQDs possess remarkable properties, making them promising for detecting and removing both organic and inorganic pollutants. Their tunable PL properties allow for accurate detection of organic pollutants even at low concentrations, while their large surface area, together with their functional groups on the surface, increase the adsorption and desorption efficiency [114]. However, a few research studies have been conducted to develop and apply CQD-based nanocomposites to detect and adsorb organic contaminants.

Several adsorption mechanisms govern the interaction between CQDs and organic contaminants, primarily including $\pi$–$\pi$ interactions, electrostatic interactions, and hydrogen bonding [115]. In 2019, Deng et al. synthesized magnetic carbon nanotubes modified with CQDS and magnetite to evaluate their effectiveness in removing carbamazepine from water. The resulting adsorbent demonstrated an adsorption capacity of 65 mg of carbamazepine per gram of adsorbent at pH 7.0, with $\pi$–$\pi$ interactions and hydrogen bonds playing similar roles in the adsorption process [116].

Nanocomposites based on CQDs are also used to adsorb and remove cationic and anionic dyes. Rabeie et al. [108] synthesized MIL88A (Fe) and GQDs nanocomposites with various morphologies to remove Acid Blue 92 synthetic dye. The spindle-type nanocomposite exhibited the highest adsorption capacity, with 617 mg/g, compared to 335 mg/g for the rod-type and 114 mg/g for the diamond-type nanocomposites. Additionally, the dye adsorption mechanism was primarily driven by $\pi$–$\pi$ and electrostatic interactions [108].

CQDs possess the essential optical characteristics for detecting organic pollutants, making them highly effective for fluorescence-based detection. In 2023, Wang et al. developed a novel fluorescent probe by combining CQDs with a molecularly imprinted polymer for detecting 4-nitrophenol in aqueous environments. The probe demonstrated high effectiveness in real samples, including tap water, wastewater, and seawater, achieving satisfactory recoveries (95.1%–107.8%), with a broad linear detection range (0–144 μmol/L) and a low detection limit of 0.41 μmol/L [117].

Applying CQDs in fluorescent sensors offers fast and accurate methods to identify pollutants at very low concentrations. This is crucial to prevent environmental damage and protect public health. However, it is essential to recognize the challenges, such as the potential toxicity of heavy metal-based quantum dots. Therefore, continuous research into safer alternatives and optimizing quantum dot stability is essential to ensure their sustainable and safe use, with a focus on carbon-based CQDs. In summary, CQDs represent a promising and innovative approach to detecting various contaminants, with significant potential to contribute to more effective and sustainable environmental solutions. The combination of technological advancements, appropriate regulations, and ongoing research could further enhance the positive impact of this technology in resource management and treatment.

## 5 Challenges and Perspectives of CQDs in Environmental Applications

In the last decades, the anthropogenic effect on the environment has been increasing due to several factors, of which we can mention the need to produce food in proportion to demographic development (according to the UN, the World population is expected to reach 9.8 billion in 2050) or the increase in the average life expectancy of human beings. Consequently, the impact of persistent pollutants on water bodies has been increasing both in quantity and quality [118]. Pollutants that generate new challenges for the development of monitoring/sensing and treatment technologies include disinfection by-products (DBPs), per- and polyfluoroalkyl substances (PFAS), pesticides and antibiotics [112, 119–121].

DBPs originate during drinking water disinfection processes, necessary to comply with drinking water regulations in different countries, and result from the reaction of disinfectants (e.g., chlorine gas, hypochlorite and chloramines) with organic matter existing in the water. Families of DBPs found in drinking water include trihalomethanes (THMs), haloacetic acids (HAAs) and haloketones (HKs) [122].

Per- and polyfluoroalkyl substances (PFAS) are present in several consumer goods and industries (from biotech to pharmaceutics), due to their amphiphilic characteristics and their high chemical and thermal stability. They can accumulate in water bodies and soils [123]. PFAS can easily enter both the water cycle and the animal and human food chain, representing a threat to the environment and human health, as they have been shown to increase the risk of many diseases, including cardiovascular, neurological, and oncological [124].

Concerning, the last two families of pollutants, antibiotics are drugs prescribed for the treatment of bacterial infections in humans and animals; however, as in pharmaceutical formulations, a significant amount of consumed antibiotic is not absorbed by organisms and are excreted in native or metabolized forms, having an environmental impact in water bodies, crops and food [125]. An important impact of antibiotics in the food chain is the increase in antimicrobial resistance, following the prolonged exposure of bacteria to these compounds, allowing them to develop protective mechanisms from the bactericidal/bacteriostatic effects of antibiotics, thereby reducing their effectiveness and aggravating health consequences [126]. Pesticides are chemical compounds used to terminate pests in agriculture and domestic life. It is a major concern for today's aquatic life and human health. Pesticides used in crop production are directly involved in the food chain and have severe effects on the environment. Nowadays nanosponges have been used for the removal of pesticides and their intermediates from the water bodies. Persistent organic pesticides have long been responsible for controlling human disease vectors and significantly increased agricultural productivity. In the last decade or so, however, the effects of pesticides on the environment, biodiversity, and human health became a major concern that needs to be addressed urgently [127].

All these pollutants have in common the fact that they are persistent, emerging and priority pollutants, that they have a harmful impact on living beings and that they are found in bodies of water in often trace quantities. This situation is quite challenging because many of the classic treatment techniques (e.g., AOPs, electrochemical and reverse osmosis) are not effective, and their detection is quite limited, generally requiring expensive hyphenated laboratory techniques.

Due to the unique characteristics of CQDs discussed above, such as their luminescence and conductive properties, as well as their high potential for functionalization, either covalently or through surface modification, there is a high potential for the development of materials with, at least, dual function: sensing and removal by adsorption. In fact, the use of functionalized CQDs allows not only to be highly sensitive, showing very low LODs towards pollutants [128]. This results from the interaction leading to detection occurring through different mechanisms, such as, fluorescence quenching, static or dynamic quenching, energy transfer and fluorescence resonance energy transfer (FRET). Additionally, CQDs can also be used in the development of electrochemical sensors [129].

However, CQDs have the limitation that for these applications they need a solid matrix. Being a drawback, it can be, in fact, an opportunity. In the previous section, we have already described the use of MIPs for the incorporation of CQDs and, in this synergetic way, increase the capabilities of both components [130]. Although the reverse might not be true, materials that have a good adsorption capacity have a high potential for detection. The development of hybrid materials that allow the simultaneous detection and removal of different pollutants is also a challenge for researchers working with CQDs. In this framework, the use of, for example, nanosponges as CQD matrices is a path with much to explore. Nanosponges can be synthesized with different biopolymers, ensuring a mesoporous structure with high adsorption capacity. Among the monomers used, we can use macrocyclic compounds such as cyclodextrins. These cyclic oligosaccharides have the possibility of being easily functionalized, allowing them to be adapted to diverse types of pollutants [131]. Additionally, its amphiphilic character that gives it the hydrophobic cavity, allows the stabilization of the photoluminescent properties of CQDs [132], allowing its reuse, and allowing these types of hybrid materials to have the aforementioned dual function. It is also expected that other macrocycles can be used as matrices for the incorporation and stabilization of CQDs. Among these, and as soon as they become more accessible to the scientific community, cucurbit[*n*]urils [133] and bambus[*n*]urils [134] are just examples of what may be envisaged in the future.

In brief, it is crucial to develop efficient solutions to detect and retain these pollutants in the water bodies, if not at their source points. CQDs may play a key role in finding solutions to reach that goal, either because of the possibility of having a photoelectrochromic response or because of their ability to contribute to the removal of pollutants by adsorption, as components of hybrid materials. We should also emphasize the possibility of developing materials based on CQDs with both functions.

## 6 Conclusions

In this Chapter, we demonstrated that CQDs have been consolidated as highly versatile nanomaterials with significant potential for environmental sensing and remediation, particularly in the detection and removal of toxic contaminants from water. Their unique combination of photoluminescent, electrical, and adsorptive properties, alongside their biocompatibility and ease of functionalization, makes them valuable candidates for developing advanced sensors and high-performance adsorbents. Sustainable synthesis routes, particularly those based on biomass and other natural precursors, have significantly enhanced the applicability of CQDs by providing environmentally friendly and cost-effective alternatives to traditional carbon sources. Additionally, techniques such as heteroatom doping, surface passivation, and integration into composites have allowed for precise tuning of their properties to target specific contaminants. Although CQDs present many advantages, certain limitations related to syn-thesis standardization, scalability, and long-term performance remain as discussed in the previous section.

Nevertheless, the rapid growth of research in this area continues to expand their range of applications and improve their performance in real-world scenarios. Taking all these aspects together, CQDs represent a promising and innovative approach to tackling global water contamination challenges. Their multifunctionality, coupled with ongoing advances in nanotechnology and green chemistry, positions them as strategic components in the development of more sustainable and efficient environmental technologies.

**Acknowledgements** The authors gratefully acknowledge financial support from Brazilian agencies Conselho Nacional de Desenvolvimento Científico e Tecnológico (CNPq) and Coordenação de Aperfeiçoamento de Pessoal de Nível Superior (CAPES). Professor Fernando Machado extends his gratitude to CNPq for his Produtividade em Pesquisa (PQ) fellowship (Process 308203/2022-4) and research projects (Processes 408413/2022-0, 443031/2023-1, and 403522/2025-0). Professor André R. Fajardo also thanks CNPq for his PQ fellowship (Process 303125/2022-5). Professor Artur J. M. Valente acknowledges financial support from the Coimbra Chemistry Centre (CQC), funded by the Fundação para a Ciência e Tecnologia (FCT, Portugal) through programs UIDB/00313/2025 and UIDP/00313/2025.

## References

1. Lin, X., Xu, J., Keller, A.A., et al.: Occurrence and risk assessment of emerging contaminants in a water reclamation and ecological reuse project. Sci. Total. Environ. **744**, 140977 (2020). https://doi.org/10.1016/j.scitotenv.2020.140977
2. Vita, N.A., Brohem, C.A., Canavez, A.D.P.M., et al.: Parameters for assessing the aquatic environmental impact of cosmetic products. Toxicol. Lett. **287**,(2018). https://doi.org/10.1016/j.toxlet.2018.01.015
3. Singh, V., Ahmed, G., Vedika, S., et al.: Toxic heavy metal ions contamination in water and their sustainable reduction by eco-friendly methods: isotherms, thermodynamics and kinetics study. Sci. Rep. **14**, 7595 (2024). https://doi.org/10.1038/s41598-024-58061-3
4. Ugbede, F.O., Aduo, B.C., Ogbonna, O.N., Ekoh, O.C.: Natural radionuclides, heavy metals and health risk assessment in surface water of Nkalagu river dam with statistical analysis. Sci. Afr. **8**, e00439 (2020). https://doi.org/10.1016/j.sciaf.2020.e00439
5. Zhang, Q., Zheng, W., Zhang, H., et al.: Occurrence, fate and distribution of emerging organic pollutants in full-scale hybrid constructed wetlands treating municipal effluents. J. Water Process. Eng. **61**, 105291 (2024). https://doi.org/10.1016/j.jwpe.2024.105291
6. Badran, A.M., Utra, U., Yussof, N.S., Bashir, M.J.K.: Advancements in adsorption techniques for sustainable water purification: a focus on lead removal. Separations **10**(11), 565 (2023). https://doi.org/10.3390/separations10110565
7. Naha, A., Antony, S., Nath, S., et al.: A hypothetical model of multi-layered cost-effective wastewater treatment plant integrating microbial fuel cell and nanofiltration technology: a comprehensive review on wastewater treatment and sustainable remediation. Environ. Pollut. **15**(323), 121274 (2023). https://doi.org/10.1016/j.envpol.2023.121274
8. Melo, B.C., Paulino, F.A.A., Cardoso, V.A., et al.: Cellulose nanowhiskers improve the methylene blue adsorption capacity of chitosan-g-poly(acrylic acid) hydrogel. Carbohydr. Polym. **181**,(2018). https://doi.org/10.1016/j.carbpol.2017.10.079
9. da Costa, J.S., Fajardo, A.R.: Polypyrrole/stearic acid-coated Luffa cylindrica for enhanced removal of sodium diclofenac from water: batch and continuous adsorption studies. J. Clean. Prod. **389**,(2023). https://doi.org/10.1016/j.jclepro.2023.136084
10. Karimi, Z., Rahbar-Kelishami, A.: Preparation of highly efficient and eco-friendly alumina magnetic hybrid nanosorbent from red mud: excellent adsorption capacity towards nitrate. J. Mol. Liq. **368**,(2022). https://doi.org/10.1016/j.molliq.2022.120751

11. Iwuozor, K.O., Ighalo, J.O., Chizitere Emenike, E. et al.: Do adsorbent pore size and specific surface area affect the kinetics of methyl orange aqueous phase adsorption? J. Chem. Lett. **2**, 1048 (2021). https://doi.org/10.22034/jchemlett.2022.327407.1048
12. Gkika, D.A., Mitropoulos, A.C., Kyzas, G.Z.: Why reuse spent adsorbents? The latest challenges and limitations. Sci. Total Environ. (2022). https://doi.org/10.1016/j.scitotenv.2022.153612
13. Pettersson, M., Sjöberg, V., Storm, N., et al.: Application of coniferous bark as sorbent material for per- and polyfluoroalkyl substances—a case study in Sweden. Environ. Pollut. **363**, 125237 (2024). https://doi.org/10.1016/j.envpol.2024.125237
14. Yang, H., He, K., Lu, D., et al.: Removal of phosphate by aluminum-modified clay in a heavily polluted lake, Southwest China: effectiveness and ecological risks. Sci. Total. Environ. **705**, 135850 (2020). https://doi.org/10.1016/j.scitotenv.2019.135850
15. Zhang, J., Liu, D., Wang, X., et al.: Efficient uranium removal from radioactive wastewater using a MOF-based adsorbent functionalized with amidoxime. J. Mol. Liq. **419**, 126759 (2025). https://doi.org/10.1016/j.molliq.2024.126759
16. Hoang, A.T., Nižetić, S., Cheng, C.K., et al.: Heavy metal removal by biomass-derived carbon nanotubes as a greener environmental remediation: a comprehensive re-view. Chemosphere **287**, 131959 (2022). https://doi.org/10.1016/j.chemosphere.2021.131959
17. Cantoni, B., Turolla, A., Wellmitz, J., et al.: Perfluoroalkyl substances (PFAS) adsorption in drinking water by granular activated carbon: influence of activated carbon and PFAS characteristics. Sci. Total. Environ. **795**, 148821 (2021). https://doi.org/10.1016/j.scitotenv.2021.148821
18. Maity, A., Pu, H., Sui, X., et al.: Scalable graphene sensor array for real-time toxins monitoring in flowing water. Nat. Commun. **14**, 4184 (2023). https://doi.org/10.1038/s41467-023-39701-0
19. Rather, J.A., De Wael, K.: Fullerene-C60 sensor for ultrahigh sensitive detection of bisphenol-A and its treatment by green technology. Sens. Actuators B Chem. **176**, 110–117 (2013). https://doi.org/10.1016/j.snb.2012.08.081
20. Khan, R., Shukla, S., Kumar, M., et al.: Progress and obstacles in employing carbon quantum dots for sustainable wastewater treatment. Environ. Res. **261**, 119671 (2024). https://doi.org/10.1016/j.envres.2024.119671
21. Xu, Y., Lan, J., Wang, B., et al.: Simple fabrication of carbon quantum dots and acti-vated carbon from waste wolfberry stems for detection and adsorption of copper ion. RSC Adv. **13**, 21199–21210 (2023). https://doi.org/10.1039/d3ra04026g
22. Praseetha, P.K., Litany, R.I.J., Alharbi, H.M., et al.: Green synthesis of highly fluorescent carbon quantum dots from almond resin for advanced theranostics in biomedical applications. Sci. Rep. **14**, 24435 (2024). https://doi.org/10.1038/s41598-024-75333-0
23. Wei, T., Ni, H., Ren, X., et al.: Fabrication of nitrogen-doped carbon dots biomass composite hydrogel for adsorption of Cu (II) in wastewater or soil and DFT simulation for adsorption mechanism. Chemosphere **361**, 142432 (2024). https://doi.org/10.1016/j.chemosphere.2024.142432
24. Abdulsatar Esmail, L., Sanaan Jabbar, H.: Violuric acid carbon dots as a highly fluorescence probe for ultra-sensitive determination of Zn (II) in tomato paste. Food Chem. **413**, 135638 (2023). https://doi.org/10.1016/j.foodchem.2023.135638
25. Martindale, B.C.M., Hutton, G.A.M., Caputo, C.A., et al.: Solar hydrogen production using carbon quan-tum dots and a molecular nickel catalyst. J. Am. Chem. Soc. **137**(18), 6018–6025 (2015). https://doi.org/10.1021/jacs.5b01650
26. Wen, F., Li, P., Yan, H., et al.: Turmeric carbon quantum dots enhanced chitosan nanocomposite films based on photodynamic inactivation technology for antibacterial food packaging. Carbohydr. Polym. **311**, 120784 (2023). https://doi.org/10.1016/j.carbpol.2023.120784
27. Soares, V.R., Silva, E.C., Gomes, C.G., et al.: Fluorescent composite beads: an advanced tool for environmental monitoring and harmful pollutants removal from water. Chemosphere **350**, 140911 (2024). https://doi.org/10.1016/j.chemosphere.2023.140911

28. Jain, S., Dilbaghi, N., Kumar Singhal, N., et al.: Carbon quantum dots@metal–organic framework based catalytic nucleic acid fluorescent system for highly sensitive and selective detection of Pb2+ in aqueous solutions. Chem. Eng. J. **457**, 141375 (2023). https://doi.org/10.1016/j.cej.2023.141375
29. Dong, Y., Ma, J., Liu, C., et al.: Ordered mesoporous silica encapsulated carbon quantum dots and its application in Fe3+ detection. Ceram. Int. **46**, 11115–11123 (2020). https://doi.org/10.1016/j.ceramint.2020.01.131
30. Zhang, M., He, H., Huang, Y., et al.: Machine learning integrated high quantum yield blue light carbon dots for real-time and on-site detection of Cr(VI) in groundwater and drinking water. Sci. Total. Environ. **904**, 166822 (2023). https://doi.org/10.1016/j.scitotenv.2023.166822
31. Xu, X., Ray, R., Gu, Y., et al.: Electrophoretic analysis and purification of fluorescent single-walled carbon nanotube fragments. J. Am. Chem. Soc. **126**, 12736–12737 (2004). https://doi.org/10.1021/ja040082h
32. Kurian, M., Paul, A.: Recent trends in the use of green sources for carbon dot synthesis–A short review. Carbon Trends **3**, 100032 (2021). https://doi.org/10.1016/j.cartre.2021.100032
33. Li, M., Chen, T., Gooding, J.J., et al.: Review of carbon and graphene quantum dots for sensing. ACS Sens. **4**, 1732–1748 (2019). https://doi.org/10.1021/acssensors.9b00514
34. Rani, U.A., Ng, L.Y., Ng, C.Y., et al.: A review of carbon quantum dots and their applications in wastewater treatment. Adv. Colloid Interface Sci. **278**, 102124 (2020). https://doi.org/10.1016/j.cis.2020.102124
35. Sun, Y.P., Zhou, B., Lin, Y., et al.: Quantum-sized carbon dots for bright and colorful photoluminescence. J. Am. Chem. Soc. **128**, 7756–7757 (2006). https://doi.org/10.1021/ja062677d
36. Shinde, D.B., Pillai, V.K.: Electrochemical preparation of luminescent graphene quantum dots from multiwalled carbon nanotubes. Chem. A Eur. J. **18**, 12522–12528 (2012). https://doi.org/10.1002/chem.201201043
37. Dong, Y., Zhou, N., Lin, X., et al.: Extraction of electro-chemiluminescent oxidized carbon quantum dots from activated carbon. Chem. Mater. **22**, 5895–5899 (2010). https://doi.org/10.1021/cm1018844
38. Huang, H., Yang, S., Li, Q., et al.: Electrochemical cut-ting in weak aqueous electrolytes: the strategy for efficient and controllable preparation of graphene quantum dots. Langmuir **34**(1), 250–258 (2018). https://doi.org/10.1021/acs.langmuir.7b03425
39. Kaczmarek, A., Hoffman, J., Morgiel, J., et al.: Luminescent carbon dots synthesized by the laser ablation of graphite in polyethylenimine and ethylenediamine. Materials **14**(4), 729 (2021). https://doi.org/10.3390/ma14040729
40. Biazar, N., Poursalehi, R., Delavari, H.: Optical and structural properties of carbon dots/$TiO_2$ nanostructures prepared via DC arc discharge in liquid. AIP Conf. Proc. **1920**, 020033 (2018). https://doi.org/10.1063/1.5018965
41. Wang, Y., Hu, A.: Carbon quantum dots: synthesis, properties and applications. J. Mater. Chem. C Mater. **2**, 6921 (2014). https://doi.org/10.1039/C4TC00988F
42. Tan, X., Li, Y., Li, X., et al.: Electrochemical synthesis of small-sized red fluorescent graphene quantum dots as a bioimaging platform. Chem. Commun. **51**, 2544–2546 (2015). https://doi.org/10.1039/C4CC09332A
43. Desmond, L.J., Phan, A.N., Gentile, P.: Critical overview on the green synthesis of carbon quantum dots and their application for cancer therapy. Environ. Sci. Nano **8**, 848–862 (2021). https://doi.org/10.1039/D1EN00017A
44. Chahal, S., Yousefi, N., Tufenkji, N.: Green synthesis of high quantum yield carbon dots from phenylalanine and citric acid: role of stoichiometry and nitrogen doping. ACS Sustain. Chem. Eng. **8**, 5566–5575 (2020). https://doi.org/10.1021/acssuschemeng.9b07463
45. Ezati, P., Rhim, J.W., Molaei, R., et al.: Preparation and characterization of B, S, and N-doped glucose carbon dots: antibacterial, antifungal, and antioxidant activity. Sustain. Mater. Technol. **32**, e00397 (2022). https://doi.org/10.1016/j.susmat.2022.e00397
46. Wu, X., Tian, F., Wang, W., et al.: Fabrication of highly fluorescent graphene quantum dots using L-glutamic acid for in vitro/in vivo imaging and sensing. J. Mater. Chem. C **1**, 4676 (2013). https://doi.org/10.1039/c3tc30820k

47. Aslan, M., Eskalen, H., Kavgaci, M.: Carbon quantum dot (CQD) nanoparticles synthesized by sucrose and urea: application as reinforcement effect on Al–Mg–Cu–Zn composite. Russ. J. Gen. Chem. **93**, 2152–2160 (2023). https://doi.org/10.1134/S1070363223080236
48. Zhu, L., Shen, D., Wu, C., et al.: State-of-the-art on the preparation, modification, and application of biomass-derived carbon quantum dots. Ind. Eng. Chem. Res. **59**, 22017–22039 (2020). https://doi.org/10.1021/acs.iecr.0c04760
49. Devi, P., Kaur, G., Thakur, A., et al.: Waste derivitized blue luminescent carbon quantum dots for selenite sensing in water. Talanta **170**, 49–55 (2017). https://doi.org/10.1016/j.talanta.2017.03.069
50. Dager, A., Uchida, T., Maekawa, T., et al.: Synthesis and characterization of mono-disperse carbon quantum dots from fennel seeds: photoluminescence analysis using machine learning. Sci. Rep. **9**, 14004 (2019). https://doi.org/10.1038/s41598-019-50397-5
51. Campbell, E., Hasan, Md.T., Gonzalez Rodriguez, R., et al.: Doped graphene quantum dots for intracellular multicolor imaging and cancer detection. ACS Biomater. Sci. Eng. **5**, 4671–4682 (2019). https://doi.org/10.1021/acsbiomaterials.9b00603
52. Ye, Z., Li, G., Lei, J., et al.: One-step and one-precursor hydrothermal synthesis of carbon dots with superior antibacterial activity. ACS Appl. Bio Mater. **3**, 7095–7102 (2020). https://doi.org/10.1021/acsabm.0c00923
53. Lu, Z., Su, T., Feng, Y., et al.: Potential application of nitrogen-doped carbon quantum dots synthesized by a solvothermal method for detecting silver ions in food packaging. Int. J. Environ. Res. Public Health **16**, 2518 (2019). https://doi.org/10.3390/ijerph16142518
54. Rong, M., Feng, Y., Wang, Y., et al.: One-pot solid phase pyrolysis synthesis of nitrogen-doped carbon dots for Fe3+ sensing and bioimaging. Sens. Actuators B Chem. **245**, 868–874 (2017). https://doi.org/10.1016/j.snb.2017.02.014
55. Xue, M., Zou, M., Zhao, J., et al.: Green preparation of fluorescent carbon dots from lychee seeds and their application for the selective detection of methylene blue and imaging in living cells. J. Mater. Chem. B **3**, 6783–6789 (2015). https://doi.org/10.1039/C5TB01073J
56. Yadav, P.K., Chandra, S., Kumar, V., et al.: Carbon quantum dots: synthesis, structure, properties, and cata-lytic applications for organic synthesis. Catalysts **13**, 422 (2023). https://doi.org/10.3390/catal13020422
57. Kou, X., Jiang, S., Park, S.J., et al.: A review: recent advances in preparations and applications of heteroatom-doped carbon quantum dots. Dalton Trans. **49**, 6915–6938 (2020). https://doi.org/10.1039/d0dt01004a
58. Manikandan, V., Lee, N.Y.: Green synthesis of carbon quantum dots and their environmental applications. Environ. Res. **212**, 113283 (2022). https://doi.org/10.1016/j.envres.2022.113283
59. Baweja, H., Jeet, K.: Economical and green synthesis of graphene and carbon quantum dots from agricultural waste. Mater. Res. Express **6**, 0850g8 (2019). https://doi.org/10.1088/2053-1591/ab28e5
60. Haryadi, H., Purnama, M.R.W., Wibowo, A.: C dots de-rived from waste of biomass and their photocatalytic activities. Indones J. Chem. **18**, 594 (2018). https://doi.org/10.22146/ijc.26652
61. Xu, Y., Fan, Y., Zhang, L., et al.: A novel enhanced fluorescence method based on multifunctional carbon dots for specific detection of Hg2+ in complex samples. Spec-trochim Acta A Mol. Biomol. Spectrosc. **220**, 117109 (2019). https://doi.org/10.1016/j.saa.2019.05.014
62. Pramanik, S., Chatterjee, S., Suresh Kumar, G., et al.: Egg-shell derived carbon dots for base pair selective DNA binding and recognition. Phys. Chem. Chem. Phys. **20**, 20476–20488 (2018). https://doi.org/10.1039/C8CP02872A
63. Wang, L., Li, W., Wu, B., et al.: Facile synthesis of fluorescent graphene quantum dots from coffee grounds for bioimaging and sensing. Chem. Eng. J. **300**, 75–82 (2016). https://doi.org/10.1016/j.cej.2016.04.123
64. de Oliveira, B.P., da Silva Abreu, F.O.M.: Carbon quantum dots synthesis from waste and by-products: perspectives and challenges. Mater. Lett. **282**, 128764 (2021). https://doi.org/10.1016/j.matlet.2020.128764
65. Wang, Y., Sun, J., He, B., et al.: Synthesis and modification of biomass derived carbon dots in ionic liquids and their application: a mini review. Green Chem. Eng. **1**, 94–108 (2020). https://doi.org/10.1016/j.gce.2020.09.010

66. Huang, Z., Lin, F., Hu, M., et al.: Carbon dots with tun-able emission, controllable size and their application for sensing hypochlorous acid. J. Lumin. **151**, 100–105 (2014). https://doi.org/10.1016/j.jlumin.2014.02.013
67. Shen, J., Shang, S., Chen, X., et al.: Facile synthesis of fluorescence carbon dots from sweet potato for Fe3+ sensing and cell imaging. Mater. Sci. Eng. C **76**, 856–864 (2017). https://doi.org/10.1016/j.msec.2017.03.178
68. Jiang, K., Sun, S., Zhang, L., et al.: Red, green, and blue luminescence by carbon dots: full-color emission tuning and multicolor cellular imaging. Angew. Chem. Int. Ed. **54**, 5360–5363 (2015). https://doi.org/10.1002/anie.201501193
69. Kim, S., Hwang, S.W., Kim, M.-K., et al.: Anomalous behaviors of visible luminescence from graphene quan-tum dots: interplay between size and shape. ACS Nano. **6**, 8203–8208 (2012). https://doi.org/10.1021/nn302878r
70. Rani, U.A., Ng, L.Y., Ng, C.Y., et al.: Development of green photocatalyst using empty fruit bunches from Elae-is guineensis for methylene blue degradation. J. Mol. Struct. **1282**, 135218 (2023). https://doi.org/10.1016/j.molstruc.2023.135218
71. Tohamy, H.S., El Sakhawy, M., Kamel, S.: Eco-friendly synthesis of carbon quantum dots as an effec-tive adsorbent. J. Fluoresc. **33**, 423–435 (2023). https://doi.org/10.1007/s10895-022-03085-z
72. Kalita, H., Mohapatra, J., Pradhan, L., et al.: Efficient synthesis of rice based graphene quantum dots and their fluorescent properties. RSC Adv. **6**, 23518–23524 (2016). https://doi.org/10.1039/C5RA25706A
73. Pan, J., Zheng, Z., Yang, J., et al.: A novel and sensitive fluorescence sensor for glutathione detection by controlling the surface passivation degree of carbon quantum dots. Talanta **166**, 1–7 (2017). https://doi.org/10.1016/j.talanta.2017.01.033
74. Liu, Y., Yong, C., Tong, B., et al.: Modification of car-bon dots derived from biomass by exogenous nitrogen doping: action mechanism and difference analysis. Opt. Mater (Amst) **134**, 113144 (2022). https://doi.org/10.1016/j.optmat.2022.113144
75. Pan, J., Sheng, Y., Zhang, J., et al.: Preparation of carbon quantum dots/$TiO_2$ nanotubes composites and their visible light catalytic applications. J. Mater. Chem. A **2**, 18082–18086 (2014). https://doi.org/10.1039/C4TA03528C
76. Yu, Y., Zhu, Z., Liu, Z., et al.: Construction of the bio-mass carbon quantum dots modified heterojunction Bi2WO6/Cu2O photocatalysis for enhancing light utilization and mechanism insight. J. Taiwan Inst. Chem. Eng. **102**, 197–201 (2019). https://doi.org/10.1016/j.jtice.2019.05.029
77. Zhang, H., Huang, H., Ming, H., et al.: Carbon quantum dots/$Ag_3PO_4$ complex photocatalysts with enhanced photocatalytic activity and stability under visible light. J. Mater. Chem. **22**, 10501 (2012). https://doi.org/10.1039/c2jm30703k
78. Hu, Z., Jiao, X.-Y., Xu, L.: The N, S co-doped carbon dots with excellent luminescent proper-ties from green tea leaf residue and its sensing of gefitinib. Microchem. J. **154**, 104588 (2020). https://doi.org/10.1016/j.microc.2019.104588
79. Long, C., Jiang, Z., Shangguan, J., et al.: Applications of carbon dots in environmental pollu-tion control: a review. Chem. Eng. J. **406**, 126848 (2021). https://doi.org/10.1016/j.cej.2020.126848
80. Zhao, S., Song, X., Chai, X., et al.: Green production of fluorescent carbon quantum dots based on pine wood and its application in the detection of Fe3+. J. Clean. Prod. **263**, 121561 (2020). https://doi.org/10.1016/j.jclepro.2020.121561
81. Atchudan, R., Jebakumar Immanuel Edison, T.N., Shanmu-gam, M., et al.: Sustainable synthesis of carbon quantum dots from banana peel waste using hydrother-mal process for in vivo bioimaging. Physica E Low Di-mens Syst. Nanostruct. **126**, 114417 (2021). https://doi.org/10.1016/j.physe.2020.114417
82. Han, C., Li, Y.-H., Qi, M.-Y., et al.: Surface/interface engineering of carbon-based materials for constructing multidimensional functional hybrids. Sol RRL **4**, 1900577 (2020). https://doi.org/10.1002/solr.201900577

83. Tadesse, A., RamaDevi, D., Hagos, M., et al.: Synthesis of nitrogen doped carbon quantum dots/magnetite nano-composites for efficient removal of methyl blue dye pollutant from contaminated water. RSC Adv. **8**, 8528–8536 (2018). https://doi.org/10.1039/C8RA00158H
84. Wang, C., Bi, L., Liu, J., et al.: Microalgae-derived carbon quantum dots mediated formation of metal sulfide nano-adsorbents with exceptional cadmium removal performance. J. Colloid Interface Sci. **629**, 994–1002 (2023). https://doi.org/10.1016/j.jcis.2022.08.188
85. Yang, M., Yan, Y., He, F., et al.: Green synthesis of fluorescent carbon dots from discarded cigarette butts as an effective fluorescence probe for sensing iron ions. Mater. Lett. **357**, 135613 (2024). https://doi.org/10.1016/j.matlet.2023.135613
86. Raji, K., Ramanan, V., Ramamurthy, P.: Facile and green synthesis of highly fluorescent nitrogen-doped carbon dots from jackfruit seeds and its applications towards the fluorimetric detection of Au 3+ ions in aqueous medium and in in vitro multicolor cell imaging. New J. Chem. **43**, 11710–11719 (2019). https://doi.org/10.1039/C9NJ02590A
87. Eskalen, H., Uruş, S., Cömertpay, S., et al.: Microwave-assisted ultra-fast synthesis of carbon quantum dots from linter: fluorescence cancer imaging and human cell growth inhibition properties. Ind. Crops Prod. **147**, 112209 (2020). https://doi.org/10.1016/j.indcrop.2020.112209
88. D.P., Singh, L., Thakur, A. et al.: Green synthesis of glowing carbon dots from Carica papaya waste pulp and their application as a labelfreechemo probe for chromium detection in water. Sens. Actuators B Chem. **283**, 363–372 (2019). https://doi.org/10.1016/j.snb.2018.12.027
89. Jiao, X.Y., Li, L., Qin, S., et al.: The synthesis of fluorescent carbon dots from mango peel and their multiple applications. Colloids Surf A Physicochem. Eng. Asp **577**, 306–314 (2019). https://doi.org/10.1016/j.colsurfa.2019.05.073
90. Desai, M.L., Jha, S., Basu, H., et al.: Acid oxidation of muskmelon fruit for the fabrication of carbon dots with specific emission colors for recognition of Hg 2+ ions and cell imaging. ACS Omega **4**, 19332–19340 (2019). https://doi.org/10.1021/acsomega.9b02730
91. Sheriff, I., Debela, S.A., Mans-Davies, A.: The listing of new persistent organic pollutants in the stockholm convention: its burden on developing countries. Environ. Sci. Policy **130**, 9–15 (2022). https://doi.org/10.1016/j.envsci.2022.01.005
92. Ding, R., Chen, Y., Wang, Q., et al.: Recent advances in quantum dots-based biosensors for antibiotics detection. J. Pharm. Anal. **12**(3), 55–364 (2022). https://doi.org/10.1016/j.jpha.2021.08.002
93. Sobiech, M., Luliński, P., Wieczorek, P.P., et al.: Quantum and carbon dots conjugated molecularly im-printed polymers as advanced nanomaterials for selective recognition of analytes in environmental, food and bio-medical applications. TrAC Trends. Anal. Chem. **142**, 116306 (2021). https://doi.org/10.1016/j.trac.2021.116306
94. Mirzaei, A., Kordrostami, Z., Shahbaz, M., et al.: Resistive-based gas sensors using quantum dots: a review. Sensors **22**, 4369 (2022). https://doi.org/10.3390/s22124369
95. Tang, J., Ma, X., Liu, J. et al.: Simultaneous determination of hydroquinone and catechol using carbon glass electrode modified with graphene quantum dots. Int. J. Electrochem. Sci. **13**, 11250–11262 (2018). https://doi.org/10.20964/2018.11.52
96. Wagner, A.M., Knipe, J.M., Orive, G., et al.: Quan-tum dots in biomedical applications. Acta Biomater. **94**, 44–63 (2019). https://doi.org/10.1016/j.actbio.2019.05.022
97. Silva, E.C., Pina, J., Pereira, R.F.P., et al.: Synthesis and characterization of fluorescent poly(α-cyclodextrin)/carbon quantum dots composite for efficient removal and detection of toluene and xylene from aqueous media. Environ. Pollut. **347**, 123778 (2024). https://doi.org/10.1016/j.envpol.2024.123778
98. Jiang, T., Liu, X., Sun, J.: UV-enhanced $NO_2$ sensor using ZnO quantum dots sensitized $SnO_2$ porous nanowires. Nanotechnology **33**, 185501 (2022). https://doi.org/10.1088/1361-6528/ac49c1
99. Jin, Y., Liu, K., Li, G., et al.: In situ reduction triggers the highly sensitive detection of pesticide by classic gold nanoparticle and quantum dots nanocomposite. Anal. Chim. Acta **1172**, 338679 (2021). https://doi.org/10.1016/j.aca.2021.338679

100. Randriamahazaka, H., Asaka, K.: Electromechanical analysis by means of complex capacitance of bucky-gel actuators based on single-walled carbon nanotubes and an ionic liquid. J. Phys. Chem. C **114**, 17982–17988 (2010). https://doi.org/10.1021/jp106232s
101. Pasandideh-Nadamani, M., Omrani, A., Sadeghi-Maleki, M.R., et al.: Application of CdS quantum dots modified carbon paste electrode for monitoring the process of acetaminophen preparation. Anal. Biochem. **502**, 36–42 (2016). https://doi.org/10.1016/j.ab.2016.02.015
102. Algarra, M., González-Calabuig, A., Radotić, K., et al.: Enhanced electrochemical response of carbon quantum dot modified electrodes. Talanta **178**, 679–685 (2018). https://doi.org/10.1016/j.talanta.2017.09.082
103. Zhao, M.X., Zhu, B.J.: The research and applications of quantum dots as nano-carriers for targeted drug delivery and cancer therapy. Nanoscale Res. Lett. **11**, 207 (2016). https://doi.org/10.1186/s11671-016-1394-9
104. Bimberg, D., Ledentsov, N.N., Grundmann, M., et al.: Luminescence properties of semiconductor quantum dots. J. Lumin. **72–74**, 34–37 (1997). https://doi.org/10.1016/S0022-2313(97)00084-7
105. Ahirwar, R.C., Mehra, S., Reddy, S.M., et al.: Progression of quantum dots confined polymeric systems for sen-sorics. Polymers (Basel) **15**, 405 (2023). https://doi.org/10.3390/polym15020405
106. Xing, X., Wang, D., Chen, Z., et al.: ZnTe quantum dots as fluorescence sensors for the detection of iron ions. J. Mater. Sci. Mater. Electron. **29**, 14192–14199 (2018). https://doi.org/10.1007/s10854-018-9552-8
107. Biranje, A., Azmi, N., Tiwari, A., et al.: Quantum dots based fluorescent probe for the selective detection of heavy metal ions. J. Fluoresc. **31**, 1241–1250 (2021). https://doi.org/10.1007/s10895-021-02755-8
108. Rabeie, B., Mahmoodi, N.M., Mahkam, M.: Morpho-logical diversity effect of graphene quantum dot/MIL88A(Fe) composites on dye and pharmaceuticals (tetracycline and doxycycline) removal. J. Environ. Chem. Eng. **10**, 108321 (2022). https://doi.org/10.1016/j.jece.2022.108321
109. Zhao, F., Repo, E., Yin, D., et al.: One-pot synthesis of trifunctional chitosan-EDTA-β-cyclodextrin polymer for simultaneous removal of metals and organic micropollutants. Sci. Rep. **7**, 15811 (2017). https://doi.org/10.1038/s41598-017-16222-7
110. Liu, S., Wu, Z., Nian, N., et al.: The speciation of heavy metal chromium in water environment by car-bon quantum dots system. Water Air Soil Pollut. **235**, 449 (2024). https://doi.org/10.1007/s11270-024-07222-1
111. Liu, Y., Chen, J., Xu, Z., et al.: Detection of multiple metal ions in water with a fluorescence sensor based on carbon quantum dots assisted by stepwise prediction and machine learning. Environ. Chem. Lett. **20**, 3415–3420 (2022). https://doi.org/10.1007/s10311-022-01475-0
112. Tshangana, C.S., Nhlengethwa, S.T., Glass, S., et al.: Technology status to treat PFAS-contaminated water and limiting factors for their effective full-scale application. NPJ. Clean Water **8**, 41 (2025). https://doi.org/10.1038/s41545-025-00457-3
113. Titchou, F.E., Zazou, H., Afanga, H. et al.: Removal of organic pollutants from wastewater by advanced oxidation processes and its combination with membrane processes. Chem. Eng. Process. Process Intensif. **169**, 108631 (2021). https://doi.org/10.1016/j.cep.2021.108631
114. Hassan Ahmed, H.E., Soylak, M.: Exploring the potential of carbon quantum dots (CQDs) as an advanced nanomaterial for effective sensing and extraction of toxic pollutants. TrAC Trends Anal. Chem. **180**, 117939 (2024). https://doi.org/10.1016/j.trac.2024.117939
115. Parambil, A.M., Rajan, S., Huang, P.-C., et al.: Carbon and graphene quantum dots based architectonics for efficient aqueous decontamination by adsorption chromatography technique—Current state and prospects. Environ. Res. **251**, 118541 (2024). https://doi.org/10.1016/j.envres.2024.118541
116. Deng, Y., Ok, Y.S., Mohan, D., et al.: Carbamazepine removal from water by carbon dot-modified magnetic carbon nanotubes. Environ. Res. **169**, 434–444 (2019). https://doi.org/10.1016/j.envres.2018.11.035

117. Wang, K., Tan, L., Zhang, Y., et al.: A molecular im-printed fluorescence sensor based on carbon quantum dots for selective detection of 4-nitrophenol in aqueous environments. Mar. Pollut. Bull. **187**, 114587 (2023). https://doi.org/10.1016/j.marpolbul.2023.114587
118. Ochs, C., Garrison, K., Saxena, P., et al.: Contamination of aquatic ecosystems by persistent organic pollutants (POPs) originating from landfills in Canada and the Unit-ed States: a rapid scoping review. Sci. Total. Environ. **924**, 171490 (2024). https://doi.org/10.1016/j.scitotenv.2024.171490
119. Kalita, I., Kamilaris, A., Havinga, P., et al.: Assessing the health impact of disinfection byproducts in drinking water. ACS ES T Water **4**(4), 1564–1578 (2024). https://doi.org/10.1021/acsestwater.3c00664
120. Zhou, W., Li, M., Achal, V.: A comprehensive review on environmental and human health impacts of chemical pesticide usage. Emerg. Contam. **11**, 100410 (2025). https://doi.org/10.1016/j.emcon.2024.100410
121. Gangar, T., Patra, S.: Antibiotic persistence and its impact on the environment. 3 Biotech **13**(12), 401 (2023). https://doi.org/10.1007/s13205-023-03806-6
122. Roque, M.I., Gomes, J., Reva, I., et al.: An Opinion on the removal of disinfection byproducts from drinking water. Water (Switzerland) **15**(9), 1724 (2023). https://doi.org/10.3390/w15091724
123. Dhore, R., Murthy, G.S.: Per/polyfluoroalkyl sub-stances production, applications and environmental impacts. Bioresour. Technol. **341**, 125808 (2021). https://doi.org/10.1016/j.biortech.2021.125808
124. eBioMedicine: Forever chemicals: the persistent effects of perfluoroalkyl and polyfluoroalkyl substances on human health. EBioMedicine **95**, 104806 (2023). https://doi.org/10.1016/j.ebiom.2023.104806
125. Gomes, M.P.: The convergence of antibiotic contamination, resistance, and climate dynamics in fresh-water ecosystems. Water (Basel) **16**, 2606 (2024). https://doi.org/10.3390/w16182606
126. Ajekiigbe, V.O., Agbo, C.E., Ogieuhi, I.J., et al.: The increasing burden of global environmental threats: role of antibiotic pollution from pharmaceutical wastes in the rise of antibiotic resistance. Discov. Public Health **22**, 120 (2025). https://doi.org/10.1186/s12982-025-00506-9
127. Jones, K.C.: Persistent organic pollutants (POPs) and related chemicals in the global environment: some personal reflections. Environ. Sci. Technol. **55**, 9400–9412 (2021). https://doi.org/10.1021/acs.est.0c08093
128. Molaei, M.J.: Principles, mechanisms, and application of carbon quantum dots in sensors: a review. Anal. Methods **12**, 1266–1287 (2020). https://doi.org/10.1039/C9AY02696G
129. Rafiq, K., Sadia, I., Abid, M.Z., et al.: Scientific insights into the quantum dots (QDs)-based electrochemical sensors for state-of-the-art applications. ACS Biomater. Sci. Eng. **10**, 7268–7313 (2024). https://doi.org/10.1021/acsbiomaterials.4c01256
130. Zhou, J., Zhang, Y., Ding, J., et al.: A More efficient method for preparing a MIP-CQDs/ZnO1-x photodegradant with highly selective adsorption and photocatalytic properties. ACS Appl. Mater. Interfaces **16**(2), 2365–2377 (2024). https://doi.org/10.1021/acsami.3c16135
131. Utzeri, G., Murtinho, D., Valente, A.J.M.: Introduction to cyclodextrin-based nanosponges. In: Nanosponges for Environmental Remediation (2023). https://doi.org/10.1007/978-3-031-41077-2
132. Silva, E.C., Gomes, C.G., Pina, J., et al.: Carbon quantum dots-containing poly(β-cyclodextrin) for simultaneous removal and detection of metal ions from water. Carbohydr. Polym. **323**, 121464 (2024). https://doi.org/10.1016/j.carbpol.2023.121464
133. Cong, H., Ni, X.L., Xiao, X., et al.: Synthesis and separation of cucurbit[n]urils and their derivatives. Org. Biomol. Chem. **14**, 4335–4364 (2016). https://doi.org/10.1039/C6OB00268D
134. Havel, V., Svec, J., Wimmerova, M., et al.: Bam-bus[n]urils: a new family of macrocyclic anion receptors. Org. Lett. **13**(15), 4000–4003 (2011). https://doi.org/10.1021/ol201515c
135. Yuan, F., Yuan, T., Sui, L., et al.: Engineering triangular carbon quantum dots with unprecedented narrow bandwidth emission for multicolored LEDs. Nat. Commun. **9**, 2249 (2018). https://doi.org/10.1038/s41467-018-04635-5

# Exploring Carbon Nanotubes for Removing Organic and Inorganic Pollutants from Wastewater

**Cristiane Ferraz de Azevedo, Alice Gonçalves Osorio, and Fernando Machado Machado**

**Abstract** Carbon nanotubes (CNTs) have been extensively used to extract pollutants from water and wastewater. These advanced materials exhibit exceptional textural properties, notably their remarkable porosity, high specific surface area, hollow and layered structures, and numerous external and internal adsorption sites. Furthermore, their intrinsic properties can be enhanced through covalent or noncovalent modifications by adding specific functional groups to their surface, thereby improving their selectivity toward particular classes of pollutants. This chapter presents the distinct synthesis routes and modification techniques for CNTs, along with insights into the key characteristics to consider when using them as nanoadsorbents for water and wastewater purification. We also bring several examples of systems where different CNTs structures are employed as adsorbents to remove inorganic and organic contaminants. Finally, we discuss the primary challenges of utilizing CNTs as adsorbents in water and wastewater treatment applications.

**Keywords** Textural properties · Nanoadsorbents · Interaction mechanisms · Synthesis · Functionalization · Nanostructures

## Abbreviations

| | |
|---|---|
| 1D | One-dimensional |
| 3D | Bulk |
| AC | Activated Carbon |

C. Ferraz de Azevedo (✉) · A. Gonçalves Osorio · F. Machado Machado
Materials Research Laboratory, Technology Development Center, Federal University of Pelotas (UFPel), Pelotas, Brazil
e-mail: cazevedo.iqg@ufpel.edu.br

A. Gonçalves Osorio
e-mail: alice.osorio@ufpel.edu.br

F. Machado Machado
e-mail: fernando.machado@ufpel.edu.br

© The Author(s), under exclusive license to Springer Nature Switzerland AG 2026 
C. P. Bergmann and F. Machado Machado (eds.), *Carbon Nanomaterials as Adsorbents for Environmental and Biological Applications*, Carbon Nanostructures,
https://doi.org/10.1007/978-3-032-08669-3_3

| | |
|---|---|
| CCVD | Catalytic chemical vapor deposition |
| CNTs | Carbon Nanotubes |
| CVD | Chemical Vapor Deposition |
| DC | Direct current |
| DWCNTs | Double-Walled Carbon Nanotubes |
| FCCVD | Catalyst chemical vapor deposition |
| HiPCO | High-pressure carbon monoxide chemical vapor deposition |
| MWCNT | Multi-Walled Carbon Nanotube |
| PLV | Pulsed laser vaporization |
| SWCNT | Single-Walled Carbon Nanotube |

## 1 Introduction

CNTs have been widely utilized as effective nanoadsorbents for removing both inorganic and organic pollutants from aqueous systems. This class of carbon allotropes exists in two primary structural configurations: single-walled nanotubes (SWNTs) [1, 2] and multi-walled nanotubes (MWNTs) [3], as illustrated in Fig. 1. One or more graphene sheets can represent these 1D systems rolled up in a concentric form [4, 5]. Such cylindrical structures can range in diameter from a few nanometers to micrometers in length [4]. These morphological differences enable distinct interactions between the carbonaceous nanoadsorbents and adsorbate molecules [4, 5].

CNTs are highly effective adsorbents for removing contaminants, toxins, and pollutants due to their large specific surface area and porous structure [6–9] high surface-to-volume ratio, hydrophobic properties [7], interconnected voids with layered architectures, an easy process for chemical activation and functionalization [6], and high chemical stability [8]. While they exhibit a strong affinity for organic

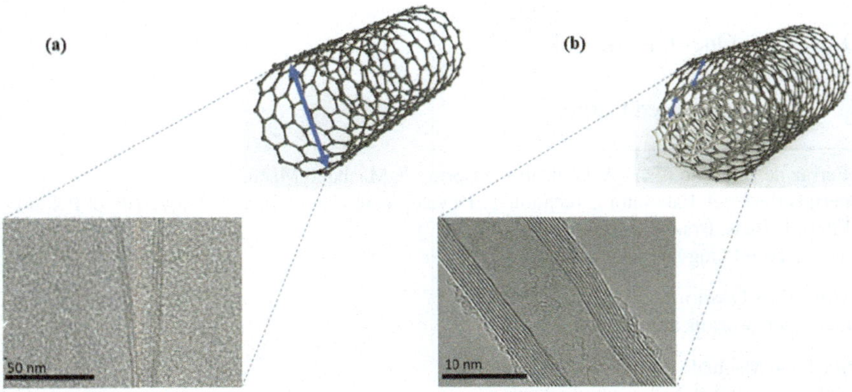


**Fig. 1** Layered structure of **a** single-walled and **b** multi-walled carbon nanotubes

contaminants, their practical applications are limited by their ultrafine particle size and tendency to aggregate. Therefore, stabilization processes that aim to prevent the formation of aggregates can be applied to optimize the application, preserving the surface characteristics [10]. Consequently, CNT-based water purification technologies hold significant potential for addressing water scarcity and improving access to clean water [9]. Current technologies utilizing CNTs for water and wastewater treatment primarily employ these materials in membranes or filters, adsorbents, electrodes, and catalysts to degrade pollutants [10].

The remarkable versatility of these carbon nanostructures stems from their exceptional properties, including high strength, unique electronic characteristics, and outstanding thermal conductivity. These attributes make them unparalleled candidates for a diverse range of applications. Metallic CNTs can transmit electric current ballistically without radiating heat, enabling the development of nanoscale electronic devices with distinctive properties. The chiral configuration of CNTs causes the electronic behavior of this nanostructure to vary from semiconductor to metallic based on the orientation of the hexagons along the CNT axis. These unique properties have been explored and better understood through various studies in literature, leading to a more detailed understanding of CNTs. This understanding has enabled the application of these materials in various devices, including supercapacitors, sensors, and water treatment systems [9].

## 2 Synthesis of CNTs

Carbon nanotubes are generally synthesized via bottom-up approaches, in which carbon atoms or clusters reorganize and self-assemble into tubular nanostructures. Methods such as arc discharge, laser ablation, and chemical vapor deposition are therefore classified as bottom-up synthesis routes, despite the use of macroscopic carbon sources.

Since multiple approaches can be applied to CNTs production, those yielding materials with fewer structural defects are preferred [9]. The number of publications addressing the synthesis of these materials has increased in recent years; however, several challenges still need to be addressed [11], such as:

(a) Developing cost-effective methods to produce high-quality nanotubes, including SWNTs, on a large scale; (b) Achieving selective synthesis that allows precise control over the structural and electronic characteristics of CNTs. High-quality nanotubes are particularly valuable for scientific and technological applications, as they should exhibit minimal structural and chemical defects along their length, typically ranging from 1 to 10 microns; (c) Ensuring proper organization of CNTs, which includes controlling their positioning and alignment on a given substrate; (d) Gaining a deeper understanding of the mechanisms underlying CNTs formation. The exact process by which CNTs grow remains unclear, and evidence suggests that multiple mechanisms may contribute to their synthesis.

For both research and industrial-scale applications, the ability to manufacture high-quality nanotubes is essential. A major limitation of currently available CNTs is their heterogeneity. The synthesis process is influenced by numerous factors, including the type of hydrocarbon used, catalyst selection, temperature, pressure, gas flow rate, deposition duration, and reactor design [11].

## *2.1 Arc Discharge Method*

CNTs were first synthesized using this technique in 1991 by Iijima [3], making it one of the most common methods for CNTs production [12]. This synthesis enables the formation of both SWNTs and MWNTs of relatively high quality [11], making it the preferred method for MWCNT production [13]. It relies on a high-voltage electric arc applied between graphite electrodes in an inert gas environment, such as helium or argon, at low pressure. The process involves heating the chamber to approximately 3727 °C while applying direct current (DC), resulting in the deposition of CNTs on the cathode [9, 14]. Key factors influencing synthesis include arc current, gas pressure, and plasma stability [9, 11, 12, 14].

The type of CNTs produced, SWCNTs or MWCNTs, depends on the inert gas used and the presence of metal catalysts such as Ni, Fe, Co, and/or Y [9]. SWCNTs are often synthesized using Ni-Y composites. Additionally, metals such as Gd, Ag, Pt, and Pd, as well as alloy mixtures like Co–Pt, Ni-Y, and Fe–Ni, can also be employed [14].

The nanotubes generated by this method usually present short structures, typically less than 1 mm in length [11, 12]. This approach is relatively straightforward to implement [12]. However, controlling their chirality remains challenging, affecting the uniformity of the produced nanotubes [11].

In this synthesis process, CNTs purification is necessary due to the presence of catalyst residues [9, 11] and the formation of byproducts [11]. Techniques such as oxidation, filtration, and acid treatment help remove impurities. Notably, MWCNTs can be synthesized using arc discharge in liquid nitrogen, offering lower energy costs [9].

The yield of this synthesis method varies depending on experimental conditions, with reported values ranging from 30% [12] to 75% [14]. A direct current of 50–100 A promotes CNTs growth with fewer structural defects compared to other synthesis methods [11, 14].

## *2.2 Laser Ablation*

Laser ablation, also known as Pulsed laser vaporization (PLV), is a bottom-up technique that vaporizes a carbon target in a controlled environment to produce CNTs [9]. This method, developed in the 1990s, was first employed to synthesize SWCNTs

[11, 12, 15]. Similar to arc discharge [12], this process vaporizes a graphite target containing a catalyst inside a quartz tube furnace at 1200 °C, while an inert gas (helium or argon) maintains a pressure of 500 Torr [11].

Introducing small amounts of transition metals, such as Ni and Co, into graphite significantly alters the resulting products, promoting the formation of SWCNTs over MWCNTs. The yield of SWCNTs depends on the type of metal catalyst used and tends to increase with furnace temperature [11, 12]. Growth ceases when catalyst particles become coated with carbon, and the resulting SWCNTs aggregate into bundles due to weak van der Waals forces [11].

Furthermore, laser ablation is influenced by several factors, including the structural and chemical composition of the target material, as well as laser parameters such as peak power, energy, wavelength, repetition rate, and buffer gas pressure. SWCNTs produced by this method exhibit high purity and quality [14]. However, the process generates subproducts such as amorphous carbon and metal nanoparticles, requiring extensive purification [15].

Despite producing high-quality nanotubes, laser ablation relies on high-purity graphite and powerful lasers [9], making it economically unfeasible for large-scale production due to its relatively low yield when compared to other methods [9, 13]. One advantage of the laser ablation method is its ability to produce high-quality SWCNTs with minimal defects and good crystallinity. Although its yield is lower than that obtained from chemical vapor deposition (CVD), it is relatively high compared to arc discharge [15]. The process enables the synthesis of nanotubes with diameters ranging from 5 to 20 nm and lengths extending from hundreds of nanometers to several micrometers [13].

### *2.3 Chemical Vapor Deposition (CVD)*

Introduced in 1996, chemical vapor deposition (CVD) is a widely used method for synthesizing nanotubes, enabling controlled growth and large-scale production [14]. It remains one of the most popular techniques for CNTs production [16, 17]. In this process, a heated substrate (700–1000 °C) is exposed to carbon-containing gases such as acetylene, methane, or ethylene, while nitrogen, argon, or hydrogen serve as carrier gases [14, 17]. The reaction occurs in a tubular reactor containing a transition metal catalyst, typically Fe, Co, or Ni [11, 17]. Once the reaction is complete, the system is cooled to room temperature, allowing the collected materials to be retrieved [18].

Nanotubes are typically synthesized with a 90% yield and high purity as carbon deposits on the nanoparticle edges, although the exact mechanism remains under investigation. This method offers two key advantages: minimal catalyst usage and controlled deposition on the substrate, leading to well-structured nanotube arrangements [14]. The saturation of metal particles with carbon promotes the formation of solid tubular carbon with a $sp^2$ structure, which is favored over other carbon arrangements, such as graphitic sheets with open edges [18].

CVD is the preferred method for CNTs production due to its simplicity, low cost, and operation at atmospheric pressure and relatively low temperatures [11, 17]. For MWCNT synthesis, ethylene or acetylene is commonly used as the carbon source, with growth occurring at temperatures between 550 and 750 °C in the presence of Fe, Ni, or Co nanoparticles as catalysts [18].

In contrast, the synthesis of SWCNTs via CVD yields structures of higher quality. Methane is typically used as the carbon source, with growth temperatures ranging from 850 to 1000 °C, and suitable catalysts are employed. These higher temperatures are essential to obtain small-diameter CNTs, as they provide the necessary energy for precise structural development [18].

Among the CVD variants, high-pressure carbon monoxide chemical vapor deposition (HiPCO) operates under elevated pressure, which influences the growth and structure of CNTs. This method is particularly effective in producing high-purity SWCNTs [19].

Another variant, catalytic chemical vapor deposition (CCVD), employs metal nanoparticles (such as Fe, Co, or Ni) as catalysts, supported on flat substrates like silicon, alumina, or carbon fibers. In this process, hydrocarbon gases like methane or ethylene decompose at high temperatures to form CNTs. The catalyst lowers the decomposition temperature, facilitating nucleation. The morphology of the resulting nanotube depends on the carbon source: methane or ethylene typically yields straight CNTs, whereas benzene or fullerene leads to curved structures. Additionally, lower temperatures favor the formation of MWCNTs, whereas higher temperatures promote the growth of SWCNTs. CCVD enables precise control over CNTs density and positioning on substrates, making it a versatile technique for various applications [19].

Floating catalyst chemical vapor deposition (FCCVD) differs from conventional CVD methods in that the catalyst nanoparticles are suspended in the gas phase rather than being placed on a substrate. This technique enhances catalyst efficiency and provides greater control over CNTs growth by allowing adjustments to key factors, including carbon source, growth temperature, catalyst precursor, and carrier gas flow. FCCVD enables a fine-tuning of CNTs properties, including the number of walls, diameter, length, and morphology. During the process, carbon sources such as ethylene and catalysts like ferrocene decompose in the reactor, producing CNTs with a variety of structures, including SWCNTs, double-walled carbon nanotubes (DWCNTs), and MWCNTs. By modifying growth parameters, CNTs can be synthesized into various forms, including sponges, arrays, or fibers, making them suitable for diverse applications [19].

Recently, continuous CVD reactors have been employed as an alternative to achieve large-scale production of CNTs, which provides feasible technology for industrial applications [20].

## 3 Functionalization of CNTs and CNTs-Based Composites

Although CNTs exhibit excellent electrical, physical, and chemical properties that enable their application in various fields, some limitations arise due to the interaction forces between carbon structures, which can lead to aggregation, low dispersion, and challenging manipulation. Since CNTs are considered highly promising materials for wastewater treatment, they can be redesigned and modified to enhance efficiency and reduce costs [10, 21], reversing these challenges.

The functionalization of CNTs minimizes their tendency to aggregate, improving their suitability for a wide range of applications. This process can be carried out using mechanical, irradiation, and physicochemical techniques. Among the physicochemical approaches, functionalization is classified as either covalent or non-covalent.

Covalent functionalization involves the attachment of chemical groups directly to the CNTs framework through a reaction, forming bonds [21–23], consequently altering the $sp^2$ hybridization to $sp^3$, partially disrupting conjugation [23, 24]. Since the tips of CNTs exhibit greater reactivity compared to their walls, functional groups are predominantly attached at the ends [24]. In contrast, non-covalent functionalization occurs when molecules adhere to the nanotube surface without altering its structure [21, 22], thereby preserving the $sp^2$ hybridization and maintaining the intrinsic properties of CNTs [24]. Non-covalent functionalization is often preferred, as it preserves the pore textural properties of CNTs [25]. Additionally, the surface of these nanomaterials can be modified by the addition of nanoparticles [22, 26, 27].

The structure of CNTs allows one or more functional groups to be added to the structure, which can improve the selectivity and stability of the resulting system [28–30]. The different chemical functional groups can be fixed to the surface of the CNTs by functionalization or purification processes [28, 29]. Purification processes can cause significant changes to the specific surface area of the CNTs and can alter the volume of micropores and mesopores and the average pore diameter. The chemical purification method, also called oxidative purification, can be carried out by applying acid solutions such as HCl, $HNO_3$ and $H_2SO_4$ [7, 31], basic compounds (such as KOH and $NH_3$) or gaseous compounds (such as $O_2$, $CO_2$, and $O_3$) [31], as well as the incorporation of heteroatoms like N, O, B, and S [32] which aim to improve the properties of the material [7] or enable selective adsorption [32].

Textural properties can also be altered when thermal purification processes are applied. Table 1 summarizes the textural parameters of different CNTs samples reported in previous studies, including specific surface area (BET), total pore volume, and average pore diameter (calculated using the Barrett-Joyner-Halenda—BJH—method). It also outlines the effects of certain purification and functionalization procedures on these properties.

Chemical functionalization techniques modify the surface by introducing hydroxyl, carbonyl, or carboxyl groups. These functional groups alter the wettability of CNTs, enhancing their affinity for polar compounds. Studies indicate that oxidized CNTs exhibit greater efficiency in dye adsorption compared to pristine ones

**Table 1** Characteristics and textural properties of SWCNT and MWCNT

| | Characteristic | SBET ($m^2 g^{-1}$) | VT pores ($cm^3 g^{-1}$) | Diameter (nm) | Syntheses | References |
|---|---|---|---|---|---|---|
| MWCNT | Pristine | 448.0 | 1.100 | 8.26 | Commercial | [53] |
| MWCNT | Pristine | 40.0 | – | 50.0 | CVD | [27] |
| SWCNT $O_3$ treated | Introduces holes and functional groups in the surface | 320.0 | 0.81 | 1.83 | Commercial | [54] |
| SWCNT $O_3/H_2O_2$ treated | Introduces holes and functional groups in the surface | 480 | 0.58 | 1.67 | Commercial | [54] |
| SWCNT HCl treated | Removing the support oxide | 387.7 | 0.66 | 6.83 | Catalytic CVD | [55] |
| SWCNT $HNO_3$ treated | –COOH Functional groups | 820.0 | 0.629 | 2.70 | Commercial | [56] |
| CNTs-SH@$Fe_3O_4$ | Functionalized and magnetic MWCNT—thiol | 200.2 | 0.427 | – | Commercial | [34] |
| SH-CNTs | Functionalized MWCNT—thiol | 163.2 | 0.435 | – | Commercial | [34] |
| $NH_2$-CNT | Modified MWCNT—$NH_2$ | 174.2 | 0.469 | – | Commercial | [34] |
| MWCNT | Pristine | 507.0 | 1.060 | 7.84 | Commercial | [57] |
| MWCNT Heat-treated | Thermal decomposition of functional groups | 550.0 | 1.150 | 7.77 | Commercial | [57] |
| $NH_2$-CNTs@$Fe_3O_4$ | Modified and magnetic MWCNT—$NH_2$ | 191.1 | 0.343 | – | Commercial | [34] |
| MWCNT $H_2SO_3$/ $KMnO_4$ treated | Functional groups (carbonyl, carboxyl, and hydroxyl) | 110.0 | – | – | Commercial | [58] |
| MWCNT KOH treated | Treatment with KOH | 785.0 | – | 2.0–4.0 | Floating catalyst method | [59] |
| MWCNT-$NH_2$ | Functionalized $NH_2$ | 112.8 | 0.650 | 8.0–15.0 | Commercial MWCNT | [60] |
| MWCNT-COOH | Functionalized COOH | 98.3 | 0.494 | 19.0 | Commercial | [61] |
| SWCNT | Pristine | 1587.0 | – | – | HiPCO | [38] |

(continued)

**Table 1** (continued)

| | Characteristic | SBET ($m^2\ g^{-1}$) | VT pores ($cm^3 g^{-1}$) | Diameter (nm) | Syntheses | References |
|---|---|---|---|---|---|---|
| SWCNT | Pristine | 577.0 | 1.15 | 7.98 | Commercial | [53] |
| SWCNT | Pristine | 580.0 | 1.30 | 1.72 | Commercial | [54] |
| SWCNT | Pristine | 370.7 | 0.41 | 5.00 | Commercial | [43] |
| MWCNT | Pristine | 1245.0 | 0.820 | 16.0–25.0 | CVD | [46] |
| MWCNT-$NH_2$ | Functionalized $NH_2$ | 143.0 | 0.602 | – | Commercial | [62] |
| MWCNT $CO_2$ treated | $CO_2$ activated | 429.0 | – | 1.2–4.0 | Commercial | [58] |

[7]. Moreover, these modifications reduce the hydrophobic nature of CNTs by introducing polar surface groups that impart a hydrophilic character, facilitating dispersion in solutions [21–23, 30]. The use of green functionalization methods enhances the process's appeal by combining environmental benefits with potential cost reductions. These approaches often incorporate agro-industrial waste as raw materials, making the synthesis more sustainable and economically viable [4].

A study by Cheng et al. [33] focuses on the production of CNTs from plastic waste, doping the material with nitrogen and iron. This approach not only provides a promising alternative for waste management but also yields a magnetic and functionalized material with a surface area of 158.71 $m^2\ g^{-1}$ and a Cr (VI) ion removal capacity of 27.47 mg $g^{-1}$ from aqueous solutions [33].

The research by Fan et al. [34] presents thiol-functionalized magnetic carbon nanotubes (CNTs-SH@$Fe_3O_4$) as an advanced adsorbent. The modification of MWCNTs aims to enhance their efficiency and selectivity for Hg(II) adsorption. Functionalization-induced nanoarchitectural changes enhance the surface area and pore volume of CNTs, leading to an adsorption efficiency exceeding 98% [34].

Besides, CNTs can be integrated into existing lower-efficiency technologies to enhance the decomposition rate and removal of contaminants in wastewater treatment systems [10].

## 4 Properties of CNTs Relevant to Wastewater Treatment

The application of CNTs in wastewater treatment is largely determined by their textural properties, surface chemistry, morphology, and structural features. These characteristics govern their adsorption performance, selectivity, and efficiency in removing contaminants. The following sections outline the key aspects relevant to their adsorption behavior.

## 4.1 Textural and Surface Chemistry Properties

The textural properties, such as surface area, pore volume, and average pore diameter, along with the presence of functional groups on their surfaces, are the main factors that determine the adsorption capacity of CNTs [5, 22, 35]. As shown in Table 1, the processes of purification or functionalization can greatly influence the textural characteristics of carbon nanotubes. These treatments may either enhance or reduce the specific surface area (BET), total pore volume, and can also lead to changes in the surface chemistry of the nanostructures.

Variations in surface area, pore volume, and average pore diameter are not only linked to differences in sample morphology but also to the specific purification or functionalization methods applied. The presence of oxygen-containing functional groups (such as OH, COOH, and C=O) on CNTs plays a significant role in determining their maximum adsorption capacity. This oxygen content can result from synthesis and purification methods, be purposefully introduced via oxidation processes using acids, plasma, or ozone, or be reduced through thermal treatment [36].

Moderate oxidation of CNTs enhances their ability to adsorb aromatic compounds through π-π interactions. On the other hand, excessive oxidation can lead to the formation of amorphous carbon, which negatively affects the availability of active sites for such interactions [37]. It has been reported that when the oxygen concentration on MWCNTs reaches approximately 18%, factors like electron polarizability, solution pH, and solute pKa become less influential. Although CNTs with lower degrees of oxidation are more responsive to environmental conditions compared to functionalized ones—which, in turn, tend to be more selective than pristine CNTs—a higher oxygen content can lead to distinct outcomes. It may reduce adsorption capacity due to a drop in surface area and fewer accessible sites, or alternatively, promote interactions such as hydrogen bonding, dispersive forces, and electrostatic attraction [36, 37].

Additionally, in some instances, residual impurities like catalyst particles and amorphous carbon can coat SWCNT bundles, affecting adsorption behavior on their outer surfaces [38].

In addition, the surface adsorption sites—often referred to as dangling bonds—can significantly influence the adsorption process due to their typically high reactivity [5]. The surface of carbon-based nanomaterials is also considered amphoteric, capable of gaining protons (becoming positively charged) or losing them (becoming negatively charged) [35]. Moreover, oxygen-containing groups can be introduced, creating additional active sites for chemical adsorption [35].

## 4.2 *Porosity and Surface Area*

Several investigations have emphasized the porous structure of carbon nanotubes [39–41]. In the case of MWCNTs, porosity can generally be categorized into two main types: inner hollow cavities of small diameter (typically between 3 and 6 nm, exhibiting a narrow size distribution) and aggregated pores (widely distributed, 20–40 nm) formed by the interaction between the CNT walls [41].

Various characterization methods have confirmed the microporous structure of SWCNTs and the mesoporous structure of MWCNTs [42]. As a result, SWCNTs typically display a higher specific surface area compared to MWCNTs. Experimental data (see Table 1) show that the surface area of as-synthesized SWCNTs varies between 370 [43] to 1587 $m^2\ g^{-1}$ [38] (micropore volume of 0.15–0.3 $cm^3\ g^{-1}$), less than the theoretically expected specific surface area for an isolated SWCNT (2600 $m^2\ g^{-1}$) [44]. The as-grown MWCNT has a surface area ranging from 180.9 $m^2\ g^{-1}$ [45] to 1306 $m^2\ g^{-1}$ [46] (mesopore volume of 0.5–2 $cm^3\ g^{-1}$) (see Table 1). The observed diameters range from 0.7 to 2.5 nm for SWCNTs and from 4 to 150 nm for MWCNTs [47].

## 4.3 *Morphology, Aggregation, and Adsorption Sites*

Due to van der Waals forces, SWCNTs tend to aggregate and form bundles, as illustrated in Fig. 2. For SWCNTs, both the tube diameter and the number of tubes within a bundle significantly influence the textural properties, as the adsorption sites in the bundle differ from those on a single nanotube [48, 49]. As depicted in Fig. 1, adsorption on an SWCNT bundle can occur at four distinct sites: (i) the grooves formed at the contact between adjacent tubes; (ii) on the outside of the bundles; (iii) the interstitial channels between the tubes in bundles; and (iv) the inside of the nanotubes (pores) with open ends [23, 40, 48, 49].

The interstitial channels between SWCNTs in bundles offer considerable potential for adsorbing small molecules, such as gases [38, 49]. In contrast, MWCNTs do not form bundles, and their adsorption sites are located in the pores that are aggregated on the inner surface of the nanotubes with open ends, as well as on the outer walls, as illustrated in Fig. 3 [41, 45]. In this context, defects in the structure should also be considered as reactive sites for adsorption. Notably, the aggregated pores in MWCNTs play a crucial role in the adsorption of larger biological contaminants, including bacteria and viruses [44].

When examining the adsorption behavior of carbon nanotubes, it is important to take into account both their inner and outer surfaces. Specifically, the transformation of a graphene sheet into a tubular structure leads to a rehybridization of the carbon orbitals into a non-planar $sp^2$ configuration, altering the surface charge in comparison to flat graphene [42]. The curvature of CNTs can lead to distinct thermodynamic

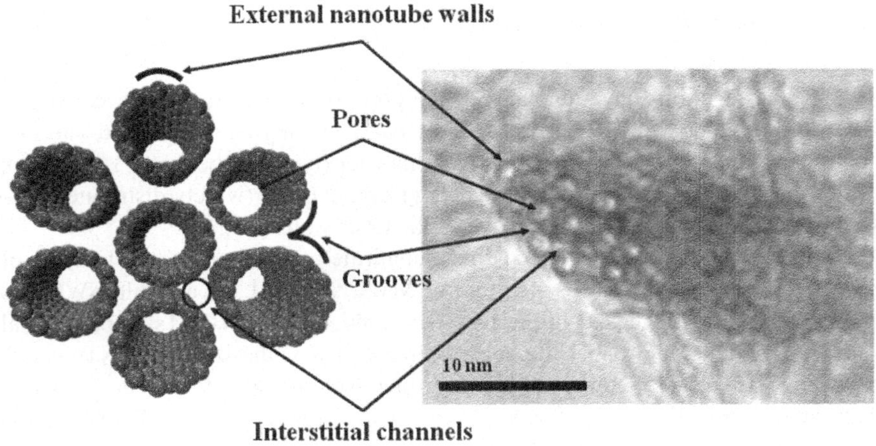


**Fig. 2** Schematic model for the possible adsorption sites of SWCNT bundle (left) and TEM image of SWCNT bundle (right). *Reprinted with permission from Machado FM, Fagan SB, da Silva IZ* et al. *(2015). Carbon Nanoadsorbents. In: Bergmann C, Machado FM (eds) Carbon Nanomaterials as Adsorbents for Environmental and Biological Applications (11–32). Carbon Nanostructures. Springer Nature* [50]

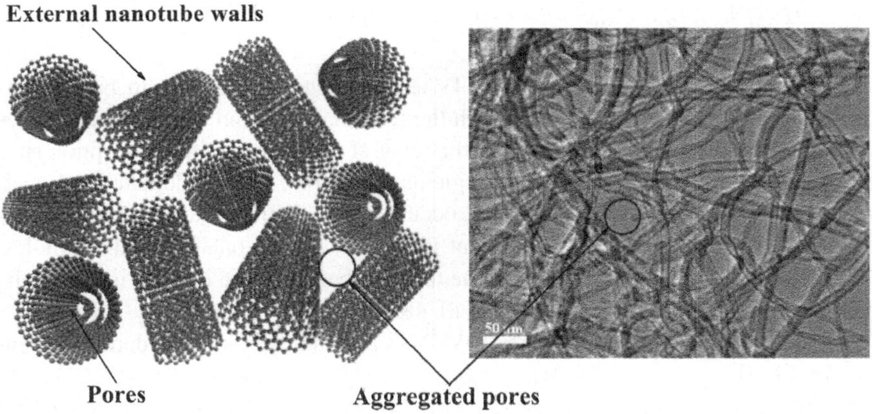


**Fig. 3** Schematic model for the pores aggregated of MWCNT (left) and TEM image of aggregated MWCNTs (right). *Reprinted with permission from Machado FM, Fagan SB, da Silva IZ* et al. *(2015). Carbon Nanoadsorbents. In: Bergmann C, Machado FM (eds) Carbon Nanomaterials as Adsorbents for Environmental and Biological Applications (11–32). Carbon Nanostructures. Springer Nature* [50]

properties during the adsorption process when compared to those observed on the flat carbon sheet [42].

### 4.4 Adsorption Mechanisms and Case Study

The structural properties of CNTs enable effective interactions with organic molecules through non-covalent forces such as hydrogen bonding, π–π stacking, electrostatic forces, van der Waals forces, hydrophobic interactions [24, 51], chemical bonding, surface complexation [24], and Lewis's acid–base interactions [24, 51].

Machado et al. [52] reported the adsorption of Alizarin Red S synthetic dye on MWCNTs and SWCNTs, highlighting the superior performance of the latter. This higher efficiency is attributed to their greater pore volume (127.8% higher) and larger specific surface area (129.5% greater) than MWCNT. The increased surface area provides more active adsorption sites, enhancing the interaction with adsorbates. Additionally, while MWCNTs possess larger pores due to agglomeration, SWCNTs tend to form bundles through van der Waals interactions. These bundles create unique adsorption sites, such as interstitial channels and groove sites, which contribute to their superior adsorption capacity [52].

## 5 Adsorption of Water Contaminants Using CNTs: Efficiency and Mechanisms

The survival of humans, animals, and plants depends on the availability of fresh water. In recent decades, there has been growing concern about the contamination of water bodies with different classes of pollutants [5, 9]. Water contaminants' qualitative and quantitative characteristics are the key parameters to select wastewater treatment options [5]. The nature of contaminants present in wastewater varies according to the source of discharge, and their toxicity becomes significant when concentrations exceed regulatory thresholds [5, 63]. They are typically classified as organic, inorganic, or biological pollutants [5, 63] and can be found in colloidal, solvated, or suspended forms [63].

This diverse and heterogeneous group of pollutants, which includes dyes, pesticides, pharmaceuticals, and heavy metals, has driven the search for efficient and innovative solutions to mitigate these problems [9]. Table 2 presents adsorption studies of different classes of contaminants using CNTs as adsorbents.

### 5.1 Inorganic Contaminants

Heavy metals are highly reactive and tend to form complex structures, resulting in significant biochemical and physical activity. They can impact ecosystems and environmental receptors and become hazardous when their concentrations exceed regulatory levels [64]. Heavy metals are toxic and non-biodegradable, leading to their bioaccumulation in living organisms and causing severe health issues [33, 47,

**Table 2** Adsorption of different classes of contaminants by CNTs

| Adsorbent | SBET ($m^2\ g^{-1}$) | Characteristic | $Q_{max}$ ($mg\ g^{-1}$) | Adsorbate | Isotherm model | References |
|---|---|---|---|---|---|---|
| MWCNT | 40 | Pristine | 59.00 | Cu(II) | Langmuir | [27] |
| MWCNT | – | Functionalized (COOH) | 118.4 | Cu(II) | Langmuir | [27] |
| MWCNT/DAP | 112.8 | Functionalized $NH_2$ | 92.71 | $CO_2$ | Freundlich | [59] |
| CNTs-SH@$Fe_3O_4$ | 200.2 | SH@$Fe_3O_4$ | 172.4 | Hg(II) | Freundlich | [34] |
| (MWCNT-COOH/ Fe-HAP | 255.7 | Composite (Fe, COOH and hydroxyapatite) | 27.65 | Benzohy-droxamic acid | Langmuir | [8] |
| CNT-COOH/$MnO_2$/$Fe_3O_4$ | 114.2 | Nanocomposite (COOH/$MnO_2$/ $Fe_3O_4$) | 103.1 | Ibuprofen | Langmuir | [67] |
| CNT-COOH/$MnO_2$/$Fe_3O_4$ | 114.2 | Nanocomposite (COOH/$MnO_2$ /$Fe_3O_4$) | 80.64 | Paracetamol | Langmuir | [67] |
| $MNO_2$-MWCNT NHs | 85.13 | Functionalized nanoheterostructures and $MnO_2$ | 8.780 | Nanoplastic | Sips | [80] |
| $MNO_2$-MWCNT NHs | 85.13 | Functionalized nanoheterostructures and $MnO_2$ | 3.720 | Bisphenol A | Sips | [80] |
| MWCNT | 180.9 | Pristine | 77.54 | Potassium diclofenac | Liu | [81] |
| SWCNT | 415.3 | Pristine | 150.4 | Potassium diclofenac | Liu | [81] |
| MWCNT | 191.5 | Pristine | 1.745 | Ciprofloxacin hydrochloride | Freundlich | [82] |
| MWCNT | 1306 | Pristine | 230.8 | Ni(II) | Langmuir | [46] |
| MWCNT | 1245 | Pristine | 201.3 | Pb(II) | Langmuir | [46] |

(continued)

**Table 2** (continued)

| Adsorbent | SBET ($m^2\ g^{-1}$) | Characteristic | $Q_{max}$ ($mg\ g^{-1}$) | Adsorbate | Isotherm model | References |
|---|---|---|---|---|---|---|
| $MnFe_2O_4$/MWCNT | 220.6 | Composite ($MnFe_2O_4$) | 494.9 | tetracycline | Langmuir | [71] |
| SWCNT | 582 | Pristine | 30.30 | Triclosan | Prausnitz-Radke | [72] |
| MWCNT-$NH_2$ | 143 | Amino functionalized | 714.0 | Acid Blue 45 | Langmuir | [62] |
| SWCNT | 712.9 | Pristine | 86.90 | Acid Blue 92 | Langmuir | [70] |

61, 64]. Sources of heavy metal contamination include industrial wastewater, river sediments, mine drainage, ash, and electronic waste [21, 63].

Due to their high solubility, stability, and mobility in aquatic environments, inadequate wastewater treatment can cause severe ecological and human health issues [64]. Among the most harmful and widespread heavy metals in aquatic organisms are chromium, copper, lead, cadmium, arsenic, manganese, and mercury [21].

The adsorption of metal ions by CNTs in aqueous solution depends on pH, ionic strength, adsorbent concentration, and contact time. pH influences the adsorbent's surface charge, making it negative at higher values and positive at lower ones, with the transition occurring at the point of zero charge ($pH_{zpc}$), affecting interactions with metal ions. Electrostatic attraction takes place between the negatively charged CNTs surface and positively charged metal ions [24].

For a comprehensive understanding of the CNTs-adsorbate interaction mechanism, several factors must be considered, including surface area, volume, pore size, nanostructure diameter and morphology, as well as the presence of functional groups. Surface area alone does not determine adsorption capacity. For example, although SWCNTs possess a smaller surface area than conventional adsorbents, such as activated carbon (AC), they exhibit significantly higher adsorption capacity [65].

CNTs interact with pollutants through physisorption or chemisorption, depending on the pollutant's characteristics. These mechanisms include hydrophobic interactions, chemical bonding, electrostatic attraction, $\pi$–$\pi$ stacking (between the $\pi$ electrons in CNTs' $sp^2$-hybridized carbon atoms and those in pollutants), Lewis acid–base interactions, Van der Waals forces, surface complexation, and hydrogen bonding [24]. The high removal efficiency presented by CNTs is related to their ease of regeneration, high selectivity in adsorptive processes, and their mechanical resistance [10].

Unmodified CNTs primarily offer a surface for the physical adsorption of metal ions. To enhance their capacity for metal ion adsorption, CNTs require modifications, such as forming composites with compatible materials or introducing specific functional groups onto their surface [24].

In the case of metal ions, physical adsorption is primarily determined by weak interactions between CNTs pores, surfaces, and the ions themselves. In this process, pore size distribution and surface area are critical factors. A higher density of micropores contributes to a larger surface area, which in turn improves the efficiency of physical adsorption [24].

Hydroxylated multi-walled carbon nanotubes (O-CNTs) and graphitized multi-walled carbon nanotubes (G-CNTs) were applied for $Pb^{2+}$ removal from aqueous solution in the study by Zhang et al. [66]. The higher $Pb^{2+}$ sorption capacity of O-CNTs (22.49 $\mu mol \cdot g^{-1} \cdot m^{-2}$) compared to G-CNTs (18.03 $\mu mol \cdot g^{-1} \cdot m^{-2}$) can be attributed to the greater availability of –OH groups on the surface of O-CNTs. Furthermore, the specific surface area is also larger for O-CNTs (85.81 $m^2/g$) than for G-CNTs (81.23 $m^2/g$) [66], which additionally contributes to the superior performance of this material.

## 5.2 Organic Contaminants

The stability of organic pollutants in the environment is attributed to the presence of cyclic structures. This category includes compounds such as pesticides, phenols, fertilizers, plasticizers, oils, pharmaceutical compounds (PHCs) [63], personal care, and detergents [67]. PHCs in water and wastewater gained attention in the 1990s. They are classified as emerging contaminants of concern because they were not initially included in water quality regulations or monitoring programs. These substances are continuously released into the environment as a result of anthropogenic activities [4]. These pollutants can accumulate in living organisms, posing a significant threat to drinking water quality, public health, and environmental balance [67].

Dyes are the leading cause of wastewater pollution [55, 68, 69], as they represent a significant class of chemicals manufactured and utilized on a large scale worldwide [70]. They are classified as stable and resistant organic pollutants, remaining inert and non-biodegradable. These characteristics represent a limitation to their degradation in contaminated waters. The existence of multiple chromophore groups in the dye structure, including -N=N-, =C=O, C–NH, –CH=N–, and C–S, contributes to their high level of hazard [68, 69].

Synthetic dyes pose the potential risk of carcinogenicity, cytotoxicity, genotoxicity, and neurotoxicity in aquatic organisms [69]. The textile industry is estimated to be responsible for approximately 20% of industrial water pollution [68]. The interaction between CNTs-based adsorbents and organic molecules largely depends on the molecular properties of the adsorbate. Multiple interaction mechanisms can occur either independently or simultaneously, making it challenging to determine the exact contribution of each. While the outer surface of CNTs contains hydrophobic regions that can engage with organic molecules, these interactions alone do not fully explain their affinity [24].

Structurally, CNTs have three carbon electrons forming sigma bonds, while a fourth valence electron is delocalized, creating a $\pi$-bond. This electronic configuration enables CNTs to interact with organic molecules through various mechanisms, including physical adsorption, $\pi$–$\pi$ interactions, Van der Waals forces, and, in some cases, electrostatic attraction. The strength of these interactions depends on the presence of functional groups and the available surface area for adsorption [24].

Balarak et al. [71] evaluated the adsorption process of Acid Blue 92 synthetic dye on SWCNTs. The primary interaction mechanisms included hydrogen bonding, dipole-induced dipole interactions, London dispersion forces, $\pi$–$\pi$ acceptor–donor interactions, and the hydrophobic effect [71].

In certain cases, when organic molecules adhere to CNTs surfaces, multilayer adsorption can occur. This process, known as surface condensation, involves initial layers interacting directly with the CNTs surface, while subsequent layers engage primarily through intermolecular forces [65].

Nie et al. [72] investigated the adsorption of propanol, a beta-blocker, onto MWCNTs. Various adsorption parameters were evaluated, and the proposed interaction mechanisms included π-π EDA interactions, hydrophobic interactions, and hydrogen bonding [72].

The adsorption of potassium diclofenac onto SWCNTs and MWCNTs was described by de Azevedo et al. [73]. According to the Liu isotherm model, the maximum adsorption capacities for SWCNTs and MWCNTs were 150.37 and 77.54 mg $g^{-1}$, respectively. The higher adsorption capacity of SWCNTs is attributed to their textural properties, as these nanotubes tend to aggregate into bundles due to surface interactions, forming yarn-like structures composed of dozens to hundreds of nanotubes. Consequently, the structure of SWCNTs offers a greater variety of adsorption sites for potassium diclofenac molecules [73].

In addition to CNTs, hybrid and composite materials such as CNTs sponges and aerogels have been explored for environmental remediation due to their high efficiency in repeated oil sorption. Hybrid and/or composite materials such as CNTs sponges [74, 75] and aerogels [76, 77] stand out as a promising option due to their high efficiency in repeated oil sorption, making them a valuable material for sustainable oil spill remediation [74, 78].

## 5.3 Biological Contaminants

Biological contaminants account for a significant portion of drinking water pollutants in terms of both quantity and likelihood of occurrence in treatment plants [25]. This group of pollutants includes microorganisms such as bacteria and viruses, natural organic matter, biological toxins, and human pathogens like Salmonella, Rotavirus, and Giardia [25].

CNTs play a crucial role in eliminating bacterial and viral microorganisms from water, making them highly valuable for environmental applications. Their antifouling properties, which inhibit bacterial adhesion and biofilm formation, have been extensively studied. Various CNTs derivatives have been explored for these purposes, demonstrating effectiveness in blocking microbial transport through filter membranes [51].

The efficiency of CNTs in removing biological contaminants is attributed to their large external surface area and the presence of aggregated pores, which provide higher adsorption volumes compared to mesopores. Unlike conventional adsorbents, such as activated carbon, which has a high surface area but a microporous structure that limits its accessibility to biological contaminants, CNTs offer multiple adsorption sites, making them significantly more effective [25].

Since biological agents in water can cause severe health problems, including irreversible consequences in extreme cases [63], the use of CNTs for their removal becomes even more essential.

Beyond adsorption applications, CNTs have also been explored for biosensing technologies. The study by Deng et al. [79] proposed the development of CNTs-based biosensors for the rapid detection of bacteria in water, demonstrating their high efficiency. It was shown that SWCNTs can successfully adsorb pure or mixed cultures of *Escherichia coli* and *Staphylococcus aureus* with varying sizes and characteristics. Their findings showed that due to its smaller size, Staphylococcus aureus diffuses 5 to 10 times faster than Escherichia coli. However, E. coli has an adsorption affinity for SWCNT aggregates that is 100 times higher.

## 6 Challenges and Perspectives of CNTs in Environmental Applications

Despite offering numerous benefits for wastewater treatment, the large-scale use of CNTs remains limited due to their high production costs when compared to other materials with the same performance [10, 16, 39, 83]. The use of CNTs and their derivatives in adsorption, filtration, or photocatalytic degradation remains restricted to laboratory-scale applications. Therefore, the development of integrated systems for large-scale deployment is essential [83].

Another challenge for large-scale implementation is the limited and often contradictory literature on CNTs toxicity. Current studies on their potential environmental impact, particularly in terrestrial and aquatic ecosystems, remain inconsistent and inconclusive. Uncertainty regarding CNTs toxicity in humans raises concerns about the safe dosage of CNTs in contaminated water and the need for complementary technologies to mitigate potential risks. Standardized toxicity tests and risk assessment protocols for both human health and the environment are essential. Evaluating the human and ecological impact in research projects involving CNTs and their nanocomposites is crucial for ensuring sustainable application [16].

Due to their high surface area-to-size ratio, CNTs tend to be more reactive and potentially more toxic than bulk materials. Their small size increases interaction with cell membranes and enhances toxin absorption [84, 85]. Factors such as the presence of impurities and structural defects [85], as well as the generation of superoxides [84] and reactive oxygen species [23, 84] also contribute to the toxicity of these materials [84, 85]. The main mechanisms of carbon nanotube toxicity include oxidative stress, membrane damage, and genotoxic effects. CNTs toxicity is associated with their ability to induce DNA damage, mutations, inflammatory responses, and oxidative stress [84, 85].

In addition to toxicity concerns, the limited biodegradability of CNTs and the absence of standardized regulatory frameworks present further challenges for their safe and widespread adoption [16]. Sustainable or green production methods can minimize the use of toxic precursors, reducing impurity levels in CNTs and mitigating their environmental impact [85].

Carbon nanotubes hold immense potential for a wide range of applications, accompanied by both challenges and opportunities. Overcoming these obstacles requires advances in several key areas. Expanding fundamental research will deepen the understanding of structure-performance relationships, providing a strong theoretical basis for practical applications. Enhancing synthesis techniques will improve yield, purity, and consistency. Refining dispersion methods will facilitate better integration into various matrices. Addressing biosafety concerns is crucial for the safe use of these in biomedical fields. Strengthening collaboration among academia, industry, and research institutions will drive innovation and technological progress. As science and technology continue to evolve, many of the current limitations will be resolved, enabling significant advancements in electronics, energy, composite materials, biomedical applications, and other fields, opening new frontiers for global technological development [39].

## 7 Conclusion

The application of carbon nanotubes in numerous technological fields is growing and promising. In wastewater treatment, CNTs play a key role in pollutant adsorption, demonstrating high efficiency in removing different classes of contaminants. This capability is attributed to their textural, structural, and morphological properties, which enable complex interactions with various chemical compounds. Additionally, the scalable application of CNTs is viable due to their high affinity, capacity, and selectivity, which can be optimized through functionalization processes. CNTs stand out as highly efficient adsorbents, surpassing conventional materials such as activated carbon in the removal of metals, organic compounds, and microorganisms. However, their large-scale application still faces significant challenges, including high production costs and the need for integrated systems for field applications. Furthermore, the lack of a comprehensive understanding regarding the toxicity of these nanomaterials raises environmental and human health concerns. Given these challenges, advances in sustainable production and structural modifications can help reduce negative impacts and make CNTs use safer and more accessible. Future research should focus not only on optimizing CNTs efficiency in pollutant removal but also on understanding their long-term environmental and health effects, ensuring their safe and sustainable application.

**Acknowledgements** The authors gratefully acknowledge the financial support provided by the Brazilian funding agency *Conselho Nacional de Desenvolvimento Científico e Tecnológico* (CNPq) and *Fundação de Amparo à Pesquisa do Estado do Rio Grande do Sul* (FAPERGS). Prof. Fernando M. Machado thanks CNPq for research grant support (Processes 308203/2022-4, 408413/2022-0, 443031/2023-1, and 403522/2025-0) and FAPERGS (Process 23/2551-0001863-2). Professor Alice G. Osorio also acknowledges CNPq for the funding of her research project (Process 402464/2023-0).

## References


1. Iijima, S., Ichihashi, T.: Single-shell carbon nanotubes of 1-nm diameter. Nature **363**, 603–605 (1993). https://doi.org/10.1038/363603a0
2. Bethune, D.S., Kiang, C.H., de Vries, M.S., et al.: Cobalt-catalysed growth of carbon nanotubes with single-atomic-layer walls. Nature **363**, 605–607 (1993). https://doi.org/10.1038/363605a0
3. Iijima, S.: Helical microtubules of graphitic carbon. Nature **354**, 56–58 (1991). https://doi.org/10.1038/354056a0
4. Duarte, E.D.V., Oliveira, M.G., Spaolonzi, M.P., et al.: Adsorption of pharmaceutical products from aqueous solutions on functionalized carbon nanotubes by conventional and green methods: a critical review. J. Clean. Prod. **372**, 133743 (2022). https://doi.org/10.1016/j.jclepro.2022.133743
5. Gupta, V.K., Saleh, T.A.: Sorption of pollutants by porous carbon, carbon nanotubes and fullerene-an overview. Environ. Sci. Pollut. Res. **20**, 2828–2843 (2013). https://doi.org/10.1007/s11356-013-1524-1
6. Mishra, Y., Mishra, V., Chattaraj, A., et al.: Carbon nanotube-wastewater treatment nexus: where are we heading to? Environ. Res. **238**, 117088 (2023). https://doi.org/10.1016/j.envres.2023.117088
7. Chávez-Guajardo, A.E., Solis-Sanchez, L.O., Medina-Llamas, M., et al.: Enhancement of Congo red adsorption using oxidated nitrogen-doped carbon nanotubes. Environ. Nanotechnol. Monit. Manag. **23**, 101037 (2025). https://doi.org/10.1016/j.enmm.2024.101037
8. Zhou, W., Zhou, X., Song, W., et al.: Removal of benzohydroxamic acid from aqueous solutions using multi-walled carbon nanotubes/iron-doped hydroxyapatite composites: synthesis, adsorption performance, and characteristics. J. Environ. Chem. Eng. **11**, 111259 (2023). https://doi.org/10.1016/j.jece.2023.111259
9. Bulla, M., Kumar, V., Devi, R., et al.: Exploring the frontiers of carbon nanotube synthesis techniques and their potential applications in supercapacitors, gas sensing, and water purification. J. Environ. Chem. Eng. **12**, 114504 (2024). https://doi.org/10.1016/j.jece.2024.114504
10. Wilson, M.E., Rukh, M.G., Ashraf, M.A.: The role of nanotechnology, based on carbon nanotubes in water and wastewater treatment. Desalin. Water Treat. **242**, 12–21 (2021). https://doi.org/10.5004/dwt.2021.27568
11. Gupta, N., Gupta, S.M., Sharma, S.K.: Carbon nanotubes: synthesis, properties and engineering applications. Carbon Lett. **29**, 419–447 (2019). https://doi.org/10.1007/s42823-019-00068-2
12. Mubarak, N.M., Abdullah, E.C., Jayakumar, N.S., et al.: An overview on methods for the production of carbon nanotubes. J. Ind. Eng. Chem. **20**, 1186–1197 (2014). https://doi.org/10.1016/j.jiec.2013.09.001
13. Shoukat, R., Khan, M.I.: Carbon nanotubes: a review on properties, synthesis methods and applications in micro and nanotechnology. Microsyst. Technol. **27**, 4183–4192 (2021). https://doi.org/10.1007/s00542-021-05211-6
14. Rathinavel, S., Priyadharshini, K., Panda, D.: A review on carbon nanotube: an overview of synthesis, properties, functionalization, characterization, and the application. Mater. Sci. Eng. B **268**, 115095 (2021). https://doi.org/10.1016/j.mseb.2021.115095
15. Taib, N.A.B., Rahman, M.R., Matin, M.M. et al.: A Review on Carbon Nanotubes (CNT): Structure, Synthesis, Purification and Properties for Modern day Applications. 1–22 (2021)
16. Mishra, S., Sundaram, B.: Efficacy and challenges of carbon nanotube in wastewater and water treatment. Environ. Nanotechnol. Monit. Manag. **19**, 100764 (2023). https://doi.org/10.1016/j.enmm.2022.100764
17. Kumar, M., Ando, Y.: Chemical vapor deposition of carbon nanotubes: a review on growth mechanism and mass production. J. Nanosci. Nanotechnol. **10**, 3739–3758 (2010). https://doi.org/10.1166/jnn.2010.2939
18. Dai, H.: Carbon nanotubes: synthesis, integration, and properties. Acc. Chem. Res. **35**, 1035–1044 (2002). https://doi.org/10.1021/ar0101640

19. Ramachandrarao, M., Khan, S.H., Abdullah, K.: Carbon nanotubes and nanofibers—reinforcement to carbon fiber composites—synthesis, characterizations and applications: a review. Compos. Part C Open Access **16**, 100551 (2025). https://doi.org/10.1016/j.jcomc.2024.100551
20. Gowrisankar, L., Manivannan, G., Karthikeyan, S.: Large-scale continuous production of carbon nanotubes-a review. J. Environ. Nanotechnol. **5**, 51–62 (2016). https://doi.org/10.13074/jent.2016.03.161186
21. Fiyadh, S.S., AlSaadi, M.A., Jaafar, W.Z., et al.: Review on heavy metal adsorption processes by carbon nanotubes. J. Clean. Prod. **230**, 783–793 (2019). https://doi.org/10.1016/j.jclepro.2019.05.154
22. Ray, S.S., Gusain, R., Kumar, N.: One-dimensional carbon nanomaterials-based adsorbents. In: Carbon Nanomaterial-Based Adsorbents for Water Purification, pp 195–224. Elsevier (2020). https://doi.org/10.1016/B978-0-12-821959-1.00009-X
23. Baratta, M., Nezhdanov, A.V., Mashin, A.I., et al.: Carbon nanotubes buckypapers: a new frontier in wastewater treatment technology. Sci. Total. Environ. **924**, 171578 (2024). https://doi.org/10.1016/j.scitotenv.2024.171578
24. Sajid, M., Asif, M., Baig, N., et al.: Carbon nanotubes-based adsorbents: properties, functionalization, interaction mechanisms, and applications in water purification. J. Water Process. Eng. **47**, 102815 (2022). https://doi.org/10.1016/j.jwpe.2022.102815
25. Upadhyayula, V.K.K., Deng, S., Mitchell, M.C., et al.: Application of carbon nanotube technology for removal of contaminants in drinking water: a review. Sci. Total. Environ. **408**, 1–13 (2009). https://doi.org/10.1016/j.scitotenv.2009.09.027
26. Parashar, M., Banerjee, D., Mitra, K., et al.: Electric field concentration on amorphous CNTs by silver nanoparticle functionalization to enhance the cold emission characteristics: an experimental verification of theoretical assumption. Phys. B Condens. Matter **700**, 416920 (2025). https://doi.org/10.1016/j.physb.2025.416920
27. Gupta, V.K., Agarwal, S., Bharti, A.K., et al.: Adsorption mechanism of functionalized multiwalled carbon nanotubes for advanced Cu (II) removal. J. Mol. Liq. **230**, 667–673 (2017). https://doi.org/10.1016/j.molliq.2017.01.083
28. Naiya, T.K., Bhattacharya, A.K., Das, S.K.: Adsorption of Cd(II) and Pb(II) from aqueous solutions on activated alumina. J. Colloid Interface Sci. **333**, 14–26 (2009). https://doi.org/10.1016/j.jcis.2009.01.003
29. Sze, M.F.F., Lee, V.K.C., McKay, G.: Simplified fixed bed column model for adsorption of organic pollutants using tapered activated carbon columns. Desalination **218**, 323–333 (2008). https://doi.org/10.1016/j.desal.2007.02.028
30. Osorio, A.G., Silveira, I.C.L., Bueno, V.L., et al.: $H_2SO_4$/$HNO_3$/HCl-Functionalization and its effect on dispersion of carbon nanotubes in aqueous media. Appl. Surf. Sci. **255**, 2485–2489 (2008). https://doi.org/10.1016/j.apsusc.2008.07.144
31. O'Connell, M.J.: Carbon Nanotubes Properties and Applications. CRC Press (2006). https://doi.org/10.1201/9781315222127
32. Lan, Y., Xu, S., Fu, Y., et al.: The selective adsorption mechanism of heteroatom-doped carbon nanotubes for heavy metals in wastewater. A DFT Study. Diam. Relat. Mater **152**, 111875 (2025). https://doi.org/10.1016/j.diamond.2024.111875
33. Cheng, S., Zeng, X., Liu, P.: One-step synthesis of magnetic N-doped carbon nanotubes derived from waste plastics for effective Cr(VI) removal. Arab. J. Chem. **17**, 105956 (2024). https://doi.org/10.1016/j.arabjc.2024.105956
34. Fan, L., Zhou, A., Zhong, L., et al.: Selective and effective adsorption of Hg(II) from aqueous solution over wide pH range by thiol functionalized magnetic carbon nanotubes. Chemosphere **226**, 405–412 (2019). https://doi.org/10.1016/j.chemosphere.2019.03.154
35. Ren, X., Li, J., Tan, X., et al.: Comparative study of graphene oxide, activated carbon and carbon nanotubes as adsorbents for copper decontamination. Dalt. Trans. **42**, 5266–5274 (2013). https://doi.org/10.1039/c3dt32969k
36. Kayiran, S.B., Lamari, F.D., Levesque, D.: Adsorption properties and structural characterization of activated carbons and nanocarbons. J. Phys. Chem. B **108**, 15211–15215 (2004). https://doi.org/10.1021/jp048169c

37. Chin, C.J.M., Shih, M.W., Tsai, H.J.: Adsorption of nonpolar benzene derivatives on single-walled carbon nanotubes. Appl. Surf. Sci. **256**, 6035–6039 (2010). https://doi.org/10.1016/j.apsusc.2010.03.115
38. Agnihotri, S., Mota, J.P.B., Rostam-Abadi, M., et al.: Structural characterization of single-walled carbon nanotube bundles by experiment and molecular simulation. Langmuir **21**, 896–904 (2005). https://doi.org/10.1021/la047662c
39. Tao, Z., Zhao, Y., Wang, Y., et al.: Recent advances in carbon nanotube technology: bridging the gap from fundamental science to wide applications. J. Carbon Res. Rev. **10**, 69 (2024). https://doi.org/10.3390/c10030069
40. Li, F., Wang, Y., Wang, D., et al.: Characterization of single-wall carbon nanotubes by N2 adsorption. Carbon NY **42**, 2375–2383 (2004). https://doi.org/10.1016/j.carbon.2004.02.025
41. Yang, Q.-H., Hou, P.-X., Bai, S., et al.: Adsorption and capillarity of nitrogen in aggregated multi-walled carbon nanotubes. Chem. Phys. Lett. **345**, 18–24 (2001). https://doi.org/10.1016/S0009-2614(01)00848-X
42. Serp, P., Corrias, M., Kalck, P.: Carbon nanotubes and nanofibers in catalysis. Appl. Catal. A Gen. **253**, 337–358 (2003). https://doi.org/10.1016/S0926-860X(03)00549-0
43. Yan, C., Zou, L., Short, R.: Single-walled carbon nanotubes and polyaniline composites for capacitive deionization. Desalination **290**, 125–129 (2012). https://doi.org/10.1016/j.desal.2012.01.017
44. Tanaike, O., Futaba, D.N., Hata, K., et al.: Supercapacitors using pure single-walled carbon nanotubes. Carbon Lett. **10**, 90–93 (2009). https://doi.org/10.5714/cl.2009.10.2.090
45. Machado, F.M., Bergmann, C.P., Fernandes, T.H.M., et al.: Adsorption of Reactive Red M-2BE dye from water solutions by multi-walled carbon nanotubes and activated carbon. J. Hazard. Mater. **192**, 1122–1131 (2011). https://doi.org/10.1016/j.jhazmat.2011.06.020
46. Egbosiuba, T.C., Abdulkareem, A.S., Tijani, J.O., et al.: Taguchi optimization design of diameter-controlled synthesis of multi walled carbon nanotubes for the adsorption of Pb(II) and Ni(II) from chemical industry wastewater. Chemosphere **266**, 128937 (2021). https://doi.org/10.1016/j.chemosphere.2020.128937
47. Bassyouni, M., Mansi, A.E., Elgabry, A., et al.: Utilization of carbon nanotubes in removal of heavy metals from wastewater: a review of the CNTs' potential and current challenges. Appl. Phys. A **126**, 38 (2020). https://doi.org/10.1007/s00339-019-3211-7
48. Ren, X., Chen, C., Nagatsu, M., et al.: Carbon nanotubes as adsorbents in environmental pollution management: a review. Chem. Eng. J. **170**, 395–410 (2011). https://doi.org/10.1016/j.cej.2010.08.045
49. Babaa, M.R., Dupont-Pavlovsky, N., McRae, E., et al.: Physical adsorption of carbon tetrachloride on as-produced and on mechanically opened single walled carbon nanotubes. Carbon NY **42**, 1549–1554 (2004). https://doi.org/10.1016/j.carbon.2004.02.004
50. Machado, F.M., Fagan, S.B., da Silva, I.Z. et al.: Carbon nanoadsorbents. In: Bergmann, C., Machado, F.M. (eds) Carbon Nanomaterials as Adsorbents for Environmental and Biological Applications (11–32). Carbon Nanostructures. Springer Nature (2015). https://doi.org/10.1007/978-3-319-18875-1_2
51. Jain, M., Khan, S.A., Pandey, A., et al.: Instructive analysis of engineered carbon materials for potential application in water and wastewater treatment. Sci. Total. Environ. **793**, 148583 (2021). https://doi.org/10.1016/j.scitotenv.2021.148583
52. Machado, F.M., Carmalin, S.A., Lima, E.C., et al.: Adsorption of Alizarin red S dye by carbon nanotubes: an experimental and theoretical investigation. J. Phys. Chem. C **120**, 18296–18306 (2016). https://doi.org/10.1021/acs.jpcc.6b03884
53. Lu, C., Liu, C.: Removal of nickel(II) from aqueous solution by carbon nanotubes. J. Chem. Technol. Biotechnol. **81**, 1932–1940 (2006). https://doi.org/10.1002/jctb.1626
54. Hemraj-Benny, T., Bandosz, T.J., Wong, S.S.: Effect of ozonolysis on the pore structure, surface chemistry, and bundling of single-walled carbon nanotubes. J. Colloid Interface Sci. **317**, 375–382 (2008). https://doi.org/10.1016/j.jcis.2007.09.064
55. Machado, F.M., Bergmann, C.P., Lima, E.C., et al.: Adsorption of reactive blue 4 dye from water solutions by carbon nanotubes: experiment and theory. Phys. Chem. Chem. Phys. **14**, 11139 (2012). https://doi.org/10.1039/c2cp41475a

56. Kang, J., Wen, J., Jayaram, S.H., et al.: Electrochemical characterization and equivalent circuit modeling of single-walled carbon nanotube (SWCNT) coated electrodes. J. Power. Sources **234**, 208–216 (2013). https://doi.org/10.1016/j.jpowsour.2013.01.058
57. Liao, Q., Sun, J., Gao, L.: Adsorption of chlorophenols by multi-walled carbon nanotubes treated with $HNO_3$ and $NH_3$. Carbon N Y **46**, 553–555 (2008). https://doi.org/10.1016/j.carbon.2007.12.009
58. Lu, C., Su, F.: Adsorption of natural organic matter by carbon nanotubes. Sep. Purif. Technol. **58**, 113–121 (2007). https://doi.org/10.1016/j.seppur.2007.07.036
59. Pillay, K., Cukrowska, E.M., Coville, N.J.: Multi-walled carbon nanotubes as adsorbents for the removal of parts per billion levels of hexavalent chromium from aqueous solution. J. Hazard. Mater. **166**, 1067–1075 (2009). https://doi.org/10.1016/j.jhazmat.2008.12.011
60. Chen, Y., Liu, C., Li, F., et al.: Pore structures of multi-walled carbon nanotubes activated by air, $CO_2$ and KOH. J. Porous Mater. **13**, 141–146 (2006). https://doi.org/10.1007/s10934-006-7017-6
61. Rahimi, K., Riahi, S., Abbasi, M., et al.: Modification of multi-walled carbon nanotubes by 1,3-diaminopropane to increase $CO_2$ adsorption capacity. J. Environ. Manage. **242**, 81–89 (2019). https://doi.org/10.1016/j.jenvman.2019.04.036
62. Sellaoui, L., Schnorr, C.E., Dhaouadi, F., et al.: Modeling the adsorption of divalent metallic cations onto multi-walled carbon nanotubes functionalized with COOH. J. Mol. Liq. **366**, 120275 (2022). https://doi.org/10.1016/j.molliq.2022.120275
63. Maleki, A., Hamesadeghi, U., Daraei, H., et al.: Amine functionalized multi-walled carbon nanotubes: Single and binary systems for high capacity dye removal. Chem. Eng. J. **313**, 826–835 (2017). https://doi.org/10.1016/j.cej.2016.10.058
64. Das, B., Voggu, R., Rout, C.S., et al.: Changes in the electronic structure and properties of graphene induced by molecular charge-transfer. Chem. Commun. **5155–5157**,(2008). https://doi.org/10.1039/b808955h
65. Pan, B., Xing, B.: Adsorption mechanisms of organic chemicals on carbon nanotubes. Environ. Sci. Technol. **42**, 9005–9013 (2008). https://doi.org/10.1021/es801777n
66. Zhang, J., Zhai, J., Zheng, H., et al.: Adsorption, desorption and coadsorption behaviors of sulfamerazine, Pb(II) and benzoic acid on carbon nanotubes and nano-silica. Sci. Total. Environ. **738**, 139685 (2020). https://doi.org/10.1016/j.scitotenv.2020.139685
67. Yang, C.K., Zhao, J., Lu, J.P.: Binding energies and electronic structures of adsorbed titanium chains on carbon nanotubes. Phys Rev B—Condens Matter Mater Phys. **66**, 414031–414034 (2002). https://doi.org/10.1103/PhysRevB.66.041403
68. Durgun, E., Dag, S., Bagci, V.M.K. et al.: Systematic study of adsorption of single atoms on a carbon nanotube. Phys. Rev. B Condens. Matter Mater Phys. **67**, 1–4 (2003). https://doi.org/10.1103/PhysRevB.67.201401
69. Loura, N., Rathee, K., Dhull, R., et al.: Carbon nanotubes for dye removal: a comprehensive study of batch and fixed-bed adsorption, toxicity, and functionalization approaches. J. Water Process. Eng. **67**, 106193 (2024). https://doi.org/10.1016/j.jwpe.2024.106193
70. An, W., Turner, C.H.: Electronic structure calculations of gas adsorption on boron-doped carbon nanotubes sensitized with tungsten. Chem. Phys. Lett. **482**, 274–280 (2009). https://doi.org/10.1016/j.cplett.2009.10.008
71. Balarak, D., Zafariyan, M., Igwegbe, C.A., et al.: Adsorption of acid blue 92 Dye from aqueous solutions by single-walled carbon nanotubes: isothermal, kinetic, and thermodynamic studies. Environ. Process **8**, 869–888 (2021). https://doi.org/10.1007/s40710-021-00505-3
72. Nie, W., Li, Y., Chen, L., et al.: Interaction between multi-walled carbon nanotubes and propranolol. Sci. Rep. **10**, 10259 (2020). https://doi.org/10.1038/s41598-020-66933-7
73. de Azevedo, C.F., Machado, F.M., de Souza, N.F., et al.: Comprehensive adsorption and spectroscopic studies on the interaction of carbon nanotubes with diclofenac anti-inflammatory. Chem. Eng. J. **454**, 140102 (2023). https://doi.org/10.1016/j.cej.2022.140102
74. Kim, C., Choi, Y.S., Lee, S.M., et al.: The effect of gas adsorption on the field emission mechanism of carbon nanotubes. J. Am. Chem. Soc. **124**, 9906–9911 (2002). https://doi.org/10.1021/ja026538h

75. Kang, D., Park, C., Park, N., et al.: Photocurrent measurements of carbon nanotube networks with physical adsorption of oxygen molecules. J. Korean Phys. Soc. **53**, 3343–3347 (2008). https://doi.org/10.3938/jkps.53.3343
76. Santucci, S., Picozzi, S., Di Gregorio, F., et al.: $NO_2$ and CO gas adsorption on carbon nanotubes: experiment and theory. J. Chem. Phys. **119**, 10904–10910 (2003). https://doi.org/10.1063/1.1619948
77. Ali, M.E., Hoque, M.E., Safdar Hossain, S.K., et al.: Nanoadsorbents for wastewater treatment: next generation biotechnological solution. Int. J. Environ. Sci. Technol. **17**, 4095–4132 (2020). https://doi.org/10.1007/s13762-020-02755-4
78. Aigbe, U.O., Osibote, O.A.: Carbon derived nanomaterials for the sorption of heavy metals from aqueous solution: a review. Environ. Nanotechnol. Monit. Manag. **16**, 100578 (2021). https://doi.org/10.1016/j.enmm.2021.100578
79. Deng, S., Upadhyayula, V.K.K., Smith, G.B., Mitchell, M.C.: Adsorption equilibrium and kinetics of microorganisms on single-wall carbon nanotubes. IEEE Sensors J. **8**, 954-962 (2008). https://doi.org/10.1109/JSEN.2008.923929
80. Lung, I., Soran, M.L., Stegarescu, A., et al.: Evaluation of CNT-COOH/$MnO_2$/$Fe_3O_4$ nanocomposite for ibuprofen and paracetamol removal from aqueous solutions. J. Hazard. Mater. **403**, 1–15 (2021). https://doi.org/10.1016/j.jhazmat.2020.123528
81. Yang, X., Ni, J.: Metal-semiconductor and semiconductor-semiconductor transitions in carbon nanotubes induced by intercalating alkali atoms. Phys. Rev. B Condens. Matter Mater. Phys. **71**, 1–5 (2005). https://doi.org/10.1103/PhysRevB.71.165438
82. Khan, F.S.A., Mubarak, N.M., Tan, Y.H., et al.: A comprehensive review on magnetic carbon nanotubes and carbon nanotube-based buckypaper for removal of heavy metals and dyes. J. Hazard. Mater. **413**, 125375 (2021). https://doi.org/10.1016/j.jhazmat.2021.125375
83. Kukkar, D., Rani, A., Kumar, V., et al.: Recent advances in carbon nanotube sponge–based sorption technologies for mitigation of marine oil spills. J. Colloid Interface Sci. **570**, 411–422 (2020). https://doi.org/10.1016/j.jcis.2020.03.006
84. Sehrawat, M., Singh, V., Rani, M., et al.: Nano-welded carbon nanotube sponges for efficient oil spill remediation. J. Clean. Prod. **467**, 142841 (2024). https://doi.org/10.1016/j.jclepro.2024.142841
85. Luo, Z., Wang, X., Yang, D., et al.: Photothermal hierarchical carbon nanotube/reduced graphene oxide microspherical aerogels with radially orientated microchannels for efficient cleanup of crude oil spills. J. Colloid Interface Sci. **570**, 61–71 (2020). https://doi.org/10.1016/j.jcis.2020.02.097

# Graphene-Based Adsorbents for Effective Removal of Organic Pollutants

**Mariana Zancan Tonel, Júlia Vaz Schultz, Solange Binotto Fagan, and Ivana Zanella da Silva**

**Abstract** Water pollution has emerged as a significant environmental concern due to the increasing expansion of industrial and agricultural activities. This issue has led to the contamination of water bodies with various toxic substances, including pesticides and pharmaceutical compounds. These pollutants often enter the food chain even at trace concentrations. However, conventional wastewater treatment methods are generally ineffective in eliminating such environmental contaminants. In recent decades, the rapid advancement of nanotechnology has had a profound impact on various fields of knowledge. Nanoscience and nanotechnology have created new opportunities for environmental remediation strategies, particularly in the treatment of contaminated effluents. Among the various nanomaterials, carbon-based ones—such as graphene—have become model systems for studying properties at the nanoscale, including a high surface area for adsorption, tunable surface chemistry, and unique electronic transport characteristics. However, pristine carbon nanomaterials are typically highly hydrophobic, which limits their applicability in aqueous environments. One promising solution involves chemical modification through surface functionalization, which significantly enhances their ability to remove pollutants from water. This chapter presents selected examples that highlight the potential of graphene, whether pristine or functionalized, to adsorb pollutants, including hormones, through physical or chemical interaction.

**Keywords** Environmental contaminants · Surface functionalization · Adsorption properties · Nanoscale materials

M. Z. Tonel · J. V. Schultz · S. B. Fagan · I. Z. da Silva (✉)
Universidade Franciscana, Santa Maria, RS, Brazil
e-mail: ivanazanella@gmail.com

© The Author(s), under exclusive license to Springer Nature Switzerland AG 2026 
C. P. Bergmann and F. Machado Machado (eds.), *Carbon Nanomaterials as Adsorbents for Environmental and Biological Applications*, Carbon Nanostructures,
https://doi.org/10.1007/978-3-032-08669-3_4

# 1 Introduction

Water is a fascinating liquid that occupies 2/3 of the surface of our planet and constitutes 70% of the human body. Despite being abundant, the amount of freshwater on the planet corresponds to only 2.5% of the total volume of water. Today, one in six people suffers from a water shortage, and by 2025, one in two people will face water stress [1]. The quality of water resources is deteriorating daily due to rapid population growth, agricultural activities, and geological and environmental changes. Recognized as one of the fundamental human rights, access to safe drinking water remains one of the most significant global challenges of the twenty-first century. In addition, the widespread use of persistent organic and xenobiotic contaminants, and their subsequent occurrence in the environment, has become increasingly critical in recent decades [2].

Emerging contaminants (ECs) are defined as synthetic or natural chemical substances that are not commonly monitored in the environment but which have the potential to enter the environment and cause adverse ecological effects and/or human health by acting as interfering agents [3]. ECs are hormonally active organic compounds that can disrupt the endocrine system of humans and/or animals even at low concentrations. Studies show that more than 700 emerging pollutants, their metabolites, and transformation products are listed as present in the European aquatic environment [4]. The interest in studying these substances is motivated by observations of abnormalities in the endocrine systems of animals exposed to organic compounds, both in laboratory settings and in the natural environment. Even at low concentrations, various exogenous, synthetic, or natural compounds have been detected in surface water samples from all continents, primarily due to anthropogenic activities [5]. Prominent classes are pharmaceutical (urban, livestock), pesticides (agriculture), disinfection by-products (urban, industrial), wood preservation, and industrial chemicals (industry) [3, 5].

ECs bind to receptors, mimicking natural hormones and thus interfere with the body's metabolism. There are two categories of ECs: natural and synthetic. Natural ECs are found in animals and humans, including genistein, coumarins, estradiol, and adrenocortical hormones [6]. Synthetic ECs are classified into various categories based on their application, chemical, and structural characteristics, such as pharmaceuticals, organochlorine pesticides, detergents and surfactants, heavy metals, and dyes [7].

Today, there are significant publications that are concerned about the impacts of ECs and their intermediate compounds on human health and the environment due to their widespread occurrence, high persistence, bioaccumulation capacity, and high toxicity [6–8]. The bioaccumulation of ECs in aquatic organisms, particularly fish, mussels, and daphnia, is a crucial criterion in risk assessment. For example, bisphenol A (BPA) is a ubiquitous endocrine disruptor that has recently been linked to adverse human health effects, precocious puberty, and sexual dysfunction [8].

Most pharmaceutical pollutants, such as pharmaceuticals (for humans and veterinary use), are organic and harm the environment and humans [9]. Organic pollutants typically exhibit a high oxygen demand and low biodegradability, in addition to a high rate of bioaccumulation along the food chain due to their lipophilic nature [10]. Due to the limited human metabolic capacity and inadequate removal, pharmaceutical and personal care product residues are commonly found in drinking water sources, sewage treatment plants, and water treatment facilities. When partially digested pharmaceuticals and personal care products are transported to sewage treatment plants, traditional sewage treatment plants are often inadequate for removing these substances [11].

Recent studies have highlighted the prevalence of pesticides and pharmaceuticals among xenobiotic compounds [12–14]. However, their removal remains a significant challenge due to their low degradability [12]. Various techniques, such as advanced oxidation, electrochemical, and photochemical processes [13], have been developed over the years; however, these methods are insufficient to eliminate all organic pollutants. Furthermore, these conventional approaches are not sustainable from either an economic or environmental perspective [14]. Nanoscience and Nanotechnology (N&N) offer promising opportunities to advance cutting-edge water supply and wastewater treatment systems in this context.

Adsorption has emerged as an effective method for the removal of specific ECs from aqueous environments [15]. This process involves mass transfer between two distinct phases—typically, a solid adsorbent and a liquid containing the target pollutants. The interaction relies on various intermolecular forces, including $\pi$–$\pi$ stacking, electrostatic attractions, van der Waals forces, and hydrophobic interactions, which enable the retention of contaminants on the adsorbent surface [16].

In most cases, physisorption governs the adsorption process, characterized by weak and reversible interactions that are predominantly driven by van der Waals forces. Chemisorption, on the other hand, entails stronger, often irreversible chemical bonding between the adsorbate and functional groups present on the surface of the adsorbent. This type of interaction tends to be more selective and depends on the chemical nature of both the adsorbent and the contaminant [17].

In recent decades, carbon nanomaterials, including carbon nanotubes, graphene, and their derivatives, have been utilized in various processes to develop green solutions, such as wastewater treatment, persistent contaminant detection, and water remediation [18, 19]. In the environmental field, graphene-based adsorbents have been explored as state-of-the-art materials in water and effluent treatment due to their highly hydrophobic surface, open-layer morphology, and high adsorption affinity for various organic contaminants [20], organic dyes [21], and metallic ions [22].

Graphene, discovered by Novoselov et al. [23], is a material composed of a single layer of carbon atoms arranged in a honeycomb structure, with a thickness of just one atom [24], making it the thinnest and strongest known material. Its high specific surface area makes it ideal for adsorption processes and surface reactions, as well as serving as a support for chemical functionalities and nanomaterials [25], which has driven the development of graphene-based nanocomposites. However, its applications are limited by a tendency to aggregate and processing difficulties [26], issues that can be overcome through functionalization, such as the use of graphene oxide (GO), which is an oxidized form of the material with a higher density of oxygen functional groups [27, 28], offering lower production.

Graphene oxide (GO) and reduced graphene oxide (rGO) have been extensively studied for various applications, particularly in biological environments, due to their hydrophilic surface groups, including hydroxyl, epoxy, and carboxylic groups, which enable them to disperse in water and physiological environments [29–41]. These functional groups not only enhance the specific properties of graphene but also help reduce its toxicity. GO and rGO are particularly effective for wastewater treatment, as they serve as excellent adsorbents for inorganic and organic pollutants due to their tunable properties, high surface area, and delocalized $\pi$-electron structure [29–33]. The adsorption of contaminants, including polycyclic aromatic hydrocarbons such as naphthalene, occurs through mechanisms such as $\pi$–$\pi$ interactions, hydrogen bonding, and electrostatic attraction [34, 35, 40]. Removing oxygen-containing groups on graphene derivatives improves the adsorption efficiency [37, 41]. As a result, GO, rGO, and their hybrids are gaining significant attention for water and sewage treatment applications due to their easy dispersion, high availability of adsorption sites, and efficient separation and collection properties [36, 38, 39].

Here, we provide an overview of recent advances in utilizing functionalized carbon nanomaterials, such as graphene and its derivatives, for water treatment and the removal of pollutants from wastewater. We critically review the main applications of functionalized nanomaterials based on their functions in wastewater treatment processes.

## 2 Graphene-Based Materials for Effective Removal of Organic Pollutants

Rapid population growth and the intensification of agricultural and industrial activities have dramatically increased the number of contaminants released into the environment. These diverse contaminants represent a significant environmental and public health problem [42]. As a result, efforts are being made to develop technologies that effectively remove pollutants from air, water, and wastewater. Among these technologies, adsorption, particularly using nanomaterials, is a rapid, cost-effective, and efficient method for removing contaminants from aquatic environments [43].

Figure 1 illustrates the exponential growth in publications related to at least one of the themes—graphene, graphene oxide, organic pollutants, or emerging contaminants—over the last decade (2014–2024). The data were obtained through a literature search in the Scopus database.

From Fig. 1, we can observe the growing importance of investigating wastewater treatment processes that utilize graphene-based materials. When dealing with emerging contaminants, the number of publications remains relatively small compared to the number of publications on pollutants. This fact demonstrates the need

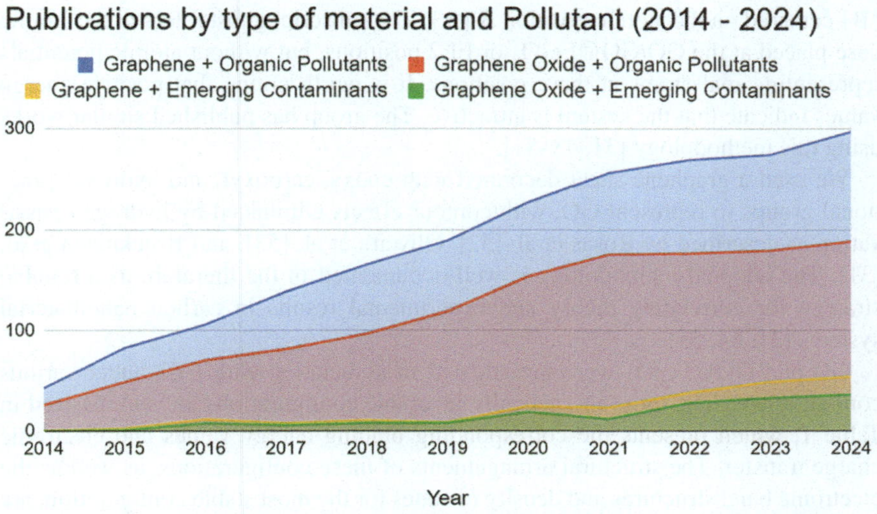


**Fig. 1** Number of publications from 2014 to 2024 involving graphene and graphene oxide in the treatment of organic pollutants and emerging contaminants. The area represents annual publication counts for each material-contaminant combination

for more research, mainly when we deal with synthetic ECs such as pharmaceuticals, hormones, organochlorine pesticides, and dyes.

Thus, in this Chapter, we will specifically evaluate the interaction of the emerging contaminant 17α-ethinylestradiol (EE2) with magnetic graphene oxide and compare it with results from the literature on the adsorption of this molecule with other nanomaterials.

In this way, we present here a systematic study of the adsorption of EE2 molecules on the surface of graphene oxide and graphene oxide functionalized with magnetite ($Fe_3O_4$). We employed the SIESTA program [44], which performs self-consistent calculations by solving the spin-polarized Kohn–Sham equations [45] using atomic numerical basis sets. We employed a double zeta basis with polarized function (DZP) to describe the pseudo-orbitals in all simulations. To represent the electronic charge in real space, we adopted a 300 Ry cutoff grid. The exchange and correlation potential were modeled by the generalized gradient approximation (GGA) with the Perdew and Zunger parametrization [46]. All atomic structures were relaxed until the residual forces were less than 0.05 eV/Å for all atoms. The periodic supercell approach used a $40 \times 40 \times 40$ Å$^3$ cell to avoid interactions between the images. The binding energy was calculated with the BSSE (Basic Superposition Error Correction) method [47], according to Eq. (1) below:

$$E = -[(Ea + b) - \left(Ea_{ghost} + b\right) - \left(Ea + b_{ghost}\right)] \quad (1)$$

system A corresponds to EE2, and system B corresponds to GO/GO@$Fe_3O_4$. They are all calculated based on their respective atomic bases, with the atomic base A

{B} centered on the atomic position B {A}. The subscript ghost refers to the atomic base placed at the GO/GO@$Fe_3O_4$ or EE2 positions, but without atomic potentials representing real atoms at these positions. It is worth noting that positive energy values indicate that the system is attractive. The group has published similar works using this methodology [31, 48–51].

We used a graphene sheet decorated with epoxy, carboxyl, and hydroxyl functional groups to represent GO, with contour effects eliminated by hydrogen passivation as described by Rosas et al. [52], Oliveira et al. [53], and Bruckmann et al. [31]. The adopted methodology is well-documented in the literature as a reliable strategy for correlating theory and experimental results in carbon nanomaterial systems [31, 54, 55].

GO and GO@$Fe_3O_4$ were investigated in association with EE2 under various configurations. The most energetically favorable configurations are summarized in Table 1, which presents the corresponding binding energy values and electronic charge transfer. The structural arrangements of these configurations, as well as the electronic band structures and density of states for the most stable configuration, are illustrated in Fig. 2.

In the GO + EE2-2 system, the most stable configuration presents a binding energy of 0.07 eV, with EE2 accepting 0.01 $e^-$, indicating that GO acts as an electron donor. In this arrangement, EE2 adopts a planar orientation relative to the GO surface. The closest interaction occurs between the oxygen atom of a hydroxyl group on GO and a hydrogen atom of EE2, with a minimum bond distance of 2.73 Å. This proximity suggests the formation of a weak hydrogen bond, contributing to the overall stabilization of the system.

**Table 1** Interaction of 17α-ethinylestradiol (EE2) with GO and GO@$Fe_3O_4$: binding energy (eV), charge transfer ($e^-$), distance (Å), and spin polarization (μB)

| Configuration | Binding energy (eV) | Charge transfer ($e^-$)* | Distance (Å) | Spin polarization (μB) |
|---|---|---|---|---|
| GO + EE2-1 | 0.06 | +0.02 | 2.91 (EE2-GO) | 1.00 |
| GO + EE2-2 | 0.07 | +0.01 | 2.73 (EE2-GO) | 1.00 |
| GO + EE2-3 | 0.04 | +0.01 | 2.75 (EE2-GO) | 1.00 |
| GO@$Fe_3O_4$ + EE2-1 | 1.56 | +0.04 (EE2)/ −0.32 ($Fe_3O_4$) | 2.42 (EE2-GO)/ 3.96 (EE2-$Fe_3O_4$) | 14.60 |
| GO@$Fe_3O_4$ + EE2-2 | 1.91 | +0.03 (EE2)/ −0.34 ($Fe_3O_4$) | 2.60 (EE2-GO)/ 2.34 (EE2-$Fe_3O_4$) | 14.68 |
| GO@$Fe_3O_4$ + EE2-3 | 2.63 | −0.20 (EE2)/ −0.26 ($Fe_3O_4$) | 2.41 (EE2-GO)/ 2.26 (EE2-$Fe_3O_4$) | 14.99 |

* A negative sign (−) indicates that the molecule donated an electronic charge, whereas a positive sign (+) signifies that it accepted a charge, acting as an electron acceptor

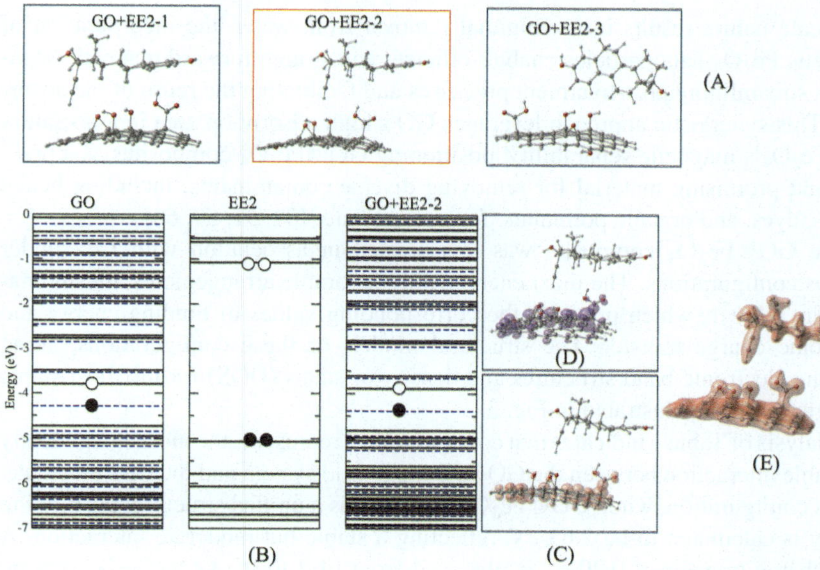


**Fig. 2** **a** Optimized configurations of EE2 interacting with GO, with the most stable configuration highlighted in red. **b** Energy levels of isolated GO and EE2, and the most stable complex (GO + EE2-2). **c** HOMO and **d** LUMO charge density plots (isosurface value: 0.001 $e^-/Å^3$), and **e** total charge density plot (isosurface value: 0.1 $e^-/Å^3$)

In Fig. 2, the dashed blue {black} lines represent the spin-up{-down} contribution. The energy band structure and charge density plots for the HOMO (highest occupied molecular orbital) and LUMO (lowest unoccupied molecular orbital) of the most stable GO + EE2-2 configuration exhibit no significant changes near the Fermi level, indicating that the electronic structure of pristine GO is largely preserved, which suggests a physical interaction between the components. The two energy levels closest to the HOMO–LUMO gap are partially filled, as evidenced by the charge density difference plots, obtained by subtracting the spin-down contribution (purple) from the spin-up contribution (orange). The HOMO is primarily associated with the spin-up state, while the LUMO corresponds to the spin-down state. Additionally, the charge densities are predominantly localized over the GO surface, further supporting the conclusion of a weak electronic interaction. Similarly, Oliveira et al. [53] reported physisorption of 17β-estradiol on GO with a binding energy of ~1.80 eV and a similar charge density distribution in the HOMO and LUMO plots, although using a Local Density Approximation (LDA) exchange–correlation functional [56] known to overestimate binding energies [57, 58].

Combining graphene oxide (GO) with $Fe_3O_4$ nanoparticles yields a highly effective and adaptable solution for environmental remediation [31, 51, 59, 60]. GO's large surface area and abundant oxygen-rich functional groups (e.g., hydroxyl, carboxyl, and epoxy) provide numerous active sites for pollutant interactions via mechanisms like hydrogen bonding, electrostatic forces, and $\pi$–$\pi$ interactions. While GO's nanoscale nature results in its minimal removal from water, the incorporation of magnetic $Fe_3O_4$ nanoparticles enables efficient and straightforward magnetic separation, streamlining post-treatment processes and facilitating the reuse of the adsorbent. This synergistic approach leverages GO's high adsorption area in association with $Fe_3O_4$'s magnetic separability, positioning GO@$Fe_3O_4$ composites as a practical and promising material for removing diverse contaminants, including heavy metals, dyes, and organic pollutants, from wastewater [31, 51, 59, 60].

The GO@$Fe_3O_4$ composite was investigated in association with EE2 under various configurations. The most energetically favorable arrangements are summarized in Table 1, which provides the corresponding values of binding energy and electronic charge transfer. The structural models of these configurations, along with the electronic band structures and density of states (DOS) for the most stable configuration, are illustrated in Fig. 3.

Analysis of Table 1 indicates that configuration 3 represents the most energetically favorable interaction between the GO@$Fe_3O_4$ hybrid system and the EE2 molecule. In this configuration, when GO@$Fe_3O_4$ is treated as a unified structure, the binding energy is calculated to be 2.63 eV, reflecting a stable but moderate interaction. A total charge transfer of 0.20 $e^-$ is observed from EE2 to the GO@$Fe_3O_4$ system, indicating that EE2 acts as an electron donor. Additionally, the $Fe_3O_4$ domain within the hybrid contributes a further 0.26 $e^-$ to the overall electronic redistribution of the

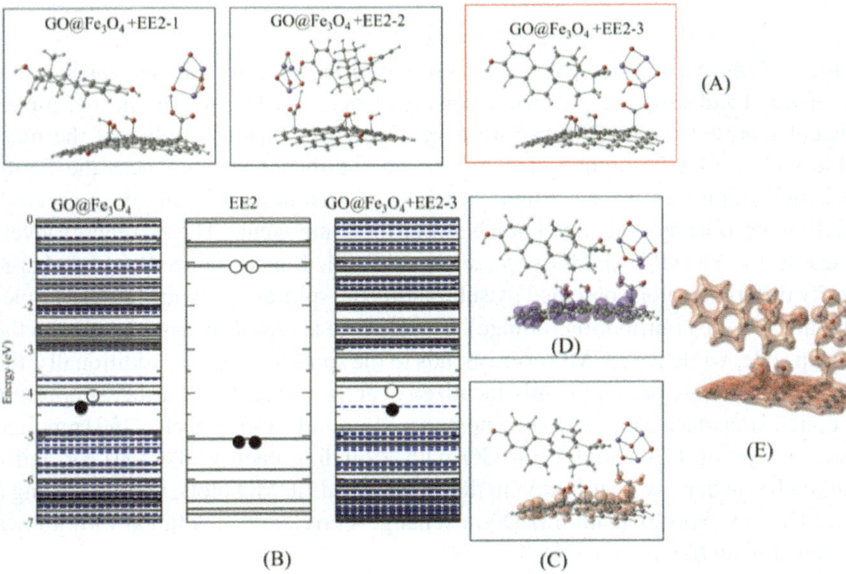


**Fig. 3** **a** Optimized configurations of EE2 interacting with GO@$Fe_3O_4$, with the most stable configuration (GO@$Fe_3O_4$ + EE2-3) highlighted in red. **b** Energy levels of isolated GO@$Fe_3O_4$ and EE2, along with the most stable complex (GO@$Fe_3O_4$ + EE2-3). **c** HOMO and **d** LUMO charge density plots (isosurface value: 0.001 $e^-/Å^3$), and **e** total charge density plot (isosurface value: 0.1 $e^-/Å^3$)

complex, suggesting an active role of the magnetic component in modulating the interaction.

The spatial arrangement of the interacting species reveals that the shortest distance between the EE2 molecule and the $Fe_3O_4$ domain is 2.26 Å, while the minimum distance between EE2 and the GO sheet is 2.41 Å. These values are indicative of non-covalent interactions, most notably hydrogen bonds, which serve as the primary stabilizing forces within the system. The proximity of these functional groups, combined with the observed charge transfer, suggests the formation of a supramolecular complex in which the electronic and geometric structures of the individual components are largely preserved.

Overall, these findings support the hypothesis that EE2 can interact favorably with GO@$Fe_3O_4$ through weak but significant interactions, without inducing major perturbations in the structural framework of the hybrid material. Such behavior is particularly relevant for applications involving adsorption, sensing, or controlled release, where moderate binding is desirable to ensure reversibility and material reusability.

Figure 3 presents the energy levels structure of the GO@$Fe_3O_4$ + EE2-3 system. When compared to the pristine GO@$Fe_3O_4$ material, notable modifications are observed near the Fermi level, indicating a significant electronic interaction between the adsorbate (EE2) and the functionalized surface. The charge density distribution reveals a distinct spatial separation of spin states: the HOMO is dominated by the spin-up contribution, predominantly localized over the GO surface, while the LUMO is characterized by a spin-down contribution mainly concentrated at the GO@$Fe_3O_4$ interface. This configuration suggests a more effective orbital coupling between EE2 and the $Fe_3O_4$ domain, favoring the formation of stable electronic interactions.

These findings indicate that the functionalization of GO with $Fe_3O_4$ nanoparticles significantly alters the electronic structure of the resulting composite, resulting in the formation of hybrid states and a redistribution of charge density. Consequently, the GO@$Fe_3O_4$ hybrid system exhibits enhanced electronic properties, thereby reinforcing its potential as an efficient adsorbent for emerging contaminants, such as EE2, particularly in environmental remediation applications.

Analyzing the total charge density plots of GO + EE2-2 and GO@$Fe_3O_4$ + EE2-3 (Figs. 2e and 3e, respectively), we observe the absence of covalent bond formation in both cases, thereby confirming physisorption as the predominant adsorption mechanism. However, the presence of the $Fe_3O_4$ component introduces magnetic properties that enhance the interaction between the adsorbent and the EE2 molecule. This increased interaction suggests that the magnetic functionalization of GO improves adsorption performance, likely due to synergistic effects between the surface chemistry and magnetic responsiveness. Additionally, the incorporation of $Fe_3O_4$ facilitates the potential for easy recovery and reuse of the composite material via external magnetic fields, making GO@$Fe_3O_4$ a promising candidate for environmental remediation applications, particularly in the removal of endocrine-disrupting compounds from aqueous media. This synergistic effect, wherein co-adsorption on the GO surface facilitates contaminant removal, is comparable to findings reported by Oliveira et al. [53], who found that the adsorption of 17-β estradiol onto GO was mediated by methanol.

Previous studies have demonstrated the potential of GO and its magnetic composite GO@$Fe_3O_4$ as efficient adsorbents for emerging contaminants such as zolpidem [61], hydrochlorothiazide [59], and furosemide [51]. While pure GO may exhibit slightly higher adsorption in some cases, the magnetic properties of GO-$Fe_3O_4$ facilitate easier recovery from aqueous solutions, rendering it more practical for real-world applications.

The adsorption performance depends on various factors, including contaminant concentration, adsorbent dosage, pH, temperature, and ionic strength. Kinetic and isotherm models have suggested chemisorption on heterogeneous surfaces, with thermodynamic data confirming a spontaneous and exothermic process.

The adsorption mechanisms are primarily governed by non-covalent interactions, including π–π stacking, n–π interactions, cation-π interactions, and hydrogen bonding. In parallel, experimental observations aligned with ab initio simulations have confirmed that tetracycline [54] adsorption onto the GO@$Fe_3O_4$ surface predominantly occurs via chemisorption. This process is strongly influenced by π–π interactions between aromatic systems and by hydrogen bonding involving functional groups on both the adsorbent and adsorbate. Furthermore, DFT calculations for furosemide [51] adsorption revealed substantial electronic interactions and significant energy involvement, reinforcing the experimentally observed adsorption behavior and indicating a thermodynamically favorable process driven by specific electronic coupling mechanisms [54].

The GO@$Fe_3O_4$ composite has demonstrated outstanding reusability, consistently maintaining high adsorption efficiency over multiple cycles. These findings underscore the synergistic interaction between graphene oxide and $Fe_3O_4$ nanoparticles, enhancing both performance and practicality. Overall, the GO@$Fe_3O_4$ nanomaterial exhibits significant potential for the adsorption of emerging contaminants by combining high efficiency, selectivity, and facile magnetic recovery. Future research and development efforts are expected to focus on scaling up these applications, further cementing the role of hybrid nanomaterials in safeguarding water quality and ecosystem health.

## 3 Challenges for the Use of Graphene-Based Materials

The properties of graphene-based nanosorbents have great application potential and may be more beneficial in several fields than other adsorbents. However, it is worth noting that carbon nanomaterials as adsorbents still face certain limitations. First, the performance of various nanotechnologies in treating natural and wastewater must be tested. Future studies should be conducted under more realistic conditions to evaluate the applicability and efficiency of multiple nanotechnologies, as well as to validate detection technologies enabled by nanomaterials. Second, the long-term effectiveness and performance of these nanotechnologies should be evaluated in water and wastewater treatment [62]. Furthermore, the cost of preparing specific carbon nanomaterials, such as carbon nanotubes and graphene, is relatively high.

On the other hand, there are some technical problems with the recyclability of some carbon nanomaterials for large-scale wastewater treatment [63]. Despite superior performance, the adoption of innovative technologies strongly depends on cost-effectiveness and the potential risk involved. Since high-purity nanomaterials are very expensive, one strategy would be to use low-purity nanomaterials without significantly compromising efficiency, as a large portion of the production cost is related to separation and purification [64]. An alternative approach to improving cost-effectiveness is the retention and reuse of nanomaterials.

Developing safe, efficient, and low-cost carbon nanocomposite adsorbents remains a significant challenge. Additionally, a deeper understanding of factors such as chemical stabilization and surface adaptations is still required to enhance its applicability in water and wastewater treatment. Graphene-based nanomaterials present unique challenges for risk assessment and management, as they are small particles rather than molecules or ions, for which existing frameworks and risk assessment protocols are not yet established. A better understanding and reduction of the risks associated with the use of nanomaterials in water and wastewater treatment will help increase public acceptance, which is crucial for the adoption of new technologies.

## 4 Conclusions

In conclusion, the increasing prevalence of emerging contaminants in aquatic environments highlights the urgent need for advanced, efficient, and sustainable water treatment technologies. Among the most promising materials explored for this purpose, graphene-based nanostructures—particularly GO and its magnetic composite GO@$Fe_3O_4$—have shown great potential due to their unique structural and electronic characteristics. Functionalization strategies further enhance their adsorption performance, stability, and reusability, addressing some of the practical limitations of conventional adsorbents. In this chapter, density functional theory simulations were employed to investigate the interactions between these materials and the endocrine-disrupting compound EE2, revealing favorable adsorption behavior mediated by non-covalent interactions and electronic coupling. These findings contribute to a deeper understanding of the physicochemical mechanisms underlying contaminant removal and reinforce the applicability of GO-based nanomaterials as versatile platforms in the development of next-generation environmental remediation technologies.

**Acknowledgements** The authors thank CENAPAD-SP (National High-Performance Computing Center in São Paulo) and UFN (Universidade Franciscana) for providing computational resources, and the Brazilian agencies CNPq (Grant Numbers: 443154/2023-6; 421701/2017-0; 309162/2021-1), CAPES (Grant Number: 88881.506898/2020-01), and FINEP (Grant Number: 01.22.0536.00) for financial support.

## Abbreviations

| | |
|---|---|
| BPA | Bisphenol A |
| BSSE | Basic Superposition Error Correction |
| DOS | Density of States |
| DZP | Double-Zeta Basis Including Polarized Orbitals |
| ECs | Emerging contaminants |
| EE2 | 17α-Ethinylestradiol |
| $Fe_3O_4$ | Magnetite |
| GGA | Generalized Gradient Approximation |
| GO | Graphene Oxide |
| GO@$Fe_3O_4$ | $Fe_3O_4$ is anchored onto GO |
| HOMO | Highest Occupied Molecular Orbital |
| LUMO | Lowest Unoccupied Molecular Orbital |
| LDA | Local Density Approximation |
| N&N | Nanoscience and Nanotechnology |
| rGO | Reduced Graphene Oxide |
| SIESTA | Spanish Initiative for Electronic Simulations with Thousands of Atoms |

## References

1. Water scarcity | International Decade for Action "Water for Life" 2005–2015. https://www.un.org/waterforlifedecade/scarcity.shtml (2018)
2. Schwarzenbach, R.P., Egli, T., Hofstetter, T.B., von Gunten, U., Wehrli, B.: Global water pollution and human health. Annu. Rev. Environ. Resour. **35**, 109–136 (2010). https://doi.org/10.1146/annurev-environ-100809-125342
3. la Farré, M., Pérez, S., Kantiani, L., Barceló, D.: Fate and toxicity of emerging pollutants, their metabolites and transformation products in the aquatic environment. TrAC Trends Anal. Chem. **27**, 991–1007 (2008). https://doi.org/10.1016/j.trac.2008.09.010
4. Norman: (n.d.). www.norman-network.net. Accessed 12 Dec 2022
5. Ma, X.Y., Li, Q., Wang, X.C., Wang, Y., Wang, D., Ngo, H.H.: Micropollutants removal and health risk reduction in a water reclamation and ecological reuse system. Water Res. **138**, 272–281 (2018). https://doi.org/10.1016/j.watres.2018.03.059
6. Campbell, C.E., Mezher, A.F., Tyszka, J.M., Nagel, B.J., Eckel, S.P., Herting, M.M.: Associations between testosterone, estradiol, and androgen receptor genotype with amygdala subregions in adolescents. Psychoneuroendocrinology **137**, 105604 (2022). https://doi.org/10.1016/j.psyneuen.2021.105604
7. Liao, Z., Zi, Y., Zhou, C., Zeng, W., Luo, W., Zeng, H., Xia, M., Luo, Z.: Recent advances in the synthesis, characterization, and application of carbon nanomaterials for the removal of endocrine-disrupting chemicals: a review. IJMS **23**, 13148 (2022). https://doi.org/10.3390/ijms232113148
8. Matuszczak, E., Komarowska, M.D., Debek, W., Hermanowicz, A.: The impact of Bisphenol A on fertility, reproductive system, and development: a review of the literature. Int. J. Endocrinol. **2019**, 1–8 (2019). https://doi.org/10.1155/2019/4068717

9. Rathi, B.S., Kumar, P.S.: Application of adsorption process for effective removal of emerging contaminants from water and wastewater. Environ. Pollut. **280**, 116995 (2021). https://doi.org/10.1016/j.envpol.2021.116995
10. Delval, F., Crini, G., Vebrel, J.: Removal of organic pollutants from aqueous solutions by adsorbents prepared from an agroalimentary by-product. Biores. Technol. **97**, 2173–2181 (2006). https://doi.org/10.1016/j.biortech.2005.09.022
11. Yang, Y., Ok, Y.S., Kim, K.-H., Kwon, E.E., Tsang, Y.F.: Occurrences and removal of pharmaceuticals and personal care products (PPCPs) in drinking water and water/sewage treatment plants: a review. Sci. Total Environ. **596–597**, 303–320 (2017). https://doi.org/10.1016/j.scitotenv.2017.04.102
12. Liu, J., Song, Y., Tang, M., Lu, Q., Zhong, G.: Enhanced dissipation of xenobiotic agrochemicals harnessing soil microbiome in the tillage-reduced rice-dominated agroecosystem. J. Hazard. Mater. **398**, 122954 (2020). https://doi.org/10.1016/j.jhazmat.2020.122954
13. Chakraborty, I., Sathe, S.M., Khuman, C.N., Ghangrekar, M.M.: Bioelectrochemically powered remediation of xenobiotic compounds and heavy metal toxicity using microbial fuel cell and microbial electrolysis cell. Mater. Sci. Energy Technol. **3**, 104–115 (2020). https://doi.org/10.1016/j.mset.2019.09.011
14. Chong, M.N., Jin, B., Chow, C.W.K., Saint, C.: Recent developments in photocatalytic water treatment technology: a review. Water Res. **44**, 2997–3027 (2010). https://doi.org/10.1016/j.watres.2010.02.039
15. Shahid, M.K., Kashif, A., Fuwad, A., Choi, Y.: Current advances in treatment technologies for removal of emerging contaminants from water—a critical review. Coord. Chem. Rev. **442**, 213993 (2021). https://doi.org/10.1016/j.ccr.2021.213993
16. Khalil, A.M.E., Memon, F.A., Tabish, T.A., Salmon, D., Zhang, S., Butler, D.: Nanostructured porous graphene for efficient removal of emerging contaminants (pharmaceuticals) from water. Chem. Eng. J. **398**, 125440 (2020). https://doi.org/10.1016/j.cej.2020.125440
17. Yu, W., Sisi, L., Haiyan, Y., Jie, L.: Progress in the functional modification of graphene/graphene oxide: a review. RSC Adv. **10**, 15328–15345 (2020). https://doi.org/10.1039/D0RA01068E
18. Bolade, O.P., Williams, A.B., Benson, N.U.: Green synthesis of iron-based nanomaterials for environmental remediation: a review. Environ. Nanotechnol. Monit. Manag. **13**, 100279 (2020). https://doi.org/10.1016/j.enmm.2019.100279
19. Ali, I., Basheer, A.A., Mbianda, X.Y., Burakov, A., Galunin, E., Burakova, I., Mkrtchyan, E., Tkachev, A., Grachev, V.: Graphene-based adsorbents for remediation of noxious pollutants from wastewater. Environ. Int. **127**, 160–180 (2019). https://doi.org/10.1016/j.envint.2019.03.029
20. Beless, B., Rifai, H.S., Rodrigues, D.F.: Efficacy of carbonaceous materials for sorbing polychlorinated biphenyls from aqueous solution. Environ. Sci. Technol. **48**, 10372–10379 (2014). https://doi.org/10.1021/es502647n
21. Sharma, P., Saikia, B.K., Das, M.R.: Removal of methyl green dye molecule from aqueous system using reduced graphene oxide as an efficient adsorbent: kinetics, isotherm and thermodynamic parameters. Colloids Surf. A **457**, 125–133 (2014). https://doi.org/10.1016/j.colsurfa.2014.05.054
22. Zhao, G., Jiang, L., He, Y., Li, J., Dong, H., Wang, X., Hu, W.: Sulfonated graphene for persistent aromatic pollutant management. Adv. Mater. **23**, 3959–3963 (2011). https://doi.org/10.1002/adma.201101007
23. Novoselov, K.S., Geim, A.K., Morozov, S.V., Jiang, D., Zhang, Y., Dubonos, S.V., Grigorieva, I.V., Firsov, A.A.: Electric field effect in atomically thin carbon films. Science **306**, 666–669 (2004). https://doi.org/10.1126/science.1102896
24. Sanchez, V.C., Jachak, A., Hurt, R.H., Kane, A.B.: Biological interactions of graphene-family nanomaterials: an interdisciplinary review. Chem. Res. Toxicol. **25**, 15–34 (2012). https://doi.org/10.1021/tx200339h
25. Compton, O.C., Nguyen, S.T.: Graphene oxide, highly reduced graphene oxide, and graphene: versatile building blocks for carbon-based materials. Small **6**, 711–723 (2010). https://doi.org/10.1002/smll.200901934

26. Loh, K.P., Bao, Q., Eda, G., Chhowalla, M.: Graphene oxide as a chemically tunable platform for optical applications. Nat. Chem **2**, 1015–1024 (2010). https://doi.org/10.1038/nchem.907
27. Dreyer, D.R., Park, S., Bielawski, C.W., Ruoff, R.S.: The chemistry of graphene oxide. Chem. Soc. Rev. **39**, 228–240 (2010). https://doi.org/10.1039/B917103G
28. Mmaduka Obodo, R., Ahmad, I., Ifeanyichukwu Ezema, F.: Introductory chapter: graphene and its applications. In: Graphene and Its Derivatives—Synthesis and Applications [Working Title]. IntechOpen (2019). https://doi.org/10.5772/intechopen.86023
29. Ghahghaey, Z., Hekmati, M., Darvish Ganji, M.: Theoretical investigation of phenol adsorption on functionalized graphene using DFT calculations for effective removal of organic contaminants from wastewater. J. Mol. Liq. **324**, 114777 (2021). https://doi.org/10.1016/j.molliq.2020.114777
30. Karimi-Maleh, H., Shafieizadeh, M., Taher, M.A., Opoku, F., Kiarii, E.M., Govender, P.P., Ranjbari, S., Rezapour, M., Orooji, Y.: The role of magnetite/graphene oxide nano-composite as a high-efficiency adsorbent for removal of phenazopyridine residues from water samples, an experimental/theoretical investigation. J. Mol. Liq. **298**, 112040 (2020). https://doi.org/10.1016/j.molliq.2019.112040
31. Bruckmann, F.d.S., Zuchetto, T., Ledur, C.M., dos Santos, C.L., da Silva, W.L., Binotto Fagan, S., Zanella da Silva, I., Bohn Rhoden, C.R.: Methylphenidate adsorption onto graphene derivatives: theory and experiment. New J. Chem. **46**, 4283–4291 (2022). https://doi.org/10.1039/D1NJ03916D
32. Shanmugham, S.R., Jegadeesan, G.B., Ponnusami, V.: Groundwater treatments using nanomaterials. In: Nanotechnology in the Beverage Industry, pp. 25–49. Elsevier (2020). https://doi.org/10.1016/B978-0-12-819941-1.00002-X
33. Yusuf, M., Khan, M.A., Otero, M., Abdullah, E.C., Hosomi, M., Terada, A., Riya, S.: Synthesis of CTAB intercalated graphene and its application for the adsorption of AR265 and AO7 dyes from water. J. Colloid Interface Sci. **493**, 51–61 (2017). https://doi.org/10.1016/j.jcis.2017.01.015
34. Shen, Y., Chen, B.: Sulfonated graphene nanosheets as a superb adsorbent for various environmental pollutants in water. Environ. Sci. Technol. **49**, 7364–7372 (2015). https://doi.org/10.1021/acs.est.5b01057
35. Sheng, G.D., Shao, D.D., Ren, X.M., Wang, X.Q., Li, J.X., Chen, Y.X., Wang, X.K.: Kinetics and thermodynamics of adsorption of ionizable aromatic compounds from aqueous solutions by as-prepared and oxidized multiwalled carbon nanotubes. J. Hazard. Mater. **178**, 505–516 (2010). https://doi.org/10.1016/j.jhazmat.2010.01.110
36. Ouyang, K., Zhu, C., Zhao, Y., Wang, L., Xie, S., Wang, Q.: Adsorption mechanism of magnetically separable $Fe_3O_4$/graphene oxide hybrids. Appl. Surf. Sci. **355**, 562–569 (2015). https://doi.org/10.1016/j.apsusc.2015.07.109
37. de Moraes, E.E., Tonel, M.Z., Fagan, S.B., Barbosa, M.C.: Density functional theory study of $\pi$-aromatic interaction of benzene, phenol, catechol, dopamine isolated dimers and adsorbed on graphene surface. J. Mol. Model. **25**, 302 (2019). https://doi.org/10.1007/s00894-019-4185-2
38. Bharath, G., Alhseinat, E., Ponpandian, N., Khan, M.A., Siddiqui, M.R., Ahmed, F., Alsharaeh, E.H.: Development of adsorption and electrosorption techniques for removal of organic and inorganic pollutants from wastewater using novel magnetite/porous graphene-based nanocomposites. Sep. Purif. Technol. **188**, 206–218 (2017). https://doi.org/10.1016/j.seppur.2017.07.024
39. Tiwari, J.N., Mahesh, K., Le, N.H., Kemp, K.C., Timilsina, R., Tiwari, R.N., Kim, K.S.: Reduced graphene oxide-based hydrogels for the efficient capture of dye pollutants from aqueous solutions. Carbon **56**, 173–182 (2013). https://doi.org/10.1016/j.carbon.2013.01.001
40. Wang, H., Yuan, X., Wu, Y., Chen, X., Leng, L., Wang, H., Li, H., Zeng, G.: Facile synthesis of polypyrrole decorated reduced graphene oxide–$Fe_3O_4$ magnetic composites and its application for the Cr(VI) removal. Chem. Eng. J. **262**, 597–606 (2015). https://doi.org/10.1016/j.cej.2014.10.020
41. Wang, J., Chen, Z., Chen, B.: Adsorption of polycyclic aromatic hydrocarbons by graphene and graphene oxide nanosheets. Environ. Sci. Technol. **48**, 4817–4825 (2014). https://doi.org/10.1021/es405227u

42. Lubchenco, J.: Entering the century of the environment: a new social contract for science. Science **279**, 491–497 (1998). https://doi.org/10.1126/science.279.5350.491
43. Ali, I., Gupta, V.K.: Advances in water treatment by adsorption technology. Nat. Protoc. **1**, 2661–2667 (2006). https://doi.org/10.1038/nprot.2006.370
44. Soler, J.M., Artacho, E., Gale, J.D., García, A., Junquera, J., Ordejón, P., Sánchez-Portal, D.: The SIESTA method for *ab initio* order-*N* materials simulation. J. Phys. Condens. Matter **14**, 2745–2779 (2002). https://doi.org/10.1088/0953-8984/14/11/302
45. Kohn, W., Sham, L.J.: Self-consistent equations including exchange and correlation effects. Phys. Rev. **140**, A1133–A1138 (1965). https://doi.org/10.1103/PhysRev.140.A1133
46. Perdew, J.P., Burke, K., Ernzerhof, M.: Generalized gradient approximation made simple. Phys. Rev. Lett. **77**, 3865–3868 (1996). https://doi.org/10.1103/PhysRevLett.77.3865
47. Boys, S.F., Bernardi, F.: The calculation of small molecular interactions by the differences of separate total energies. Some procedures with reduced errors. Mol. Phys. **19**, 553–566 (1970). https://doi.org/10.1080/00268977000101561
48. Bruckmann, F.d.S., Rossato Viana, A., Tonel, M.Z., Fagan, S.B., Garcia, W.J.d.S., de Oliveira, A.H., Dorneles, L.S., Roberto Mortari, S., da Silva, W.L., da Silva, I.Z., Rhoden, C.R.B.: Influence of magnetite incorporation into chitosan on the adsorption of methotrexate and in vitro cytotoxicity. Environ. Sci. Pollut. Res. **29**, 70413–70434 (2022). https://doi.org/10.1007/s11356-022-20786-x
49. Tonel, M.Z., Martins, M.O., Zanella, I., Pontes, R.B., Fagan, S.B.: A first-principles study of the interaction of doxorubicin with graphene. Comput. Theor. Chem. **1115**, 270–275 (2017). https://doi.org/10.1016/j.comptc.2017.07.004
50. Saini, S.S., Copello, G.J., Fagan, S.B., Tonel, M.Z.: Comparison of three cyclodextrins to optimize bisphenol A extraction from source water: computational, spectroscopic, and analytical studies. J Sep. Sci. **46**, 2300012 (2023). https://doi.org/10.1002/jssc.202300012
51. Vargas, G.O., Schnorr, C., Nunes, F.B., Da Rosa Salles, T., Tonel, M.Z., Fagan, S.B., Zanella Da Silva, I., Silva, L.F.O., Mortari, S.R., Dotto, G.L., Rhoden, C.R.B.: Highly furosemide uptake employing magnetic graphene oxide: DFT modeling combined to experimental approach. J. Mol. Liq. **379**, 121652 (2023). https://doi.org/10.1016/j.molliq.2023.121652
52. Hernández Rosas, J.J., Ramírez Gutiérrez, R.E., Escobedo-Morales, A., Chigo Anota, E.: First principles calculations of the electronic and chemical properties of graphene, graphane, and graphene oxide. J. Mol. Model **17**, 1133–1139 (2011). https://doi.org/10.1007/s00894-010-0818-1
53. De Oliveira, P.V., Zanella, I., Bulhões, L.O.S., Fagan, S.B.: Adsorption of 17 β- β-estradiol in graphene oxide through the competing methanol co-solvent: experimental and computational analysis. J. Mol. Liq. **321**, 114738 (2021). https://doi.org/10.1016/j.molliq.2020.114738
54. da Silva Bruckmann, F., Mafra Ledur, C., Zanella da Silva, I., Luiz Dotto, G., Rodrigo Bohn Rhoden, C.: A DFT theoretical and experimental study about tetracycline adsorption onto magnetic graphene oxide. J. Mol. Liq. **353**, 118837 (2022). https://doi.org/10.1016/j.molliq.2022.118837
55. Jauris, I.M., Matos, C.F., Saucier, C., Lima, E.C., Zarbin, A.J.G., Fagan, S.B., Machado, F.M., Zanella, I.: Adsorption of sodium diclofenac on graphene: a combined experimental and theoretical study. Phys. Chem. Chem. Phys. **18**, 1526–1536 (2016). https://doi.org/10.1039/C5CP05940B
56. Perdew, J.P., Zunger, A.: Self-interaction correction to density-functional approximations for many-electron systems. Phys. Rev. B **23**, 5048–5079 (1981). https://doi.org/10.1103/PhysRevB.23.5048
57. Sun, J., Remsing, R.C., Zhang, Y., Sun, Z., Ruzsinszky, A., Peng, H., Yang, Z., Paul, A., Waghmare, U., Wu, X., Klein, M.L., Perdew, J.P.: Accurate first-principles structures and energies of diversely bonded systems from an efficient density functional. Nat. Chem. **8**, 831–836 (2016). https://doi.org/10.1038/nchem.2535
58. Grimme, S.: Semiempirical GGA-type density functional constructed with a long-range dispersion correction. J. Comput. Chem. **27**, 1787–1799 (2006). https://doi.org/10.1002/jcc.20495

59. Rhoden, C.R.B., Bruckmann, F.D.S., Salles, T.D.R., Kaufmann Junior, C.G., Mortari, S.R.: Study from the influence of magnetite onto removal of hydrochlorothiazide from aqueous solutions applying magnetic graphene oxide. J. Water Process Eng. **43**, 102262 (2021). https://doi.org/10.1016/j.jwpe.2021.102262
60. Nunes, F.B., Da Silva Bruckmann, F., Viana, A.R., Da Rosa Salles, T., Zancanaro, L.V., Rhoden, D.S.B., Franco, C., Schuch, A.P., Dotto, G.L., Silva, L.F.O., Ramos, C.G., Rhoden, C.R.B.: Removal of selective serotonin reuptake inhibitor using magnetic graphene oxide derivatives: adsorption study in low drug concentration using HPLC quantification, in vitro safety, and phytotoxicity. J. Environ. Chem. Eng. **12**, 112336 (2024). https://doi.org/10.1016/j.jece.2024.112336
61. Barbieri, I.A., Oliveira, M.L.S., Bruckmann, F.S., Salles, T.R., Zancanaro, L.V., Silva, L.F.O., Dotto, G.L., Lima, E.C., Naushad, Mu., Bohn Rhoden, C.R.: Effective removal of hypnotic drug from the aqueous medium through adsorption on graphene oxide magnetic derivatives. J. Mol. Liq. **393**, 123657 (2024). https://doi.org/10.1016/j.molliq.2023.123657
62. Qu, X., Alvarez, P.J.J., Li, Q.: Applications of nanotechnology in water and wastewater treatment. Water Res. **47**, 3931–3946 (2013). https://doi.org/10.1016/j.watres.2012.09.058
63. Li, F., Jiang, X., Zhao, J., Zhang, S.: Graphene oxide: a promising nanomaterial for energy and environmental applications. Nano Energy **16**, 488–515 (2015). https://doi.org/10.1016/j.nanoen.2015.07.014
64. Qu, X., Brame, J., Li, Q., Alvarez, P.J.J.: Nanotechnology for a safe and sustainable water supply: enabling integrated water treatment and reuse. Acc. Chem. Res. **46**, 834–843 (2013). https://doi.org/10.1021/ar300029v

# 3D Graphene-Based Adsorbents: Synthesis, Characterization, and Applications in Water Treatment

**Mayara Bitencourt Leão and Carolina Ferreira de Matos Jauris**

**Abstract** Graphene-based materials have received significant attention due to their exceptional properties in water treatment applications. Among these, three-dimensional graphene offers distinct advantages over its two-dimensional form, particularly in addressing nanosafety concerns, which have specific implications for environmental and health-related applications. This chapter explores the synthesis, characterization, and applications of 3D graphene-based adsorbents, focusing on their unique structural features, high porosity, and tunable properties that ensure outstanding performance in water purification. It presents the main methods for synthesizing 3D graphene-based materials, examining their impact on structural and chemical properties, alongside challenges in characterizing these complex macrostructures and their interactions with contaminants. Applications of 3D graphene in water treatment are also discussed, with an emphasis on adsorption strategies, including hierarchical structuring and surface functionalization, to enhance efficiency. Key challenges, such as developing scalable and eco-friendly synthesis methods and accessing the intricate features of 3D graphene, are analyzed. Finally, conclusions and perspectives are provided, highlighting pathways to optimize material design and uncover interaction mechanisms to create efficient, sustainable solutions tailored to water treatment needs.

**Keywords** 3D-graphene · Adsorption · Water treatment · Contaminants

## Abbreviations

0D  Zero-dimensional

M. B. Leão
Universidade Federal do Pampa, Bagé, Brazil
e-mail: mayaraleao@unipampa.edu.br

C. F. de Matos Jauris (✉)
Universidade Federal de Santa Maria, Santa Maria, Brazil
e-mail: carolina.matos@ufsm.br

© The Author(s), under exclusive license to Springer Nature Switzerland AG 2026

C. P. Bergmann and F. Machado Machado (eds.), *Carbon Nanomaterials as Adsorbents for Environmental and Biological Applications*, Carbon Nanostructures,
https://doi.org/10.1007/978-3-032-08669-3_5

| | |
|---|---|
| 1D | One-dimensional |
| 2D | Two-dimensional |
| 3D | Three-dimensional |
| BET | Brunauer–Emmett–Teller |
| CVD | Chemical Vapor Deposition |
| EDX | Energy Dispersive X-ray Spectroscopy |
| FTIR | Fourier Transform Infrared Spectroscopy |
| GO | Graphene Oxide |
| Gr-O | Graphite Oxide |
| ID/IG | Ratio of the intensity of the D-band to the intensity of the G-band |
| PE-CVD | Plasma-enhanced Chemical Vapor Deposition |
| $Q_{max}$ | Adsorptive Capacity |
| rGO | Reduced Graphene Oxide |
| SEM | Scanning Electron Microscopy |
| SSA | Specific Surface Area |
| TGA | Thermogravimetric Analysis |
| UV-Vis | Ultraviolet–Visible Spectroscopy |
| XPS | X-ray Photoelectron Spectroscopy |
| XRD | X-Ray Diffraction |

## 1 Introduction

Carbon is the basis element of all organic chemistry and the raw material of life. Due to the flexibility of their bonding, carbon-based systems show an unlimited number of different structures, with a wide range of properties. These properties are largely the result of the dimensionality of these structures [1]. Due to its dimensionality, regarding synthetic structures composed of carbon, we find fullerene (0D) carbon nanotubes (1D) and graphene or reduced graphene oxide (2D). Graphene is an atomic monolayer material, composed of carbon, tightly compacted in a two-dimensional hexagonal structure, and is a basic component for graphitic materials of all other dimensions [2]. In the hexagonal carbon network, the $sp^2$ hybridization leads to a trigonal planar structure with the formation of a bond between carbon atoms that are separated by 1.42 Å [1, 3]. Among the main properties of graphene are your high intrinsic mobility [4, 5], large specific surface area, and high optical transmittance [6–8]. Due to its differentiated properties, graphene has the potential for a large number of applications, such as sensing [9], energy storage [10], and environmental remediation [11, 12], however, some applications, like water treatment, may be limited so that graphene needs to be chemically or structurally modified with the to change its functionality and expand its applications.

One of the ways to optimize some applications of graphene is through the synthesis of three-dimensional structures (3D), where 2D graphene sheets interconnect to form a 3D network. These structures have new physical and chemical properties when compared to individual building blocks and bulk material, presenting extremely low density, high porosity, and large surface area [13–17]. One of the main advantages of this type of structure is that the 3D network prevents the aggregation of graphene sheets, improving the performance of the material and offsetting the main disadvantage of using 2D graphene in environmental applications [13, 18]. These 2D layers in well-organized and interconnected 3D structures can retain the exceptional properties of 2D graphene materials, and also be used without the worry of restacking [8, 19], increasing the adsorption graphene capacity. In addition, the morphology material makes it convenient for manipulation, avoiding the release of graphene nanoparticles into the environment [20]. Therefore, the three-dimensional network prevents aggregation and improves the performance of the bulk material, compensating for the main disadvantage of lower-dimensional materials. Furthermore, the integrated morphology of a graphene-based monolith facilitates its manipulation and application [21]. With increasing investigations involving 3D graphene-based macrostructures, it has become evident that they possess new collective physicochemical properties that are remarkably different from individual building blocks, expanding their application capabilities [21, 22].

3D graphene mainly consists of the self-assembly/interconnection of two or more graphene sheets, forming different types of 3D graphene nanostructures, such as sponges, hydrogels, aerogels, and xerogels. These materials offer, in addition to the advantages already mentioned, high porosity and available surface area, excellent chemical and mechanical stabilities, and the possibility of chemical modification of the surface [23]. In recent years, three-dimensional graphene and its derivatives have been considered one of the most effective carbon-based materials for wastewater treatment due to their unique physicochemical properties [24], being developed for the removal of various types of pollutants from water [23]. Due to their excellent properties, relevant and desirable high adsorptive capacities have been found for different classes of contaminants, such as metals, dyes, organic solvents, oils, pesticides, and pharmaceuticals [20, 25, 26].

Before discussing graphene-based 3D materials, we must introduce some terms that are important and guide the development of this chapter. Specifically, the term *composition* refers to the chemical constitution of the material; *structure* refers to the arrangement of atoms, molecules, or ions that form the material. *Synthesis* refers to the methods and chemical substances required for producing a specific material, while *processing* refers to the transformation of synthesized materials into goods for use and consumption, ensuring properties suitable for each intended purpose [27]. For all materials, it is important to keep in mind that we must know about their processing, their structure, their properties, and their performance, knowing that the structure of a material will depend on how it is processed, as well as the performance will be a function of its properties [28]. For this reason, in this text we will address these four stages: we will learn about the synthesis methods (processing), which will influence the structure obtained. The determination of this structure, together with the properties of the materials obtained, is described by the characterizations. Finally, we will address the performance of these materials in their applications for water purification, considering and relating the four fundamental components of materials science.

## 2 Graphene-Based 3D Materials

Three-dimensional (3D) graphene has gained popularity due to its multifunctional properties. To better understand what a 3D graphene-based material is, it is important to first define graphene. Graphene is a 2D nanocrystalline material consisting of a single layer of $sp^2$-hybridized carbon atoms arranged in a hexagonal honeycomb lattice. Naturally, graphene serves as the building block of graphite, where graphene sheets are held together by weak van der Waals interactions. Compared to other carbon materials, graphene boasts the largest specific surface area[29], approximately 2600 $m^2/g$ [10, 30]. This value is significantly higher than the surface area of graphite, which is around 10 $m^2/g$ [31].

3D graphene is based on the formation of a network structure from 2D graphene, using different and varied synthesis techniques [32], since 2D graphene materials are ideal building blocks for the synthesis of 3D graphene-based materials, compared to 1D and 0D materials [33]. Despite the large number of studies published in recent years, the specific definition of three-dimensional graphene materials is often unclear. Sometimes, there is even confusion about whether graphite is a 3D graphene material; however, it is defined that 3D graphene materials are non-graphite 3D structured graphene materials [19]. 3D graphene includes various structures, such as 3D graphene with organized patterns and random patterns, irregular 3D graphene films, graphene sponges, foams, and 3D graphene fibers [32], among others. The goal of producing such materials, therefore, is to optimize the use and performance of graphene materials by arranging them in three dimensions rather than stacking them similarly to graphite. The latter option leads to a significant loss of surface area and other properties of few-layer graphene. Thus, a 3D graphene material should have up to 10 layers of graphene and maintain the general properties of graphene-based materials [19]. In general, a porous graphene structure, in the form of a monolith, has many interconnected, curved, low-layer graphene sheets. In other cases, such as 3D graphene films, these materials are a single low-layer graphene sheet with a distorted structure [19, 32].

Based on their shapes, 3D graphene-based materials can be classified as graphene foams and sponges, namely: organogels, hydrogels, aerogels, and xerogels. In common, these three-dimensional graphene networks have microscopic structures rich in pores and are usually obtained through the self-assembly process, as will be seen in the next section [19–21]. The porosity of macroscopic 3D graphene structures is the result of the bending, curvature, and wrinkling of the graphene sheets that compose it [19]. This structure, depending on the type of material, method, and synthesis conditions used, can present specific surface areas that have great variation, from 0.025 to approximately 1350 $m^2/g$ [26]. At the same time, these same materials have densities below 0.1 $g/cm^3$ [19].

As mentioned earlier, the main 3D graphene materials can be divided into two classes: graphene foams and graphene sponges. Foams can be defined as an interconnected network of graphene or reduced graphene, with a morphology created by the introduction of gas bubbles in the production process. Sponges, which can be divided into organogels, hydrogels, aerogels, and xerogels, are lightweight, consisting of an interconnected flexible network with excellent porosity [26]. Despite their similarity, foams have a more rigid and interconnected three-dimensional structure, while sponges exhibit greater flexibility, irregular porosity, and a superior compression capacity. Among sponges, hydrogels are materials that have their pores filled with water. They are obtained, in most cases, through simple cylindrical reactors, and their macroscopic structure depends on the size, shape, and volume of the flask used. These materials generally have homogeneous pore structures internally and externally, and the number and size of the pores can be controlled by changing the concentration of precursor or reducing agent. In addition, the size of the pores can be changed by compression [34, 35]. In turn, organogels are wet sol-gels equivalent to hydrogels, but their pores are filled with organic solvents instead of water. Aerogels can be prepared by removing the solvents in the pores of the wet gels under supercritical conditions. Since no phase boundaries are observed in the supercritical region, capillary stress is eliminated during the drying process, and a significantly reduced pore destruction can be caused by the capillary effect. Finally, xerogels are also the result of drying wet gels, but they undergo an unrestricted drying process, which results in the compaction of the structure. Briefly, in the drying of hydrogels and organogels, the liquid is replaced by air in aerogels, using techniques that keep their volume intact; on the other hand, the liquid phase present in a gel is eliminated by evaporation, in ambient or oven conditions, to form a xerogel. A comparison of the synthesis procedures of aerogels and xerogels indicates that xerogels have a greater tendency to crack during their synthesis. Aerogels have structured pores, low density, and highly branched structures [21, 36]. For cryogel preparation, the solvent within the pores is initially frozen and then sublimated under a vacuum. During freeze-drying, the solvent passes through two-phase boundaries, typically generating high capillary stress. Attempts have been made to use this method for aerogel synthesis; however, due to solvent crystallization within the pores, the final products often exhibit defects, such as cracks or even a powdery form with larger pores. A schematic showing the distinct characteristics of these materials is shown in Fig. 1.

Another important point that optimizes the application of these materials is the assessment of toxicological risk. In a previously conducted survey, it was observed that the toxicity of 0D, 1D, and 2D carbon nanomaterials does not vary according to dimensionality, that is, it is not possible to observe a direct relationship between dimensionality and toxicity. However, it was observed that 3D graphene materials tend to have lower toxicity than 0D, 1D, and 2D materials. This reduction in toxicity is attributed to the stability and compactness of the 3D structures so that the three-dimensional structure of graphene makes it less bioavailable and less biodegradable [37]. Such characteristics may also be related to the synthesis methods used to obtain

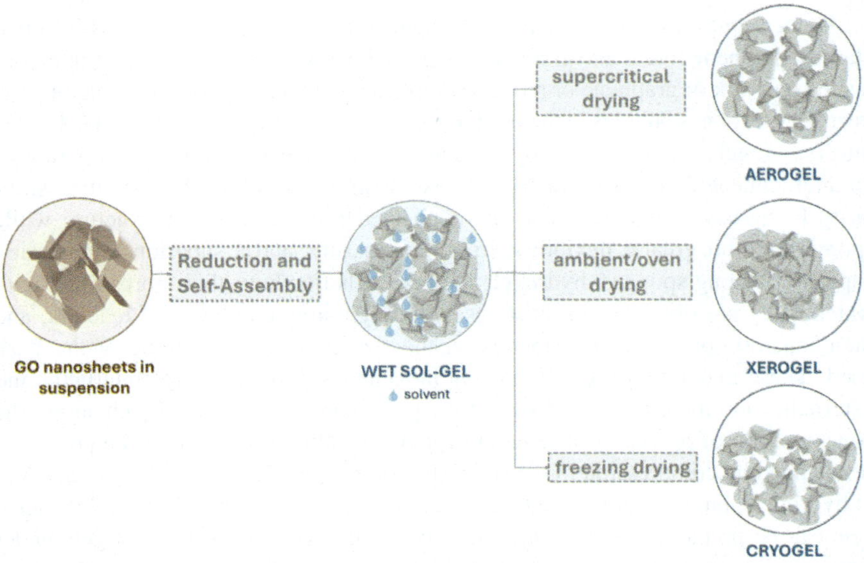


**Fig. 1** The steps of the graphene oxide reduction process and the resulting products after applying in the wet sol–gel: ambient pressure drying (Xerogel) supercritical drying (Aerogel) and freeze-drying (Cryogel)

these materials, which sometimes use only water as a solvent—as is the case with hydrogels.

## 3 Obtaining 3D Graphene-Based Nanomaterials

Before detailing the synthesis of graphene-based three-dimensional materials, it is important to define what each of the precursor materials used in their synthesis are. Pristine graphene has atomic thickness and is composed only of carbon atoms with $sp^2$ hybridization. Its production through mechanical exfoliation allows the production of materials with unique physical and chemical properties, such as high purity, high electronic conductivity, and few defects [38]. Despite its excellent properties, two main problems are related to the use of pristine graphene: the first problem is related to the greater difficulty in obtaining it compared to other graphene-based materials; another very important problem is its known hydrophobicity, resulting from the dominant presence of carbon atoms. One way to solve these problems, while expanding the applications of graphene, is the synthesis and application of graphene-based materials [39].

One of these graphene-based materials is graphene oxide (GO). It is obtained from the chemical oxidation of graphite to obtain graphite oxide (Gr-O). This graphite oxide is then exfoliated to obtain graphene oxide. Chemically, the stacking of graphene sheets, interacting through van der Waals forces, and ideally formed only by carbon, forms graphite. When performing the chemical oxidation of graphite using a strong oxidizing agent, oxygenated groups are added to the basal plane of the graphite layers. These groups can be epoxy, hydroxyl, carboxyl, and carbonyl. This chemical oxidation of graphite is usually performed using the different modified Hummers methods [40]. Then, this graphite oxide obtained can undergo a mechanical exfoliation step that will allow the obtaining of graphene oxide, and purification steps, such as centrifugation, can be performed additionally. In addition to its simpler production, the use of graphene oxide significantly increases the polarity of its predecessor. This characteristic provides excellent dispersibility in many solvents, especially in water. Furthermore, the presence of these oxygenated groups can provide reactive sites for chemical modification or functionalization, which allows the development of materials based on graphene oxide. A negative point that must be considered is that the oxygenated groups make the material electrically insulating. In addition, the material becomes quite dispersible in water, which may not be the desired effect. From graphene oxide, reduced graphene oxide (rGO) can be obtained through chemical, thermal, hydrothermal, or thermochemical reduction. Since rGO still carries residual functional groups, it has good dispersion in many solvents due to its amphiphilic character. Furthermore, it is relatively easy to control the degree of oxidation and solubility of rGO by controlling the reduction process (concentration and type of reducing agent, temperature, and time). The more reduced it is, the more the properties of this chemically reduced graphene resemble those of pristine graphene [32, 38, 39].

From the materials mentioned, it is possible to obtain materials based on three-dimensional graphene. The formation of these structures occurs through $\pi$–$\pi$ interactions and van der Waals forces between the graphene layers, which allow them to self-assemble with each other, forming an interconnected network. The manufacturing methods for these materials can be divided into two strategies: template-directed synthesis and template-free assembly of graphene oxide. Compared to template-directed synthesis, template-free assembly has the easiest synthesis, presents higher yield, greater ease of large-scale production, and lower cost [38]. Among the template-free methods, the main ones can be mentioned: chemical vapor deposition (CVD), hydrothermal method, and self-assembly. Another widely used method is 3D printing. The first 3D materials based on graphene were developed in 2009 by Vickery et al. [41] and Wang et al. [42], who developed graphene-based aerogels through freeze-drying. These and other methods have been developed and optimized over time, with an explosion of new graphene-based 3D materials and methods in recent years [43]. Each method has its advantages and disadvantages, as will be discussed below [44].

## *3.1 Chemical Vapor Deposition (CVD)*

Chemical vapor deposition (CVD) and plasma-enhanced CVD (PE-CVD) are widely adopted techniques for fabricating 3D graphene structures. Some methods for

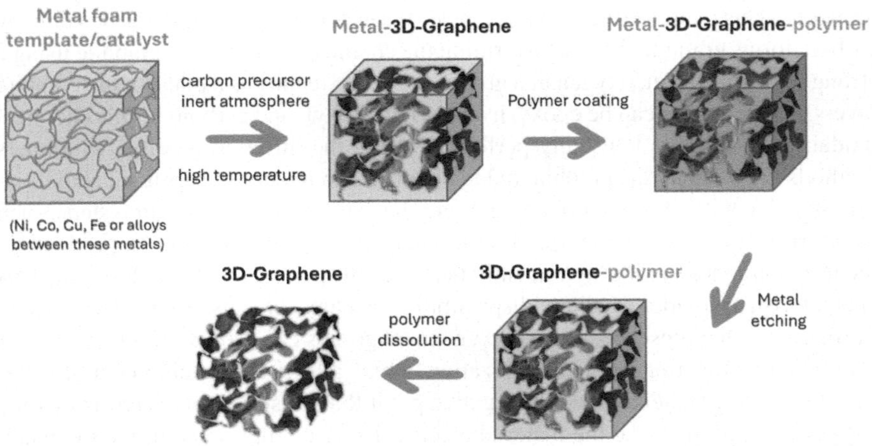


**Fig. 2** Schematic representation of the production of 3D-graphene by chemical vapor deposition method using a metal foam as a template/catalyst

preparing these materials have used hydrocarbons as precursors. At the same time, a three-dimensional substrate is used as a mold and helps to develop high-quality three-dimensional structures [32]. In these cases, nickel foam can be used as a substrate and a carbon source (such as ethanol, for example); At the end of the process, the nickel foam is removed, leaving independent material. As described in the literature, the materials obtained by this method are macroporous and extremely thin and light, while being formed only by graphene of one or few layers [45, 46], being ideal for application in the adsorption of contaminants. In contrast to in-plane graphene and 3D graphene-based nanostructures constructed from dispersed graphene oxide sheets, this material tends to have a more stable structure, but with a random microstructure and disordered pore distribution [32]. Figure 2 presents a schematic representation of the production of 3D-graphene using CVD.

## 3.2 *Hydrothermal and Thermochemical Methods*

The reduction of graphene oxide to obtain 3D graphene is one of the preferred methods in the literature. This is due to its simple and adjustable production method, allowing the production of a variety of material types and shapes. The assembly of materials from graphene oxide generally consists of two steps, the construction of the monolith and the reduction of GO in an autoclave (reactor or conventional biological sterilization autoclave [35] Fig. 3).

The two processes can occur simultaneously or successively [19, 32]. Its use in place of pristine graphene, in addition to the synthetic reasons already mentioned above, is preferable due to the oxygenated groups that result in a highly hydrophilic

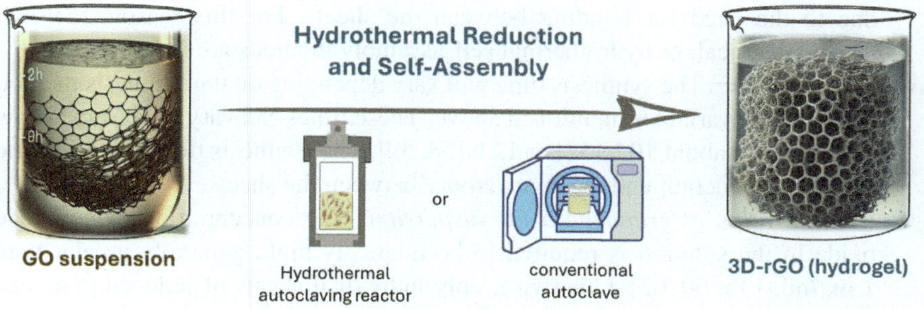


**Fig. 3** Schematic representation of the production of 3D graphene through the thermal/hydrothermal reduction of graphene oxide in an autoclave

surface, preventing the irreversible agglomeration of hydrophobic graphene [21] and returning to the initial graphitic structure. A very important factor in this type of synthesis is the concentration of graphene oxide used. Increasing the concentration of GO is effective in increasing the attraction between the sheets [35, 42].

In this strategy, hydrogels (or organogels) can be obtained by self-assembly of GO, where the dispersed GO solution acts as a precursor and self-assembles under specific conditions [38]. These conditions can be quite varied, referring to the reducing agent used and its concentration; synthesis temperature; synthesis time; pressure; and concentration of the GO dispersion. Since many factors and synthetic parameters are involved, the material of interest must be carefully studied to ensure reproducibility and fine-tuning of adsorptive properties, enabling the modulation of the analyte and tailoring the environment for the material's application [47].

i. *Type of reducing agent:* Various reducing agents have been used to obtain reduced graphene oxide from graphene oxide. The main ones include borohydrides (such as $NaBH_4$), aluminum hydride ($LiAlH_4$), hydrohalic acids (such as HI and HBr), sulfur compounds (such as $CH_4N_2O_2S$), reducing agents containing nitrogen (such as hydrazine), oxygen (such as methyl and ethyl alcohols and ascorbic acid), sulfur (such as $NaHSO_3$), acid metals, alkali metals, amino acids, plant extracts, microorganisms, proteins and hormones [48–52].
ii. *Reducing agent concentration:* Regardless of the reducing agent used, the variation in its concentration will have direct and evident effects on the degree of reduction of the reduced graphene oxide sheets. In addition, the morphology may also be altered. This occurs due to the removal of oxygenated groups that also alter the size of the sample, as well as the degree of oxidation and compaction of the three-dimensional structures based on graphene [25, 35, 52–55].
iii. *Synthesis temperature:* The synthesis temperature may vary according to the other factors mentioned in this topic. One of them, for example, is the reducing agent used, the synthesis time, and the pressure. Although some studies show the possibility of reducing graphene oxide at room temperature [56], the vast majority of methods for obtaining materials based on 3D graphene use heating

due to the need for bonding between the sheets. For this reason, thermal, thermochemical, or hydrothermal self-assembly methods are used.

iv. *Synthesis time:* The synthesis time will vary depending on the synthesis method used and the variables mentioned above. These times can vary [52] from as low as 2 h [35] to about 10 h [57] or 12 h [58, 59]. The synthesis time will affect the degree of reduction and the connectivity between the sheets.

v. *Concentration of graphene oxide dispersion:* The concentration of graphene oxide in the solution is required to be relatively high, generally greater than 1 mg/mL [35, 60–62]. Otherwise, only individual sheets of reduced graphene oxide will be obtained, due to the distance between neighboring sheets [19].

In this technique, graphene hydrogel is first obtained by chemical, thermochemical, or hydrothermal reduction. Usually, the epoxy and hydroxyl groups are more easily eliminated; ester, carboxyl, and carbonyl groups located at the edges or in defective areas are removed with greater difficulty. At the same time, curvatures, wrinkles, and defects appear in the reduced graphene oxide sheets. Such structures contribute to the porous structure of the 3D graphene product. Sequentially or concomitantly, external forces generated by centrifugation, evaporation, and pressure facilitate the bonding interaction between the sheets and assist in the self-assembly of the sheets. While the functional groups are removed, at the same time there is a decrease in the repulsive force between sheets. In this way, the closest sheets interconnect with each other, resulting in an interconnected 3D network. Due to the limited mobility of rGO sheets, simultaneous to the presence of solvent molecules between the sheets, parallel re-stacking is quite unlikely [19].

After obtaining it, it can be transformed into an aerogel, if freeze-drying is carried out. If unrestricted drying of the material is carried out, then xerogel will be obtained. Obtaining an aerogel will leave abundant micro and nanopores in the final material, with high porosity, large surface area, and extremely low apparent density [38]. On the other hand, obtaining xerogels generates a material with compacted pores and usually with low sorption capacities [53]. Furthermore, there is the possibility of using the hydrogels themselves as adsorbent materials, which have shown promise in their applications [25, 35, 54, 55].

## 3.3 3D Printing Techniques

Three-dimensional graphene-based materials can also be developed using conventional polymer 3D printing techniques. Although additive manufacturing is not directly applicable to pure carbon due to the difficulties in melting or polymerizing this material [43, 63], pure carbon 3D printing is still in the early stages of research. Since carbon can be synthesized in combination with other compounds, different 3D printing techniques have been explored for this purpose [63]. Since the manufacture of 3D graphene-based materials relies on carbon-based precursors, only techniques involving plastics or resins can be used to create these materials. Among the applicable techniques, Power Bed Fusion, Material Extrusion, and Vat Photopolymerization (Fig. 4) stand out, each with its advantages, limitations, and areas of application, adapted to the type of material and the purpose of the process [64]. Power Bed Fusion employs high-energy sources to fuse graphene-based powders layer by layer. Material Extrusion involves depositing a graphene-polymer composite through a nozzle to create intricate designs. Vat Photopolymerization uses light to solidify graphene-infused liquid resins in precise patterns, enabling the fabrication of highly detailed and functional graphene components.

Due to the methods used, there must be subsequent steps to 3D printing to ensure the integrity of the material obtained. Thus, the combination of the flexibility of 3D printing with carbonization makes it possible to create customized carbon materials in a process that involves multiple steps [63]. New 3D printing techniques have enabled major advances in the production of graphene-based materials, with emphasis on direct ink printing, material extrusion, and material jetting techniques. These techniques use precursors such as graphene [65, 66], and graphene oxide [67, 68]. Complementary processes, such as freeze-drying, carbonization, and chemical treatments, are essential to adjust the final properties of materials produced by 3D printing, expanding their application possibilities and improving the viability of their large-scale production.

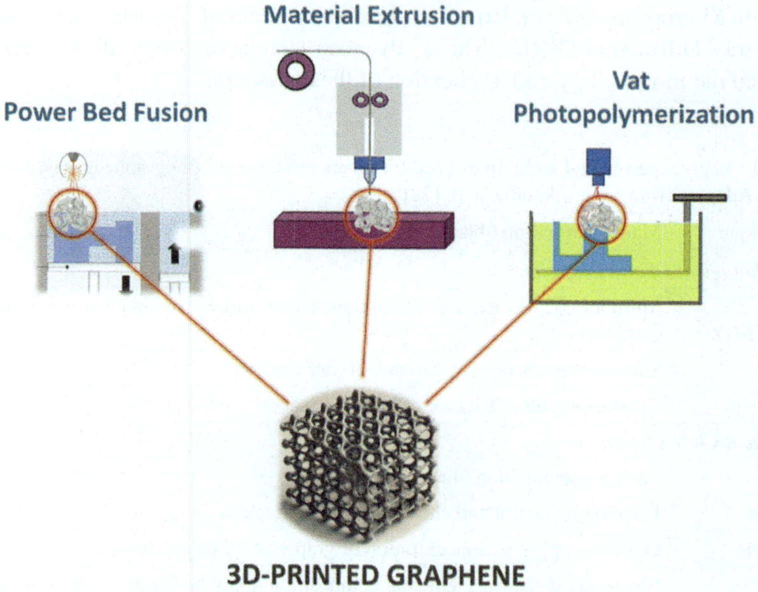


**Fig. 4** Schematic representation showing the main techniques for obtaining 3D-printed graphene structures: Power Bed Fusion, Material Extrusion, and Vat Photopolymerization

# 4 Morphological and Chemical Characterization Techniques for Nanomaterials

To understand the adsorption mechanisms of compounds on materials, it is necessary to understand the chemical composition, including the presence of functional groups, and the shape of the materials (i.e., surface area and pores) [26]. In the literature, some characterization techniques are most used to unravel the adsorption mechanisms of compounds in 3D graphene-based materials, which are described below and summarized in Table 1.

## *4.1 Morphological Characterization*

The study of the morphology of a material allows us to correlate the nano-level with the macro-level. This correlation requires morphological characterization and an understanding of its relationships with the synthesis and application of materials. With advances in the development of new characterization techniques, it is possible to analyze the morphology of materials with appropriate precision, work with high magnification, and reduce the time required for measurements [69]. The main techniques for characterizing the morphology of 3D graphene materials are Scanning Electron Microscopy (SEM), Brunauer–Emmett–Teller (BET) surface area analysis, and X-ray Diffraction (XRD). These advanced techniques often allow correlation between the morphology and application of these materials.

**Table 1** Main experimental techniques used for characterization of 3D graphene-based nanomaterials. Adapted from Mourdikoudis et al. [72]

| Technique | Main information obtained |
|---|---|
| *Morphological Characterization* | |
| SEM/ SEM–EDX | Morphology, precision in lateral dimensions, quick examination–elemental composition |
| BET | Surface area (Brunauer–Emmett–Teller method) |
| XRD | Crystal structure, distance of graphene sheets |
| *Chemical Characterization* | |
| FTIR | Surface composition, ligand binding |
| Raman | Detailed structural and chemical characterization |
| UV–Vis | Optical properties, concentration of graphene oxide precursor |
| XPS | Electronic structure, elemental composition, oxidation states, ligand binding (surface-sensitive) |
| TGA | Mass and composition of materials |

### 4.1.1 Scanning Electron Microscopy (SEM)

Microscopy is a technique that allows the visualization of structures that cannot be seen with the naked eye. Its main purpose is to form an image of the material to be characterized. Micrographs allow the visualization of the structures present on its surface. Techniques capable of improving the resolution capacity by magnifying (increasing) the images are used to visualize a sample. The Scanning Electron Microscope (SEM), which uses electrons as an excitation source, is used to observe and modify the surface of the sample. For observation, the SEM is used to capture and interpret some signals emitted during the interaction of the electron beam with the sample. These signals include electrons (secondary and backscattered), X-rays, light (ultraviolet, visible, and infrared), heat, electrons conducted through the sample, and electrons absorbed by the sample. In this way, it is possible to analyze the surface morphology, structural organization, and chemical composition of the materials [70]. The chemical composition can be analyzed through Energy Dispersive X-ray Spectroscopy (EDX), which identifies the chemical elements and their proportions in the sample. For materials based on 3D graphene, it is possible to apply SEM to understand the formation of the three-dimensional network and visualization of cavities and pores. Thus, it is possible to infer, for example, whether the structure is more or less compact.

### 4.1.2 BET Method

The Brunauer–Emmett–Teller (BET) theory explains the physical adsorption of gas molecules on a solid surface, allowing the use of the analysis technique to measure the specific surface area of materials. This theory applies to multilayer adsorption systems and generally uses gases that do not react chemically with the surfaces of materials, i.e., do not act as adsorbates. With these inert gases, it is possible to quantify the specific surface area and the pores size distribution. Usually, the most used gas for surface analysis by the BET method is nitrogen.

It is important to point out that the determination of the surface area of 3D graphene is controversial due to several factors that affect both characterization and interpretation of results. First, the highly porous and complex three-dimensional structure of 3D graphene makes precise measurement challenging, as different methods access distinct parts of the surface. For instance, gas adsorption techniques may underestimate the surface area by failing to access smaller or closed pores, while electron microscopy techniques (SEM/TEM) may provide only a visual and partial estimate of the structure.

Another important factor is the lack of standardization in defining what constitutes a surface area in the context of 3D-graphene. Some works consider only the accessible area for adsorption, while others include both the internal and external area calculated theoretically. This inconsistency makes it difficult to compare different works. Lastly, experimental conditions, such as temperature and pressure during measurements, can also influence results, especially in porous materials.

Furthermore, a material that presents a large surface area, when applied in the adsorption of a chemical compound, is expected to be efficient in its use. This behavior is expected to be due to the presence of pores (from micropores to macropores) that should help to increase the sorption capacity. However, despite the claim that good adsorption is directly related to the high surface area of the material, we often see that such a correlation is not observed. As described by Leão et al. [26], more than the surface area, for the discussion of the adsorptive capacities for different materials or analytes, the experimental conditions used must be considered.

### 4.1.3 X-Ray Diffraction (XRD)

In X-ray diffraction (XRD), atoms scatter X-ray waves, primarily using the atoms' electrons. X-ray radiation has a wavelength that is on the order of the interatomic spacing in crystalline solids and is used to determine the crystal domain size and structure of nanomaterials [71]. XRD methods are efficient in elucidating the 3D structure in terms of interplanar distance, allowing the monitoring of the variation of interplanar distances according to the synthesis method and evaluated steps. From this, and considering Bragg's Law, it is possible to calculate the variation of interplanar distances between the different materials derived from graphite, including graphene oxide, reduced graphene oxide and three-dimensional reduced graphene oxide.

## *4.2 Chemical Characterization*

Chemical characterization of materials consists of an analytical process for determining the chemical composition of a sample, aiming to identify its properties. Among the many techniques used in the characterization of three-dimensional materials based on graphene, the main ones are Fourier Transform Infrared Spectroscopy (FTIR), Raman Spectroscopy, Ultraviolet/Visible Spectroscopy (UV–Vis), X-ray photoelectron Spectroscopy (XPS) and Thermogravimetric Analysis (TGA). These analyses will be briefly described below.

### 4.2.1 Fourier Transform Infrared Spectroscopy (FTIR)

FTIR spectroscopy is a spectral method that works based on the vibration and rotation of atoms. In this technique, molecules with a dipole moment can be detected by fingerprint [73]. This occurs because almost all compounds with covalent bonds, whether organic or inorganic, absorb various frequencies of electromagnetic radiation in the infrared region of the electromagnetic spectrum. This absorption is measured in terms of wavenumbers, typically ranging from 4000 to 400 $cm^{-1}$.[74].

Characterization by FTIR allows the qualitative and semi-quantitative determination of the functional groups present in the materials. In oxidized graphene materials, it is possible to observe the presence of a variety of intense bands that can be attributed to the oxygenated groups usually found in these materials. When reduced, a decrease in the intensity of these bands is observed according to the degree of reduction produced—and this will be influenced by the concentration of the reducing agent, time, and reaction temperature. Its applicability concerning adsorption in materials based on 3D graphene is related to the confrontation of the polarity of the materials and the adsorbed compounds.

### 4.2.2 Raman Spectroscopy

Characterization by Raman spectroscopy, as a complement to FTIR, allows the analysis, prediction and explanation of the interactions that occur between the material and the analytes. Raman spectra are easily compared with previously published studies, since carbon allotropes present their fingerprints in Raman spectroscopy, with the presence of at least one of the so-called D, G and D' bands, around 1350 $cm^{-1}$, 1580 $cm^{-1}$ and 1620 $cm^{-1}$, respectively.

Obtaining graphene-based materials from graphite oxidation produces pronounced effects on the Raman spectrum, especially with the ratio of the intensities of the D and G bands. The intensity ratio of these bands ($I_D/I_G$) can be used to identify the level of disorder in the structure, providing insight into the reduction process through the removal of oxygen functional groups that create imperfections within the carbon structure and the degree of connection between the graphene sheets. This value is determined from the deconvolution of the data to obtain the area under each of the bands, which allows for obtaining the ratio of the areas of the D and G bands. A higher intensity ratio of the D and G bands ($I_D/I_G$) may be related to a greater capacity to sorb polar compounds, since it represents a greater number of defects, such as the presence of oxygenated functional groups. At the same time, materials with a lower $I_D/I_G$ ratio can predict a better interaction with more nonpolar compounds, since aromatic rings without the presence of defects can present better sorption due to $\pi$-$\pi$ interactions.

Recent studies have focused on the spectroscopic evaluation of graphene-based three-dimensional materials using FTIR and Raman techniques, both before and after adsorption. Spectroscopic differences, combined with theoretical calculations, can provide valuable insights into the adsorption mechanisms of contaminants on these 3D materials [24, 53, 54].

### 4.2.3 X-ray Photoelectron Spectroscopy (XPS)

X-ray photoelectron spectroscopy (XPS) is one of the most sensitive, informative, and accessible surface analysis techniques, and can be more advantageous over FTIR.

[71]. The use of XPS to characterize three-dimensional graphene materials has limitations due to its shallow analysis depth of 1–10 nm, which prevents the evaluation of internal layers. The porous and complex structure of 3D graphene complicates the collection of emitted electrons, leading to low resolution and inconsistent signals, as well as mixed chemical information from internal and external surfaces. Issues such as charging effects in less conductive areas, difficulties in accurately quantifying elements due to the three-dimensional geometry, and challenges in sample preparation also affect the results. Therefore, XPS is most effective when combined with other techniques for a more comprehensive characterization.

### 4.2.4 Thermogravimetric Analysis (TGA)

Thermogravimetric analysis (TGA) is a thermal analysis method in which changes in the physical and chemical properties of materials are measured as a function of increasing temperature while maintaining a constant heating rate. TGA can provide information about materials undergoing vaporization, sublimation, sorption and desorption, dehydration, and decomposition. The technique is commonly used to determine the characteristics of materials that exhibit mass loss or gain due to decomposition, oxidation or loss of volatiles such as moisture [71]. These applications are directly related to the chemical characterization of the materials of interest, since they can represent, as a function of temperature, the amount of solvent present in the sample, as well as the percentage of oxygenated groups. Finally, the degradation temperature of the material studied can be determined. This characterization can be complementary to others such as FTIR. While FTIR provides semi-quantitative information on the oxygenated functional groups, with TGA it is possible to determine the amount present, through the degradation temperature of these groups.

## 5 3D Graphene-Based Nanomaterials as Adsorbents

3D graphene-based materials are multifunctional materials, being applied in electrocatalysis [75], photocatalysis [76], supercapacitors [77–79], solar cells [80], energy storage [81, 82], sensing [83], biotechnology [84] and in the adsorption of contaminants in water [26, 85–87], among other diverse applications. The structural characteristics of 3D graphene-based materials include low density, high porosity and high surface areas. The combination of a porous structure and the large specific surface area of these materials has attracted great attention in the areas of sorption of oils, dyes, metals, pesticides, drugs and endocrine disruptors [20, 26]. Regarding the dimensionality of materials, comparing the efficiency and sorption mechanisms of 3D materials with 1D and 2D materials is not trivial, and many factors must be considered. Although 3D materials are significantly better than 1D materials in drug sorption, when compared with 2D materials, in some cases they may not present better sorption capacities, but present shorter sorption times [55]. Graphene oxide dispersions or two-dimensional graphene materials have demonstrated high efficiency and wide use in adsorption, but their practical application is restricted due to the difficulty of separation and recovery, as well as the potential health risks arising from their nanotoxic effect [35, 37, 88]. These results may be related to the intraparticle diffusion mechanism.

In pesticide adsorption, we observed that a wide range of analytes were investigated, but with few repetitions between materials. For a "simple" graphene-based material, only paraquat has been studied so far [89]. On the other hand, important contaminants such as atrazine, chlorpyrifos, glyphosate, tebuconazole, among many others, were investigated for materials based on 3D graphene and decorated with iron oxide nanoparticles, glucamine, carbon nanotubes and 3D graphene/$Fe_3O_4$ and 3D graphene/cellulose nanocomposites [90–95]. Adsorptive capacities as low as 0.31 mg/g were found for the adsorption of ametryn on a 3D graphene/cellulose composite [91], but adsorptive capacities as high as 546 mg/g were obtained in the adsorption of glyphosate on carboxylated carbon nanotubes-graphene oxide aerogels [94]. In few studies the surface areas of the materials tested in the removal of pesticides from water were presented [89, 90, 93]. Still on the removal of emerging contaminants, drug adsorption studies have shown a significant increase in recent years. For this class of contaminants, higher adsorption capacities have been found for different materials. Of note is the use of 3D reduced graphene oxide/nano-$Fe_3O_4$ hybrid hydrogel, which achieved the capacity to remove 2113 mg/g of tetracycline. The same material was also able to remove 920 mg/g of ciplofloxacin from water [57]. For these analytes, studies are also found in which there is a comparison between three-dimensional, two-dimensional and one-dimensional materials with different degrees of oxidation for the adsorption of atenolol [55] and three-dimensional with different degrees of oxidation in the removal of paracetamol [25]. Both studies found better performance in the removal of atenolol, a highly hydrophilic drug, in more oxidized materials. In the verification of dimensionality, much more outstanding performances were observed for three-dimensional materials considering the time and adsorption capacity [55].

Endocrine disruptors, which can interact with the endocrine system, generating effects such as blocking or mimicking natural hormones, are also important contaminants to be studied for water removal. Among the most studied endocrine disruptors are Bisphenol A [96–101], widely found in everyday polymers, phenol [100, 102] and hydroquinone [102, 103]. The adsorption of bisphenol A has been described along with the specific surface area on materials such as graphene hydrogel-AgBr@rGO, with adsorptive capacity ($Q_{max}$) of 80.1 mg/g and specific surface area (SSA) of approximately 200 $m^2/g$ [96]; magnesium ascorbyl phosphate graphene-based monolith ($Q_{max}$ = 324 mg/g and SSA = 294.3 $m^2/g$) [97]; rGO-β-cyclodextrin aerogel ($Q_{max}$ = 346 mg/g and SSA de 172 $m^2/g$) [98]; 3D foam-like graphene oxide ($Q_{max}$ = 420.9 mg/g and SSA = 974.8 $m^2/g$) [100]; $Ag_3PO_4$-graphene hydrogel ($Q_{max}$ = 15 mg/g and SSA = 325,3 $m^2/g$) [101]. Note that for these materials, even with the same analyte in common, the materials present great variation in the adsorptive capacity–specific surface area relationship. For example, the materials of Mu et al. [101] and Fang et al. [97] have very similar surface areas, but the adsorptive capacity of the second is about 21 times greater than that of the first. This result may be related, for example, to the longer contact time (18 h versus 0.13 h) and/or the adsorbent dosage (5 times lower in the material with greater adsorptive capacity). Therefore, it is evident that the comparison between materials alone is quite important, but experimental standardization seems to be more relevant to allow direct comparisons between materials. The removal of hydrocarbons by adsorption is usually represented in units of g/g, due to the great capacity of 3D graphene materials to adsorb these contaminants. In general, the most investigated adsorbates have been toluene, acetone and ethanol, with adsorptive capacities of up to 180 g/g in a 3D-GO [104], cyclohexane and ethyl acetate, with sorption capacities of up to 239.7 g/g in 3D-rGO-$Fe_3O_4$ [105].

Dye adsorption is widely used as a model due to the ease of quantifying dyes using more accessible techniques. In addition, many of them can be considered environmentally relevant contaminants. Undoubtedly, the most evaluated dyes for adsorption are methylene blue, rhodamine B, methyl orange and safranin. Methylene blue is widely used as a model for adsorption, having been tested on 3D-rGO with different degrees of oxidation and presenting sorption capacities ranging from 227.5 mg/g to 1034.7 mg/g. In this same work, the relationship between degree of oxidation, pore sizes, cluster formation and adsorption was revealed through theoretical molecular dynamics calculations [54]. The same behavior was observed for the same materials in the adsorption of safranin [35]. Other materials composed only of graphene-based materials showed excellent adsorptive capacities for methylene blue [106–108] and methyl violet [107]. Finally, for metals, the large number of studies carried out with Cd (II), Cr (IV), Cu (II), Hg (II) and Pb (II) is evident. As for the other analytes, the surface area does not seem to be the factor with the greatest influence on the adsorptive capacity for the different materials tested. For example, for U (VI), similar adsorptive capacities are observed for materials with surface areas that vary approximately 54 thousand times [109, 110]. For lead, the opposite was observed. Materials with very similar areas had a 2.3-fold variation in adsorption capacity [111, 112]. The same behavior was observed for Cu (II): materials with very similar areas, but with a variation of about 12 times in the adsorption capacity [113, 114].

These results make it clear that a much deeper relationship than just composition-surface area-adsorptive capacity is observed for adsorption studies for different classes of contaminants. Contact time, adsorbent dosage, pH range, and initial concentration can be equally or even more critical than the physical and morphological characteristics of the materials. For instance, the tested pH range should align positively with the pH at the material's point of zero charge. The hydrophobicity index, in turn, is directly linked to the adsorption efficiency for organic compounds with varying polarities. Similarly, using a lower adsorbent dosage might lead to an overestimation of the adsorptive capacity, as the $Q_{max}$ values are inversely proportional to the mass of the adsorbent. Additionally, in most studies, a higher initial analyte concentration corresponds to a greater adsorptive capacity. While this trend may seem intuitive, it is not necessarily proportional in all scenarios. By modulating the porosity, specific surface area and degree of oxidation of the structure, these materials can be applied in different scenarios [34, 115]. Less oxidized materials are commonly considered as competitive and efficient sorbents for the removal of oil and inorganic compounds in water, with higher sorption capacity compared to other types of sorbents. On the other hand, materials containing abundant oxygenated groups exhibit excellent sorption capacity for organic contaminants [116]. Furthermore, 3D graphene presents an architecture with pores of varying sizes, including micropores, mesopores and macropores. While micropores and mesopores provide high surface area, macropores offer easy access to the sorption surface [85]. A point that is rarely discussed in the literature refers to the influence of pore sizes in the adsorbent, due to the formation of adsorbate clusters. The average pore diameter of 3D materials is required to be very large, allowing the analyte to diffuse through them. Correspondingly, the pore volume also has a notable effect on the diffusion process. Therefore, the change in the pore structure is important for obtaining higher adsorptive capacities [23]. Some theoretical studies have shown that the distance between layers is a determining factor for the adsorption of ions and molecules in graphene-based membranes so there is an increase in adsorption for larger distances. This happens because as the positions between layers are increased, there is more space for molecules to form clusters [53]. In addition to molecular dynamics simulations showing that with increasing pore size, there is an increase in the amount of molecules inside the pore, they also show that there is a decrease in the interaction of the adsorbate with the external walls of the material. This can be observed due to the formation of clusters, which prevent molecules from entering pores of reduced size [54]. Thus, clustering can be considered an important driving force for contaminant adsorption in these systems. In addition, three-dimensional graphene has high stability in aqueous environments, adapting to a wide variety of adverse conditions, such as strong and salty aqueous solutions and even organic solvents such as dimethylformamide and cyclohexane [117]. These characteristics indicate that the use of three-dimensional macrostructures based on graphene has great potential in the water purification process, both for organic and inorganic contaminants.

The adsorption of organic contaminants on 3D graphene-based materials has been extensively studied, especially pesticides, pharmaceuticals, endocrine-disrupting chemicals, hydrocarbons, and dyes, which have gained prominence in recent years when compared to inorganic contaminants. For these contaminants, functionalized materials are usually used, since the groups (mainly containing O or N) of the contaminants are more easily attracted to the oxygenated groups of the materials. The polarity of the material is also an important point to be considered: the more polar a contaminant, the more easily it will be adsorbed by a functionalized material. For hydrocarbons, we can mention that an important factor is the contact angle, as smaller contact angles provide better adsorptive capacities [26]. Theoretical studies have shown that the functionalization of carbon nanomaterials strongly affects all interactions between organic compounds and adsorbents based on three-dimensional graphene. This functionalization allows the formation of hydrogen bonds and decreases the hydrophobicity of the material's surface, increasing its solubility in aqueous solution [25, 54]. These statements are valid for both two-dimensional and three-dimensional graphene-based materials [118, 119]. An advantage of three-dimensional materials over others lies in the already exhaustively reported characteristics of 3D materials: high porosity, structural stability, and probably a high specific surface area.

As for the surface area, a review carried out by Leão et al. [26] details and correlates the adsorptive capacities of graphene-based materials, functionalized and/or decorated or not. In this review, which addressed and reviewed dozens of studies on this topic, a significant lack of experimental standardization in adsorption studies was found; in addition, some of the studies present partial information, not including essential data such as the mass of the material and volume of the analyte used, which allow the determination of the adsorbent dosage. In some cases, adsorptive capacities are determined without the presentation of kinetic and isothermal studies. In addition, many material properties are ignored or disregarded. Among the main ones, we can mention the pH of the point of zero charge, hydrophobicity index; presence of acidic and basic functional groups on the surface of the adsorbent, as well as the pore structure, since more compact structures can prevent particles of larger diameters from penetrating the interior of the solid. For this reason, contrary to what is stated in most published works, a material that has a high specific surface area will not necessarily have a good adsorptive capacity. Due to the great chemical and morphological variety of these materials, their adsorption capacities for different pollutants vary from units to thousands of mg/g [120]. These variations can even occur for the same material being applied to remove different contaminants [100, 104, 111, 112, 121] (Fig. 5).

Adsorption on 3D graphene-based materials is a physicochemical process [122]. During adsorption, a chemical partitioning process occurs. The analyte, previously solubilized in water, undergoes this partitioning process and is then adsorbed as a solid particle. At the same time, if the adsorption is weak, a desorption and re-adsorption process may occur simultaneously. For this reason, the mutual effect between the adsorbents and the contaminants is essential to improve the adsorption performance [122]. Regarding the behavior as a function of time, adsorption in 3D graphene-based materials generally occurs first on the external surface (instantaneous adsorption) and subsequently follows the intraparticle diffusion process [122, 123]. Adsorption mechanisms are often not elucidated in publications, despite being more frequently presented in recent years. However, it is possible to predict which material-analyte interactions can be found in these materials. While pore filling is mainly responsible for the adsorption of organic pollutants, such as oils and organic solvents, $\pi$–$\pi$ interactions involve the non-covalent attraction between aromatic rings of organic pollutants and graphene sorbents. In other words, hydrophobic interaction occurs between non-polar organic molecules with the hydrophobic binding sites of graphene sorbents. There is also a good interaction between these structures and inorganic pollutants, such as metal ions. Another important adsorption mechanism, especially the adsorption of polar compounds on functionalized materials, is hydrogen bonding, which is a dipole–dipole interaction between a hydrogen atom covalently bonded to highly electronegative atoms in functionalized graphene sorbents with another electronegative atom carrying a lone pair. Finally, electrostatic interactions can significantly contribute to the adsorption of positively or negatively charged pollutants, including cationic heavy metal ions and ionic organic dyes, onto the

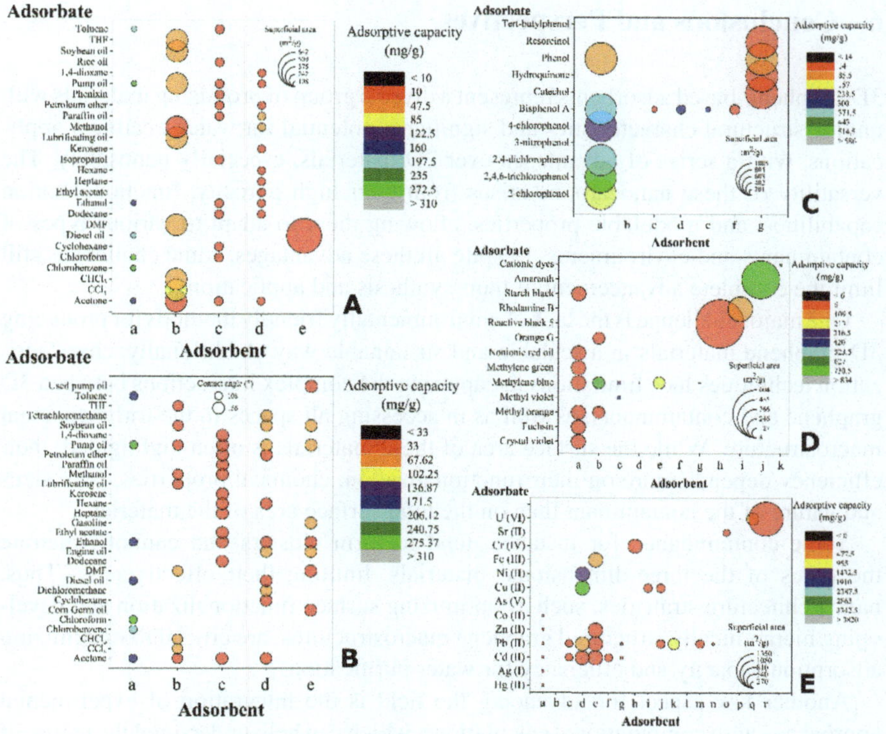


**Fig. 5** Specific surface area versus adsorption capacity for each material as a function of the analyte investigated for **a** hydrocarbons and **b** contact angle versus adsorption capacity for each material (**b**). Specific surface area versus adsorption capacity for each material as a function of **C** endocrine disruptors, **d** dyes, and **e** metals. Adapted from Leão et al. [26]

binding sites of graphene sorbents. Finally, the presence of defects on the graphene surface, such as edges and vacancies, can aid in the adsorption process, as well as the interconnected pore structure that provides more spaces to immobilize and trap water pollutants within the micropores [26, 124, 125].

Some recent theoretical–experimental studies [25, 53, 54] have shown that, when it comes to three-dimensional materials, two key factors are essential for efficient adsorption: (i) The nature of the contaminants, which, in certain media, may form aggregates large enough to prevent them from penetrating the material's pores, thereby reducing the adsorptive capacity. (ii) The degree and type of functionalization of graphene sheets, which can result in selectivity and enhanced adsorption capacity. In this sense, systematic planning of synthesis and complete characterization of 3D materials can help in understanding the adsorption mechanisms, with all these steps being interconnected according to the four components of materials science: processing, structure, properties and performance [28].

## 6 Conclusions and Perspectives

3D-graphene-based adsorbents represent a diverse group of promising materials with unique structural characteristics and significant potential for water treatment applications, with a series of advantages over 2D materials, especially nanosafety. The versatility of these nanocarbons arises from their high porosity, functionalization capabilities, and modulable properties, allowing them to adapt to various types of contaminants and environments. Despite all these advantages, some challenges still limit the complete advancement of their synthesis and application.

One major challenge is the lack of environmentally friendly methods for producing 3D-graphene materials in a scalable and sustainable way. Additionally, characterization techniques face limitations in capturing the complex interactions between 3D graphene and contaminants, as well as in accessing all spaces in the tridimensional macrostructure. While the surface area of these materials is often highlighted, their efficiency depends more on their functionalization, chemical properties, pore sizes and nature of the contaminant than on the total surface area of the material.

Some contaminants, for instance, tend to form clusters that cannot penetrate the pores of the three-dimensional materials, limiting their effectiveness. Thus, nanoarchitecture strategies, such as optimizing surface functionalization and developing hierarchically structured graphene macrostructures, are crucial to maximizing adsorption capacity and efficiency for water purification.

Another key aspect for advancing the field is the integration of experimental approaches with computational calculations, which can help understand the nature of the interactions between materials and contaminants and optimize adsorbent design. Furthermore, combined with these theoretical calculations, the use of spectroscopic techniques, such as FTIR and Raman, has proven to be essential for elucidating interaction mechanisms.

Future efforts should be focused on greener synthesis routes, nanoarchitecture strategies, and interdisciplinary approaches integrating experiments with theoretical modeling. Together, these advances could enable the development of highly efficient and adaptable three-dimensional graphene-based adsorbents tailored to the specific demands of water treatment and contribute to innovative and environmentally responsible solutions.

**Acknowledgements** The authors acknowledge funding from Brazilian agencies CNPq, CAPES, FAPERGS, National Institute of Science and Technology of carbon nanomaterials (INCT—Nanocarbon) and National Institute of Science and Technology of Nanomaterials for Life (INCT—NanoVida).

## References

1. Neto, A.C., Guinea, F., Peres, N.M., et al.: The electronic properties of graphene. Rev. Mod. Phys. **81**, 109 (2009)

2. Geim, A.K., Novoselov, K.S.: The rise of graphene, In: nanoscience and technology: a collection of reviews from nature journals. World Scientific, 11–19 (2010)
3. Tan, C., Cao, X., Wu, X.-J., et al.: Recent advances in ultrathin two-dimensional nanomaterials. Chem. Rev. **117**, 6225–6331 (2017)
4. Bolotin, K.I., Sikes, K.J., Jiang, Z., et al.: Ultrahigh electron mobility in suspended graphene. Solid State Commun. **146**, 351–355 (2008)
5. Morozov, S., Novoselov, K., Katsnelson, M., et al.: Giant intrinsic carrier mobilities in graphene and its bilayer. Phys. Rev. Lett. **100**, 016602 (2008)
6. Li, X., Zhu, Y., Cai, W., et al.: Transfer of large-area graphene films for high-performance transparent conductive electrodes. Nano Lett. **9**, 4359–4363 (2009)
7. Cai, W., Zhu, Y., Li, X., et al.: Large area few-layer graphene/graphite films as transparent thin conducting electrodes. Appl. Phys. Lett. **95**, 123115 (2009)
8. Nazeer, M., Akbar, S., Zulfiqar, S., et al.: CuO@ 3D graphene modified glassy carbon electrode towards the detection of Orange II and Rhodamine B. Mater. Sci. Eng. B **310**, 117770 (2024)
9. Duoc, P.N.D., Binh, N.H., Van Hau, T. et al.: A novel electrochemical sensor based on double-walled carbon nanotubes and graphene hybrid thin film for arsenic (V) detection. J. Hazardous Mater., 123185 (2020)
10. Ahmad, F., Zahid, M., Jamil, H., et al.: Advances in graphene-based electrode materials for high-performance supercapacitors: a review. J. Energy Storage **72**, 108731 (2023)
11. Peng, W., Li, H., Liu, Y., et al.: A review on heavy metal ions adsorption from water by graphene oxide and its composites. J. Mol. Liq. **230**, 496–504 (2017)
12. Yu, J.-G., Yu, L.-Y., Yang, H., et al.: Graphene nanosheets as novel adsorbents in adsorption, preconcentration and removal of gases, organic compounds and metal ions. Sci. Total Environ. **502**, 70–79 (2015)
13. Shen, Y., Fang, Q., Chen, B.: Environmental applications of three-dimensional graphene-based macrostructures: adsorption, transformation, and detection. Environ. Sci. Technol. **49**, 67–84 (2015)
14. Wu, Y., Zhu, J., Huang, L.: A review of three-dimensional graphene-based materials: synthesis and applications to energy conversion/storage and environment. Carbon **143**, 610–640 (2019)
15. Chen, L., Han, Q., Li, W., et al.: Three-dimensional graphene-based adsorbents in sewage disposal: a review. Environ. Sci. Pollut. Res. **25**, 25840–25861 (2018)
16. Chen, Z., Jin, L., Hao, W. et al.: Synthesis and applications of three-dimensional graphene network structures. Materials Today Nano (2019)
17. Rish, S.K., Wang, R., Tahmasebi, A., et al.: Catalytic synthesis of 3D graphene nanostructures from biomass-based activated carbon with excellent lithium storage performance. Diam. Relat. Mater. **139**, 110305 (2023)
18. Mauter, M.S., Elimelech, M.: Environmental applications of carbon-based nanomaterials. Environ. Sci. Technol. **42**, 5843–5859 (2008)
19. Sun, Z., Fang, S., Hu, Y.H.: 3D graphene materials: from understanding to design and synthesis control. Chem. Rev. **120**, 10336–10453 (2020)
20. Yousefi, N., Lu, X., Elimelech, M., et al.: Environmental performance of graphene-based 3D macrostructures. Nat. Nanotechnol. **14**, 107–119 (2019)
21. Fang, Q., Shen, Y., Chen, B.: Synthesis, decoration and properties of three-dimensional graphene-based macrostructures: a review. Chem. Eng. J. **264**, 753–771 (2015)
22. Joseph, K.M., Dangel, G.R., Shanov, V.: Modified 3D graphene for sensing and electrochemical capacitor applications. Nanomaterials **14**, 108 (2024)
23. Sahoo, P.K., Nayak, G., Gupta, J., et al.: 3D graphene-based materials and their potential application on wastewater management. Environ. Qual. Manage. **33**, 205–220 (2023)
24. Yin, Z., Cui, C., Chen, H., et al.: The application of carbon nanotube/graphene-based nanomaterials in wastewater treatment. Small **16**, 1902301 (2020)
25. de Matos, C.F., Leão, M.B., Vendrame, L.F., et al.: Unlocking the paracetamol adsorption mechanism in graphene tridimensional-based materials: an experimental-theoretical approach. Frontiers in Carbon **3**, 1305183 (2024)

26. Leão, M.B., Bordin, J.R., Matos, C.F.: Specific surface area versus adsorptive capacity: an application view of 3D graphene-based materials for the removal of emerging water pollutants. Water Air Soil Pollut. **234**, 136 (2023)
27. Askeland, D.R., Wright, W.J.: Ciência e engenharia dos materiais, Cengage Learning (2023)
28. Willian, D., Callister, J., Rethwisch, D.: Ciência e engenharia de materiais: uma introdução, in, LTC (2012)
29. Chen, R., Zhou, X.: Recent advances in 2D graphene reinforced metal matrix composites. Nanotechnology **33**, 062003 (2021)
30. Rao, C., Sood, A., Voggu, R., et al.: Some novel attributes of graphene. J. Phys. Chem. Lett. **1**, 572–580 (2010)
31. Segundo, J., Vilar, E.O.: Grafeno: Uma revisão sobre propriedades, mecanismos de produção e potenciais aplicações em sistemas energéticos. Revista eletrônica de materiais e processos **11**, 54–57 (2016)
32. Singh, M., Dawsey, T., Gupta, R.K.: Recent advancement in 3D graphene for metal-sulfur batteries. J. Energy Storage **73**, 109059 (2023)
33. Ding, M., Li, C.: Recent advances in simple preparation of 3D graphene aerogels based on 2D graphene materials. Frontiers Chem. **10**, 815463 (2022)
34. Zhi, D., Li, T., Li, J., et al.: A review of three-dimensional graphene-based aerogels: synthesis, structure and application for microwave absorption. Compos. B Eng. **211**, 108642 (2021)
35. Leão, M.B., Rosa, P.C., Dalla Corte, C.L., et al.: Eco-friendly, non-toxic and super adsorbent hydrogels based on graphene. Mater. Chem. Phys. **288**, 126408 (2022)
36. Tran, N., Choi, H.W., Tran, Q.N.: A review of green aerogel-and xerogel-based electrodes for supercapacitors. Polymers **16**, 2848 (2024)
37. da Rosa, P.C.C., Leão, M.B., Dalla Corte, C.L., et al.: Evaluation of the carbon nanostructures toxicity as a function of their dimensionality using model organisms: a review. Water Air Soil Pollut. **232**, 1–15 (2021)
38. Zhang, Y., Xia, X., Liu, B., et al.: Multiscale graphene-based materials for applications in sodium ion batteries. Adv. Energy Mater. **9**, 1803342 (2019)
39. Razaq, A., Bibi, F., Zheng, X., et al.: Review on graphene-, graphene oxide-, reduced graphene oxide-based flexible composites: From fabrication to applications. Materials **15**, 1012 (2022)
40. Hummers, W.S., Jr., Offeman, R.E.: Preparation of graphitic oxide. J. Am. Chem. Soc. **80**, 1339–1339 (1958)
41. Vickery, J.L., Patil, A.J., Mann, S.: Fabrication of graphene–polymer nanocomposites with higher-order three-dimensional architectures. Adv. Mater. **21**, 2180–2184 (2009)
42. Wang, J., Ellsworth, M.: Graphene aerogels. ECS Trans. **19**, 241 (2009)
43. Yu, W., Zhao, W., Liu, X.: Pulsed laser welding of macroscopic 3D graphene materials. Mate. Horizons **10**, 5597–5606 (2023)
44. Shah, N.R.A.M., Yunus, R.M., Rosman, N.N., et al.: Current progress on 3D graphene-based photocatalysts: from synthesis to photocatalytic hydrogen production. Int. J. Hydrogen Energy **46**, 9324–9340 (2021)
45. Dong, X., Chen, J., Ma, Y., et al.: Superhydrophobic and superoleophilic hybrid foam of graphene and carbon nanotube for selective removal of oils or organic solvents from the surface of water. Chem. Commun. **48**, 10660–10662 (2012)
46. Chen, Z., Ren, W., Gao, L., et al.: Three-dimensional flexible and conductive interconnected graphene networks grown by chemical vapour deposition. Nat. Mater. **10**, 424–428 (2011)
47. Pilato, S., Moffa, S., Siani, G., et al.: 3D graphene oxide-polyethylenimine scaffolds for cardiac tissue engineering. ACS Appl. Mater. Interfaces. **15**, 14077–14088 (2023)
48. Chua, C.K., Pumera, M.: Chemical reduction of graphene oxide: a synthetic chemistry viewpoint. Chem. Soc. Rev. **43**, 291–312 (2014)
49. Palomba, M., Carotenuto, G., Longo, A.: A brief review: the use of L-ascorbic acid as a green reducing agent of graphene oxide. Materials **15**, 6456 (2022)
50. Ismail, Z.: Green reduction of graphene oxide by plant extracts: a short review. Ceram. Int. **45**, 23857–23868 (2019)

51. De Silva, K., Huang, H.-H., Joshi, R., et al.: Chemical reduction of graphene oxide using green reductants. Carbon **119**, 190–199 (2017)
52. Guex, L.G., Sacchi, B., Peuvot, K.F., et al.: Experimental review: chemical reduction of graphene oxide (GO) to reduced graphene oxide (rGO) by aqueous chemistry. Nanoscale **9**, 9562–9571 (2017)
53. Côrtes, P.R., Bitencourt Leão, M., Reis, G.L.R. et al.: Unraveling the Phosphorus Adsorption Mechanisms in Three-dimensional Reduced Graphene Oxide Materials. Langmuir (2024)
54. Leão, M.B., Vendrame, L., Jauris, I.M., et al.: Combining multi-scale simulations and experiments to unveil the adsorption of methylene blue in graphene tridimensional-based materials. Molec. Syst. Design Eng. **8**, 666–680 (2023)
55. Leao, M.B., Fernandes, A.N., de Matos, F., Jauris, C.: Nanocarbon dimensionality and oxidation degree influence the sorption of atenolol: implications for water treatment. ACS Appl. Nano Mater. **7**, 8307–8317 (2024)
56. Zhang, J., Yang, H., Shen, G., et al.: Reduction of graphene oxide via L-ascorbic acid. Chem. Commun. **46**, 1112–1114 (2010)
57. Shan, D., Deng, S., Jiang, C., et al.: Hydrophilic and strengthened 3D reduced graphene oxide/nano-Fe 3 O 4 hybrid hydrogel for enhanced adsorption and catalytic oxidation of typical pharmaceuticals. Environ. Sci. Nano **5**, 1650–1660 (2018)
58. Gao, H., Sun, Y., Zhou, J., et al.: Mussel-inspired synthesis of polydopamine-functionalized graphene hydrogel as reusable adsorbents for water purification. ACS Appl. Mater. Interfaces. **5**, 425–432 (2013)
59. Ma, J., Yang, M., Yu, F., et al.: Water-enhanced removal of ciprofloxacin from water by porous graphene hydrogel. Sci. Rep. **5**, 13578 (2015)
60. Chen, W., Xiao, P., Chen, H., et al.: Polymeric graphene bulk materials with a 3D cross-linked monolithic graphene network. Adv. Mater. **31**, 1802403 (2019)
61. Xu, Y., Sheng, K., Li, C., et al.: Self-assembled graphene hydrogel via a one-step hydrothermal process. ACS Nano **4**, 4324–4330 (2010)
62. Singh, R., Ullah, S., Rao, N., et al.: Synthesis of three-dimensional reduced-graphene oxide from graphene oxide. J. Nanomater. **2022**, 8731429 (2022)
63. Blyweert, P., Nicolas, V., Fierro, V., et al.: 3D printing of carbon-based materials: A review. Carbon **183**, 449–485 (2021)
64. Romero, C., Liu, Z., Wei, Z. et al.: A review on hierarchical porous carbon derived from various 3D printing techniques. Nanoscale (2024)
65. Yuan, S., Fan, W., Wang, D., et al.: 3D printed carbon aerogel microlattices for customizable supercapacitors with high areal capacitance. J. Mater. Chem. A **9**, 423–432 (2021)
66. Lin, Y., Liu, F., Casano, G. et al.: Pristine Graphene Aerogels by Room-Temperature Freeze Gelation (2016)
67. Zhu, C., Han, T.Y.-J., Duoss, E.B., et al.: Highly compressible 3D periodic graphene aerogel microlattices. Nat. Commun. **6**, 6962 (2015)
68. Zhang, Q., Zhang, F., Medarametla, S.P., et al.: 3D printing of graphene aerogels. Small **12**, 1702–1708 (2016)
69. Popelka, A., Zavahir, S., Habib, S.: Morphology analysis. In: Polymer Science and Innovative Applications. Elsevier, pp. 21–68 (2020)
70. Da Róz, A.L., Leite, F.D.L., Ferreira, M. et al.: Técnicas de nanocaracterização: princípios e aplicações (2015)
71. Nasrollahzadeh, M., Atarod, M., Sajjadi, M., et al.: Plant-mediated green synthesis of nanostructures: mechanisms, characterization, and applications. Interface Sci. Technol. Elsevier **2019**, 199–322 (2019)
72. Mourdikoudis, S., Pallares, R.M., Thanh, N.T.: Characterization techniques for nanoparticles: comparison and complementarity upon studying nanoparticle properties. Nanoscale **10**, 12871–12934 (2018)
73. Gong, Y., Chen, X., Wu, W.: Application of fourier transform infrared (FTIR) spectroscopy in sample preparation: Material characterization and mechanism investigation. Adv. Sample Preparation **11**, 100122 (2024)

74. Pavia, D.L., Lampman, G.M., Kriz, G.S., et al.: Introdução à espectroscopia: Tradução da 4ª edição norteamericana. Cengage Learning, São Paulo (2010)
75. Ji, X., Zhang, X., Zhang, X.: Three-dimensional graphene-based nanomaterials as electrocatalysts for oxygen reduction reaction. J. Nanomater. **2015**, 357196 (2015)
76. Zhang, F., Liu, J., Hu, L., et al.: Recent progress of three-dimensional graphene-based composites for photocatalysis. Gels **10**, 626 (2024)
77. Yang, Z., Chabi, S., Xia, Y., et al.: Preparation of 3D graphene-based architectures and their applications in supercapacitors. Progress Nat. Sci. Mater. Int. **25**, 554–562 (2015)
78. Zaka, A., Hayat, K., Mittal, V.: Recent trends in the use of three-dimensional graphene structures for supercapacitors. ACS Appl. Electron. Mater. **3**, 574–596 (2021)
79. Zeng, J., Xu, C., Gao, T., et al.: Porous monoliths of 3D graphene for electric double-layer supercapacitors. Carbon Energy **3**, 193–224 (2021)
80. Su, H., Hu, Y.H.: 3D graphene: synthesis, properties, and solar cell applications. Chem. Commun. **59**, 6660–6673 (2023)
81. Li, G., Huang, B., Pan, Z., et al.: Advances in three-dimensional graphene-based materials: configurations, preparation and application in secondary metal (Li, Na, K, Mg, Al)-ion batteries. Energy Environ. Sci. **12**, 2030–2053 (2019)
82. Wang, Z., Gao, H., Zhang, Q., et al.: Recent advances in 3D graphene architectures and their composites for energy storage applications. Small **15**, 1803858 (2019)
83. Ding, H., Wei, Y., Wu, Z., et al.: Recent advances in gas and humidity sensors based on 3D structured and porous graphene and its derivatives. Acs Mater. Lett. **2**, 1381–1411 (2020)
84. Thiyagarajan, P.: A review on three-dimensional graphene: synthesis, electronic and biotechnology applications-The Unknown Riddles. IET Nanobiotechnol. **15**, 348–357 (2021)
85. Lai, K.C., Lee, L.Y., Hiew, B.Y.Z., et al.: Environmental application of three-dimensional graphene materials as adsorbents for dyes and heavy metals: review on ice-templating method and adsorption mechanisms. J. Environ. Sci. **79**, 174–199 (2019)
86. Liu, H., Qiu, H.: Recent advances of 3D graphene-based adsorbents for sample preparation of water pollutants: A review. Chem. Eng. J. **393**, 124691 (2020)
87. Sun, H., Lin, Y., Takeshi, H., et al.: Synthesis of 3D graphene-based materials and their applications for removing dyes and heavy metals. Environ. Sci. Pollut. Res. **28**, 52625–52650 (2021)
88. Yu, Z., Wei, L., Lu, L., et al.: Structural manipulation of 3D graphene-based macrostructures for water purification. Gels **8**, 622 (2022)
89. Huang, Y., Li, C., Lin, Z.: EDTA-induced self-assembly of 3D graphene and its superior adsorption ability for paraquat using a teabag. ACS Appl. Mater. Interfaces **6**, 19766–19773 (2014)
90. Andrade, M.B., Santos, T.R., Fernandes Silva, M., et al.: Graphene oxide impregnated with iron oxide nanoparticles for the removal of atrazine from the aqueous medium. Sep. Sci. Technol. **54**, 2653–2670 (2019)
91. Zhang, C., Zhang, R.Z., Ma, Y.Q., et al.: Preparation of cellulose/graphene composite and its applications for triazine pesticides adsorption from water. ACS Sustain. Chem. Eng. **3**, 396–405 (2015)
92. Nodeh, H.R., Kamboh, M.A., Ibrahim, W.A.W., et al.: Equilibrium, kinetic and thermodynamic study of pesticides removal from water using novel glucamine-calix [4] arene functionalized magnetic graphene oxide. Environ. Sci. Process Impacts **21**, 714–726 (2019)
93. Santos, T.R., Andrade, M.B., Silva, M.F., et al.: Development of α-and γ-$Fe_2O_3$ decorated graphene oxides for glyphosate removal from water. Environ. Technol. **40**, 1118–1137 (2019)
94. Liu, H., Wang, X., Ding, C., et al.: Carboxylated carbon nanotubes-graphene oxide aerogels as ultralight and renewable high performance adsorbents for efficient adsorption of glyphosate. Environ. Chem. **17**, 6–16 (2019)
95. Wang, Z., Zhang, J., Hu, B., et al.: Graphene/$Fe_3O_4$ nanocomposite for effective removal of ten triazole fungicides from water solution: Tebuconazole as an example for investigation of the adsorption mechanism by experimental and molecular docking study. J. Taiwan Inst. Chem. Eng. **95**, 635–642 (2019)

96. Chen, F., An, W., Liu, L., et al.: Highly efficient removal of bisphenol A by a three-dimensional graphene hydrogel-AgBr@ rGO exhibiting adsorption/photocatalysis synergy. Appl. Catal. B **217**, 65–80 (2017)
97. Fang, Z., Hu, Y., Wu, X., et al.: A novel magnesium ascorbyl phosphate graphene-based monolith and its superior adsorption capability for bisphenol A. Chem. Eng. J. **334**, 948–956 (2018)
98. Sun, Z., Zhao, L., Liu, C., et al.: A novel 3D adsorbent of reduced graphene oxide-β-cyclodextrin aerogel coupled hardness with softness for efficient removal of bisphenol A. Chem. Eng. J. **372**, 896–904 (2019)
99. Zhang, Y., Cui, W., An, W., et al.: Combination of photoelectrocatalysis and adsorption for removal of bisphenol A over $TiO_2$-graphene hydrogel with 3D network structure. Appl. Catal. B **221**, 36–46 (2018)
100. Wang, W., Gong, Q., Chen, Z., et al.: Adsorption and competition investigation of phenolic compounds on the solid-liquid interface of three-dimensional foam-like graphene oxide. Chem. Eng. J. **378**, 122085 (2019)
101. Mu, C., Zhang, Y., Cui, W., et al.: Removal of bisphenol A over a separation free 3D Ag3PO4-graphene hydrogel via an adsorption-photocatalysis synergy. Appl. Catal. B **212**, 41–49 (2017)
102. Wang, X., Lu, M., Wang, H., et al.: Three-dimensional graphene aerogels–mesoporous silica frameworks for superior adsorption capability of phenols. Sep. Purif. Technol. **153**, 7–13 (2015)
103. Liao, C., Zhao, X.-R., Jiang, X.-Y., et al.: Hydrothermal fabrication of novel three-dimensional graphene oxide-pentaerythritol composites with abundant oxygen-containing groups as efficient adsorbents. Microchem. J. **152**, 104288 (2020)
104. Park, H.S., Kang, S.O.: Sorption behavior of slightly reduced, three-dimensionally macroporous graphene oxides for physical loading of oils and organic solvents. Carbon Lett. (Carbon Lett.) **18**, 24–29 (2016)
105. Wang, Y., Xie, W., Liu, H., et al.: Hyperelastic magnetic reduced graphene oxide three-dimensional framework with superb oil and organic solvent adsorption capability. Adv. Compos. Hybrid Mater. **3**, 473–484 (2020)
106. Liu, F., Chung, S., Oh, G., et al.: Three-dimensional graphene oxide nanostructure for fast and efficient water-soluble dye removal. ACS Appl. Mater. Interfaces. **4**, 922–927 (2012)
107. Deng, J., Lei, B., He, A., et al.: Toward 3D graphene oxide gels based adsorbents for high-efficient water treatment via the promotion of biopolymers. J. Hazard. Mater. **263**, 467–478 (2013)
108. Ma, Y.-X., Li, X., Shao, W.-J., et al.: Fabrication of 3d porous polyvinyl alcohol/sodium alginate/graphene oxide spherical composites for the adsorption of methylene blue. J. Nanosci. Nanotechnol. **20**, 2205–2213 (2020)
109. Amini, A., Khajeh, M., Oveisi, A.R., et al.: A porous multifunctional and magnetic layered graphene oxide/3D mesoporous MOF nanocomposite for rapid adsorption of uranium (VI) from aqueous solutions. J. Ind. Eng. Chem. **93**, 322–332 (2021)
110. Wang, X., Li, R., Liu, J., et al.: Melamine modified graphene hydrogels for the removal of uranium (VI) from aqueous solution. New J. Chem. **41**, 10899–10907 (2017)
111. Li, W., Gao, S., Wu, L., et al.: High-density three-dimension graphene macroscopic objects for high-capacity removal of heavy metal ions. Sci. Rep. **3**, 2125 (2013)
112. Lei, Y., Chen, F., Luo, Y., et al.: Synthesis of three-dimensional graphene oxide foam for the removal of heavy metal ions. Chem. Phys. Lett. **593**, 122–127 (2014)
113. Yu, B., Xu, J., Liu, J.-H., et al.: Adsorption behavior of copper ions on graphene oxide–chitosan aerogel. J. Environ. Chem. Eng. **1**, 1044–1050 (2013)
114. Zhan, W., Gao, L., Fu, X., et al.: Green synthesis of amino-functionalized carbon nanotube-graphene hybrid aerogels for high performance heavy metal ions removal. Appl. Surf. Sci. **467**, 1122–1133 (2019)
115. Chen, Y., Chen, L., Bai, H., et al.: Graphene oxide–chitosan composite hydrogels as broad-spectrum adsorbents for water purification. J. Mater. Chem. A **1**, 1992–2001 (2013)

116. Jing, J., Qian, X., Si, Y., et al.: Recent advances in the synthesis and application of three-dimensional graphene-based aerogels. Molecules **27**, 924 (2022)
117. Wang, Y., Zhang, P., Liu, C.F., et al.: A facile and green method to fabricate graphene-based multifunctional hydrogels for miniature-scale water purification. RSC Adv. **3**, 9240–9246 (2013)
118. Jauris, I., Matos, C., Zarbin, A., et al.: Adsorption of anti-inflammatory nimesulide by graphene materials: a combined theoretical and experimental study. Phys. Chem. Chem. Phys. **19**, 22099–22110 (2017)
119. Jauris, I., Matos, C., Saucier, C., et al.: Adsorption of sodium diclofenac on graphene: a combined experimental and theoretical study. Phys. Chem. Chem. Phys. **18**, 1526–1536 (2016)
120. Yu, Z., Wei, L., Lu, L. et al.: Structural Manipulation of 3D Graphene-Based Macrostructures for Water Purification. Gels 2022, 8, 622, in, s Note: MDPI stays neutral with regard to jurisdictional claims in published … (2022)
121. Zheng, X., Xiong, X., Yang, J., et al.: A strong and compressible three dimensional graphene/polyurushiol composite for efficient water cleanup. Chem. Eng. J. **333**, 153–161 (2018)
122. Weng, D., Song, L., Li, W., et al.: Review on synthesis of three-dimensional graphene skeletons and their absorption performance for oily wastewater. Environ. Sci. Pollut. Res. **28**, 16–34 (2021)
123. Soltani, R., Marjani, A., Shirazian, S.: Facile one-pot synthesis of thiol-functionalized mesoporous silica submicrospheres for Tl (I) adsorption: isotherm, kinetic and thermodynamic studies. J. Hazardous Mater. **371**, 146–155 (2019)
124. Yap, P.L., Auyoong, Y.L., Hassan, K. et al.: Multithiol functionalized graphene bio-sponge via photoinitiated thiol-ene click chemistry for efficient heavy metal ions adsorption. Chem. Eng. J., 124965 (2020)
125. Huo, J., Yu, G., Wang, J.: Selective adsorption of cesium (I) from water by Prussian blue analogues anchored on 3D reduced graphene oxide aerogel. Sci. Total Environ. **761**, 143286 (2021)

# Theoretical Fundamentals for the Batch Contact Adsorption Method

**Eder Claudio Lima, Yones Dehmani, Mohamed Abdel Salam, Glaydson Simões dos Reis, and Fernando Machado Machado**

**Abstract** Although batch adsorption studies are prevalent in literature, there are many significant errors related to physical chemistry concepts that users of adsorption may not fully understand. This paper will define adsorption, chemisorption, and physisorption. Following this, adsorption equilibrium will be presented using several isotherm models with nonlinear fitting. The determination of each equilibrium constant in L mg$^{-1}$ will also be illustrated. These equilibrium constants need to be converted to L mol$^{-1}$ before calculating the dimensionless thermodynamic equilibrium constant. Once this constant is obtained, it can be applied in both linearized and nonlinear van't Hoff equations to estimate changes in adsorption enthalpy and entropy. The statistical evaluation of the models will utilize determination coefficient ($R^2$), adjusted determination coefficient ($R^2_{adj}$), and Bayesian Information Criterion (BIC) values. Furthermore, various kinetic models—including those based on chemical reactions, empirical kinetic models, and diffusive models—will be explored. The

E. C. Lima (✉)
Institute of Chemistry, Federal University of Rio Grande do Sul (UFRGS), Porto Alegre, RS, Brazil
e-mail: profederlima@gmail.com; eder.lima@ufrgs.br

Y. Dehmani
Interfaces, Confinement, Matériaux et Nanostructures (ICMN), CNRS-Université d'Orléans, UMR 7374, Orléans Cedex 2, France

M. A. Salam
Department of Chemistry, College of Science and Technology, Florida A&M University, Tallahassee, Florida 32307, USA
e-mail: mohamed.salam@famu.edu

G. S. dos Reis
Laboratory of Industrial Chemistry and Reaction Engineering, Faculty of Science and Engineering, Åbo Akademi University, Åbo/Turku, Finland
e-mail: glaydson.simoes.dos.reis@slu.se

F. Machado Machado
Centro de Desenvolvimento Tecnológico, Universidade Federal de Pelotas (UFPEL), Pelotas, RS, Brazil
e-mail: fernando.machado@ufpel.edu.br

© The Author(s), under exclusive license to Springer Nature Switzerland AG 2026 
C. P. Bergmann and F. Machado Machado (eds.), *Carbon Nanomaterials as Adsorbents for Environmental and Biological Applications*, Carbon Nanostructures,
https://doi.org/10.1007/978-3-032-08669-3_6

authors will employ the concepts of half-adsorption time ($t_{0.5}$) and near-equilibrium time ($t_{0.95}$) to compare the kinetic parameters of the different models. Additionally, comparisons between linear and nonlinear pseudo-first-order and pseudo-second-order kinetics models, as well as the Langmuir isotherm model, will be made. The results indicate that it is advisable to use nonlinear fittings for both isotherms and the kinetics of adsorption.

**Keywords** Nonlinear equilibrium and kinetic adsorption models · The correct thermodynamic equilibrium constant for thermodynamic calculations · Statistical evaluation of adsorption models

## 1 Introduction

Adsorption refers to the process in which a sorbing species (adsorbate) present in a fluid phase adheres to the surface of a condensed phase (solid or liquid). Typically, adsorbents are solids. In this chapter, we will focus on solid-phase adsorbents. The adsorbate exists in a fluidic phase, either as a solute dissolved in a solvent or as a gas. When the sorbing species penetrates into the adsorbent, this process is termed absorption. Distinguishing between adsorption and absorption can sometimes be difficult, so the term sorption is often preferred.

$$Sorption = Adsorption + Absorption$$

Thus, the term adsorbent may also be referred to as a sorbent, while adsorbate is alternatively known as sorbate. Various materials serve as adsorbents, including silica, alumina, carbon-based substances (such as activated carbons, biochars and activated biochars, carbon nanotubes, graphene-family materials, and polymeric resins), along with plant-derived (agricultural and forestry byproducts) or microorganism-derived adsorbents (like algae, bacteria, fungi, and yeast). When the biomass is inactive (for example, dead algae, fungi, or agricultural residues like fruit peels, shells, seeds, or wood sawdust), it is termed a biosorbent. Conversely, if the biomass is alive, this process of species retention is referred to as bioaccumulation, which should not be confused with biosorption.

Adsorption can occur as either physical (physisorption) or chemical (chemisorption). Both types can be irreversible or reversible. Contrary to misconceptions in earlier literature, chemisorption does not exclusively indicate irreversible adsorption, just as physical adsorption does not solely imply reversibility. These misunderstandings have circulated within adsorption studies. Typically, the adsorption of organic compounds onto carbon-based adsorbents is seen as physisorption, while the adsorption of inorganic substances, like heavy metals, is generally classified as chemisorption. This serves as a key guideline for those working with adsorption.

Estimating the enthalpy of adsorption (ΔH°ads) for the adsorbate and adsorbent aids in determining whether the adsorption process involves chemisorption or physisorption [1]. The energies of interactions, including covalent chemical bonds and other interactions between two species, are presented in Table 1 [2].

Nevertheless, the enthalpy of adsorption should not be viewed as the sole indicator of chemisorption and physisorption. Other factors, including the adsorbate's chemical affinity for the adsorbent's surface, must be taken into account, along with confirmation from various analytical techniques such as Raman Spectroscopy, X-ray Photoelectron Spectroscopy (XPS), and Scanning Electron Microscopy with Energy Dispersive X-ray Spectroscopy (SEM–EDS) for metallic species. Despite the use of these advanced techniques, differentiating between physisorption and chemisorption remains challenging due to the difficulty in quantifying small amounts of adsorbate on the adsorbent's surface. Therefore, it is best to adhere to the rule of thumb: "Cationic metallic ions typically form complexes with adsorbents featuring the following functional groups: OH, NH, SH; whereas organic molecules are generally physisorbed onto carbon-based materials and other inorganic solids".

At a consistent temperature, an adsorption isotherm outlines the relationship between the quantity of adsorbate retained by the adsorbent ($q_e$) and the concentration of adsorbate that remains in solution after equilibrium ($C_e$). The parameters derived from adsorption equilibrium models offer valuable insights into surface characteristics, adsorption mechanisms, and the interactions between adsorbent and adsorbate. Numerous equations exist to describe the adsorption equilibrium of an adsorbate on an adsorbent. Langmuir is the most frequently cited in the literature [3]. Other isotherm models, including the Freundlich isotherm [4], Sips isotherm [5], Liu isotherm [6], and Redlich-Peterson isotherm [7], are also extensively discussed in scholarly work.

Studying adsorption kinetics is crucial for treating aqueous effluents as it offers insights into the adsorption mechanism. Various kinetic models have been developed to derive intrinsic kinetic adsorption constants. This chapter will cover kinetic adsorption models related to chemical reactions (including the pseudo-first-order

**Table 1** Interactions between species [1, 2]

| Type of interaction | The magnitude of energy (kJ $mol^{-1}$) |
|---|---|
| Covalent bond | 100–800[a] |
| Ionic bond with the formation of a crystalline lattice | 100–400[a] |
| Electrostatic attraction | 30–80[b] |
| Ion–dipole | 5–60[b] |
| Dipole–dipole | 0.5–15[b] |
| Ion-induced dipole | 0.4–4[b] |
| Dispersion | 4–40[b] |
| Hydrogen bond | 4–60[b] |

[a] Chemisorption. [b] Physisorption

equation [8], pseudo-second-order equation [9], and general order equation [10]) as well as empirical models such as the fractal-like-pseudo-first-order and fractal-like-pseudo-second-order model [11] and the Elovich chemisorption model [12].

## 2 Equilibrium Isotherm Models

### 2.1 Langmuir Isotherm Model

The Langmuir [3] isotherm is based on the following assumptions:

- Adsorbates are chemically adsorbed at a fixed number of well-defined sites;
- The adsorbent attains a maximum saturation of the active sites;
- A monolayer of the adsorbate is formed over the surface of the adsorbent when it gets saturated;
- Each site can hold only one adsorbate species;
- All sites are energetically equivalent, and
- Interactions between the adsorbate species do not exist.

The Langmuir isotherm equation is depicted by Eq. (1):

$$q_e = \frac{Q_{max} \cdot K_L \cdot C_e}{1 + K_L \cdot C_e} \tag{1}$$

where $q_e$ is the amount of adsorbate adsorbed at the equilibrium (mg $g^{-1}$), $C_e$ is the supernatant adsorbate concentration at the equilibrium (mg $L^{-1}$), $K_L$ is the Langmuir equilibrium constant (L $mg^{-1}$), and $Q_{max}$ is the maximum adsorption capacity of the adsorbent (mg $g^{-1}$) assuming a monolayer of adsorbate uptake by the adsorbent.

This equation has several limitations regarding its principles, such as assuming all active adsorbent sites have the same energy. Such a condition could only occur if the adsorbent is a polymeric resin or organofunctionalized silica with a single functional group. In contrast, biomaterials, composite materials, biochars, and activated carbons typically contain multiple active sites (functional groups), leading to varying interaction energies with the adsorbing species. Furthermore, interactions among sorbing species do not occur in isolation; they also engage with the solvent and other entities in the medium, as is expected [1]. A key advantage of the Langmuir model is its ability to predict the saturation of the adsorbent. This attribute explains the model's prevalence for adsorbents with multiple active sites under high concentrations of sorbing species, where it is unreasonable to assume that there are no interactions between the sorbing species themselves and between non-adsorbed species and those attached to the adsorbent's surface.

As a convention, in this chapter, $q$ means sorption capacity expressed in mg $g^{-1}$, and it is an experimental value unless otherwise stated. $Q_{max}$ is the maximum sorption capacity obtained by the best isotherm model fitted. This parameter is a theoretical value that was based on an equilibrium isotherm. Several isotherm models have a predicted $Q_{max}$. Readers should not confuse $q_e$ at the saturation with $Q_{max}$. The values of $q_e$ are the sorption capacity experimentally measured at the equilibrium. When these values are obtained at the saturation of the adsorbent, they are not the $Q_{max}$. This consideration is a common mistake in the adsorption literature [1].

To illustrate the Langmuir model, consider the data on the adsorption of captopril pharmaceutical onto activated carbon derived from *Butia catarinensis* [13]. In this chapter section, we will focus exclusively on nonlinear equations. Subsequent sections will explain this choice and discuss the primary issues regarding the linearization of equilibrium and kinetic adsorption models. Thus, Fig. 1 presents the nonlinear Langmuir isotherm plot for the adsorption of captopril pharmaceutical onto activated carbon derived from *Butia catarinensis* (Fig. 2) [13].

If the user intends to use the Langmuir equilibrium constant for calculating thermodynamic parameters of adsorption, which will be detailed later, the following equation must be considered:

$$K_e = K_L \tag{2}$$

where $K_e$ is the equilibrium constant that presents the unit of L mg$^{-1}$, and that could be used in the thermodynamic calculations that will be described furtherly.

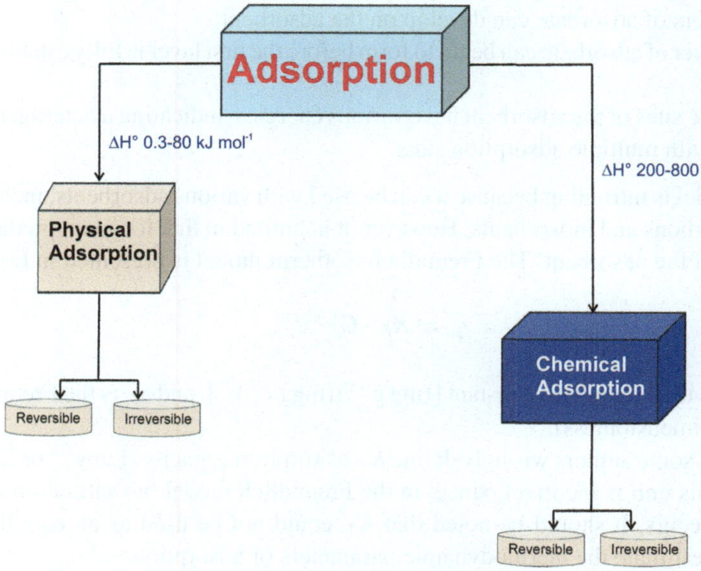


**Fig. 1** Types of adsorption

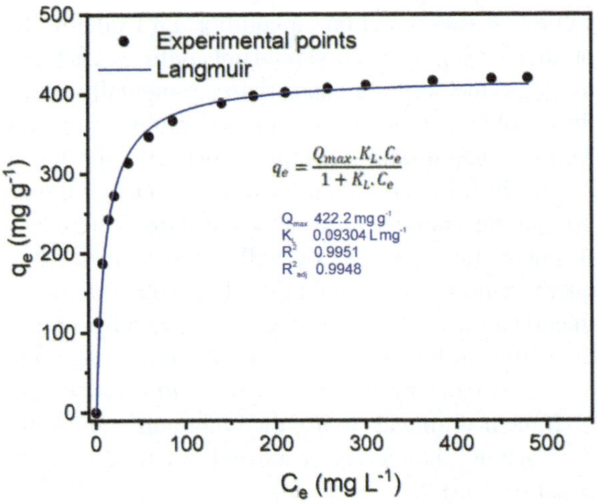


**Fig. 2** Langmuir adsorption equilibrium isotherm of captopril pharmaceutical onto activated carbon derived from *Butia catarinensis* at 10 °C and pH 7.0

## 2.2 *Freundlich Isotherm Model*

The Freundlich isotherm model has the following premises [4]:

- Adsorption increases exponentially without limit as adsorbent levels rise;
- Multilayers of adsorbate can develop on the adsorbent;
- A new layer of adsorbate can begin to form before the first layer is fully established; and
- The active sites of the adsorbent have various energies, indicating a heterogeneous surface with multiple adsorption sites.

This model is intriguing because it can be used with various adsorbents, including activated carbons and biosorbents. However, it is limited in that it cannot predict the saturation of the adsorbent. The Freundlich isotherm model is presented in Eq. (3):

$$q_e = K_F \cdot C_e^{1/n_F} \tag{3}$$

where $K_F$ is the Freundlich constant $[(\text{mg g}^{-1})(\text{mg L}^{-1})^{n_F}]$, and $n_F$ is the Freundlich exponent (dimensionless).

Note that some authors wrongly define $K_F$ as sorption capacity (L mg$^{-1}$ or L g$^{-1}$). However, this unit is incorrect, since, in the Freundlich model, no saturation of the adsorbent occurs. It should be noted that $K_F$ could not be used as an equilibrium constant to estimate the thermodynamic parameters of adsorption.

Figure 3 shows the nonlinearized Freundlich isotherm plot for the adsorption of captopril pharmaceutical onto activated carbon derived from *Butia catarinensis* [13].

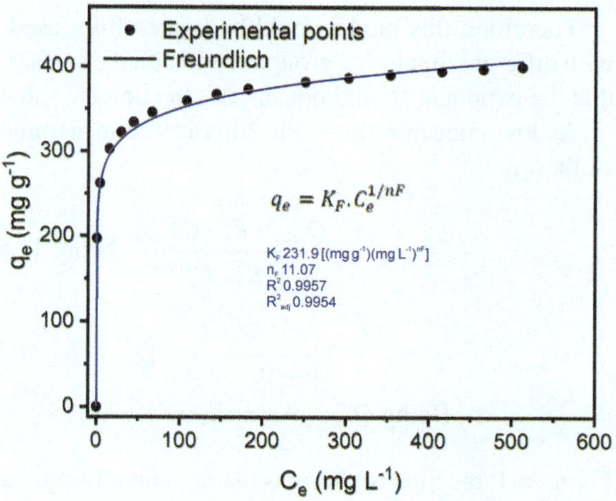


**Fig. 3** Freundlich adsorption equilibrium isotherm of captopril pharmaceutical onto activated carbon derived from *Butia catarinensis* at 25 °C and pH 7.0

## 2.3 *Langmuir–Freundlich-Like Isotherm Models*

### 2.3.1 Sips Isotherm Model

In 1948, Sips [5] proposed the isotherm model, later named the Langmuir–Freundlich model. Note, which considers the multi-layer adsorption, the interactions among adsorbates, and the varying energies of the adsorbent's activated sites. It is important to note that Sips did not originally coin this term. Any isotherm model that merges the Langmuir and Freundlich approaches can be referred to as a Langmuir–Freundlich-like model.

$$q_e = \frac{Q_{max} \cdot K_s \cdot C_e^{1/n_s}}{1 + K_s \cdot C_e^{1/n_s}} \quad \text{where } 0 < \frac{1}{n_s} \leq 1 \tag{4}$$

In Eq. (4), $K_s$ represents the affinity between the adsorbate and the adsorbent, commonly referred to as the Sips equilibrium constant $(\mathrm{L\,mg^{-1}})^{1/\mathrm{n_s}}$,, wherein ns denotes the Sips exponent (dimensionless). Moreover, $Q_{max}$ signifies the theoretical maximum adsorption capacity according to the Sips model (mg g$^{-1}$). It is defined that $1/n_s$ should be less than or equal to 1, as previously documented in the literature [5]. Nevertheless, a limitation is observed in the original Sips model; numerous authors fail to address this constraint and incorrectly apply a n greater than 1 to $1/n_s$. This misinterpretation is a prevalent error noted within the field of adsorption studies. As seen above, the Sips model will have the following conditions:

- The saturation of the adsorbent is predicted ($Q_{max}$);
- Multi-layer adsorption of the adsorbate on the adsorbent may occur;
- Initiation of the second layer is possible without completing the first layer;
- The active sites of the adsorbents may possess varying energies; and
- Interactions exist between the sorbing species.

Therefore, this model could be successfully used for heterogeneous adsorbents with different functional groups present on the surface. However, there is a constraint that the exponent should not surpass the unitary value [5].

At low concentrations, the Sips isotherm assumes the form of the Freundlich isotherm:

$$q_e = \frac{Q_{max} \cdot K_s \cdot C_e^{1/n_s}}{1 + K_s \cdot C_e^{1/n_s}} \quad \text{being } 1 \gg \mathrm{K_s} \cdot \mathrm{C_e^{1/n_s}} \tag{5}$$

$$q_e = \frac{Q_{max} \cdot K_s \cdot C_e^{1/n_s}}{1 + 0} = Q_{max} \cdot K_L \cdot C_e^{1/n_s} = K_F \cdot C_e^{1/n_s} \tag{6}$$
$$\text{Being } Q_{max} \cdot K_L = K_F$$

For $n_s = 1$, the Sips isotherm is mathematically equivalent to the Langmuir isotherm.

$$q_e = \frac{Q_{max} \cdot K_s \cdot C_e^{1/n_s}}{1 + K_s \cdot C_e^{1/n_s}} = \frac{Q_{max} \cdot K_s \cdot C_e^1}{1 + K_s \cdot C_e^1} = \frac{Q_{max} \cdot K_L \cdot C_e}{1 + K_L \cdot C_e} \tag{7}$$

Figure 4 presents the adsorption isotherm of Reactive Red 194 dye onto Fe-Lignin complex adsorbent using the Sips isotherm model [14].

If the user wants to use the Sips equilibrium constant ($K_s$) for the calculation of the thermodynamic parameters of adsorption, it should be considered:

$$K_e = K_S^{ns} \tag{8}$$

using Eq. (8), the $K_e$ has the unit of L mg$^{-1}$.

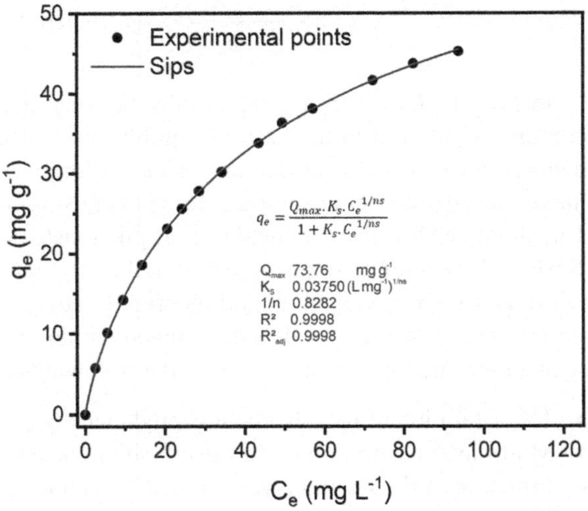


**Fig. 4** Sips adsorption equilibrium isotherm of RR-194 dye using Fe-Ligning çomplex at 25 °C and pH 2.0

### 2.3.2 Liu Isotherm Model

The Liu model [6] is classified as a "Langmuir–Freundlich Type Isotherm," representing an empirical model. However, the assumptions regarding infinite adsorption stem from the Freundlich model, while the monolayer assumptions inherent to the Langmuir model are excluded. The Liu isotherm model possesses the following characteristics:

- The saturation of the adsorbent is predicted ($Q_{max}$);
- Adsorbates may form multiple layers on the adsorbent;
- The initiation of the second layer can occur without the completion of the first layer;
- The activated sites of the adsorbents may possess varying energy levels.
- Interactions among the sorbing species in solution may occur.

The exponent of Liu's isotherm has no constraint as the Sips model. Equation (9) defines the Liu isotherm model.

$$q_e = \frac{Q_{max} \cdot (K_g \cdot C_e)^{n_L}}{1 + (K_g \cdot C_e)^{n_L}} \tag{9}$$

where $K_g$ represents the Liu equilibrium constant (L mg$^{-1}$); $n_L$ denotes the dimensionless exponent of the Liu equation, and $Q_{max}$ indicates the maximum adsorption capacity of the adsorbent (mg g$^{-1}$). In contrast to the Sips isotherm, $n_L$ may assume any positive value. Figure 5 shows the nonlinearized Liu isotherm plot for the adsorption of hydroquinone onto pine activated carbon at 10 °C [15].

When using the equilibrium constant of the Liu isotherm, it should be considered:

$$K_e = K_g \tag{10}$$

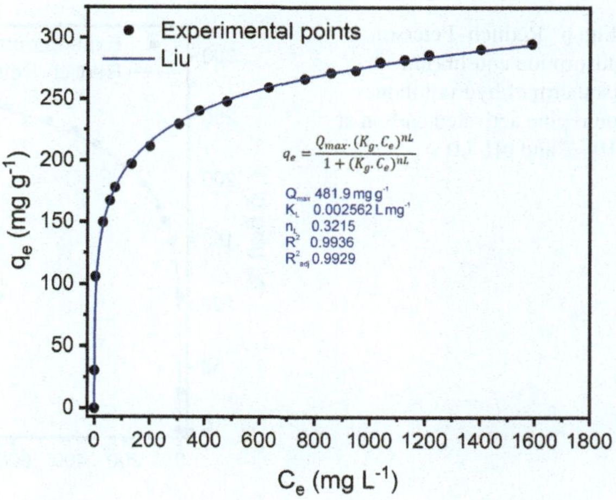


**Fig. 5** Liu adsorption equilibrium isotherm of hydroquinone onto pine activated carbon at 10 °C and pH 7.0

In Eq. (10), the unit of $K_e$ is L mg$^{-1}$, which could be used in the thermodynamic calculations.

### 2.3.3 Redlich–Peterson Isotherm Model

This isotherm model is also a "*Langmuir–Freudlich Type Isotherm*," which is an empirical equation, shown in Eq. (11) [7].

$$q_e = \frac{K_{RP} \cdot C_e}{1 + a_{RP} \cdot C_e^{\beta}} \quad \text{where } 0 < \beta \leq 1 \tag{11}$$

In this context, $a_{RP}$ represents the affinity constant of the Redlich–Peterson model, which is associated with the equilibrium constant (L mg$^{-1}$)$^{\beta}$. $K_{RP}$ (L g$^{-1}$) denotes a constant incorporating both the maximum sorption capacity and the affinity between the adsorbent and adsorbate. The symbol $\beta$ signifies the exponent of the Redlich–Peterson isotherm, which is dimensionless. By definition, the permissible range for $\beta$ is $0 < \beta < 1$. The equation becomes a Langmuir isotherm when $\beta = 1$.

$$q_e = \frac{K_{RP} \cdot C_e}{1 + a_{RP} \cdot C_e^1} = \frac{Q_{max} \cdot a_{RP} \cdot C_e}{1 + a_{RP} \cdot C_e}; \quad where\ K_{RP} = Q_{max} \cdot a_{RP} \tag{12}$$

For low concentrations, this equation becomes a linear isotherm

$$q_e = \frac{K_{RP} \cdot C_e}{1 + a_{RP} \cdot C_e} = \frac{K_{RP} \cdot C_e}{1 + 0} = K_{RP} \cdot C_e \tag{13}$$

Figure 6 presents the isotherm of hydroquinone adsorption onto pine activated carbon at 10 °C and pH 7.0 using the Redlich–Peterson model [15].

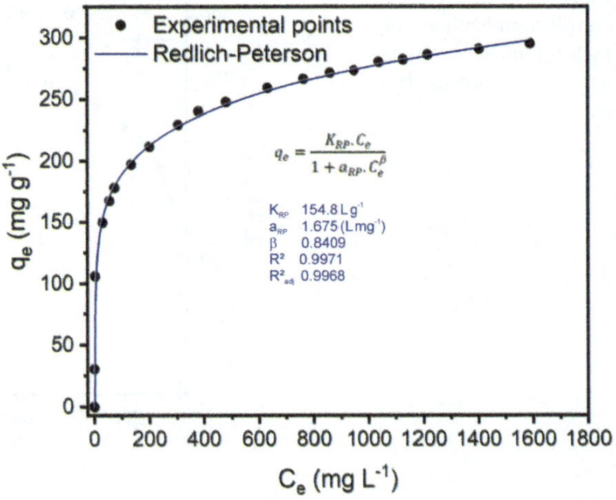


**Fig. 6** Redlich–Peterson adsorption equilibrium isotherm of hydroquinone onto pine activated carbon at 10 °C and pH 7.0

To calculate the standard enthalpy change (ΔH°) and the standard entropy change (ΔS°) of adsorption using the van't Hoff equation with the Redlich-Peterson constant of equilibrium, the following considerations are essential to consider:

$$K_e = \sqrt[\beta]{a_{RP}} \tag{14}$$

the unit of $K_e$ in this expression is L mg$^{-1}$, and it could be used in the thermodynamic calculations.

## 2.4 Other Langmuir–Freundlich-Type Isotherm Models

In addition to the most commonly used adsorption equilibrium models previously presented, other isotherm models can be applied to describe how an adsorbent absorbs a sorbing species [1, 16, 17]. One example is the Hill isotherm model:

$$q_e = \frac{Q_{max} \cdot C_e^{n_H}}{K_H + C_e^{n_H}} \tag{15}$$

where $K_H$ is the Hill affinity constant $\left[(\text{mgL}^{-1})^{\text{n}_\text{H}}\right]$, $Q_{max}$ is the maximum sorption capacity (mg g$^{-1}$), and $n_H$ is the exponent of the Hill isotherm that is dimensionless. Note that $\frac{1}{K_H}$ is the equilibrium constant $\left[(\text{L mg}^{-1})^{\text{n}_\text{H}}\right]$, according to the Hill isotherm model.

For obtaining the $K_e$ from the Hill, it is necessary to make the following consideration:

$$K_e = \frac{1}{\sqrt[n_H]{K_H}} \tag{16}$$

This value of $K_e$ will be practically the value of the equilibrium constant obtained by the Liu isotherm ($K_g$). Therefore, it is essential to state that Liu's isotherm is a derivation of Hill's isotherm. The unit of $K_e$ in Eq. (16) is L mg$^{-1}$.

Figure 7 shows the isotherm of hydroquinone adsorption onto pine activated carbon at 10 °C and pH 7.0 [15].

Another example is the Khan isotherm model, which is described in Eq. (17).

$$q_e = \frac{Q_{max} \cdot K_K \cdot C_e}{(1 + K_K \cdot C_e)^{n_K}} \tag{17}$$

where $Q_{max}$ is the maximum sorption capacity (mg g$^{-1}$) according to the Khan isotherm, $K_K$ is the Khan equilibrium constant expressed in L mg$^{-1}$, and $n_K$ is the dimensionless exponent of the Khan isotherm equation.

Figure 8 presents the Khan isotherm of hydroquinone adsorption onto pine activated carbon at 10 °C and pH 7.0 [15].

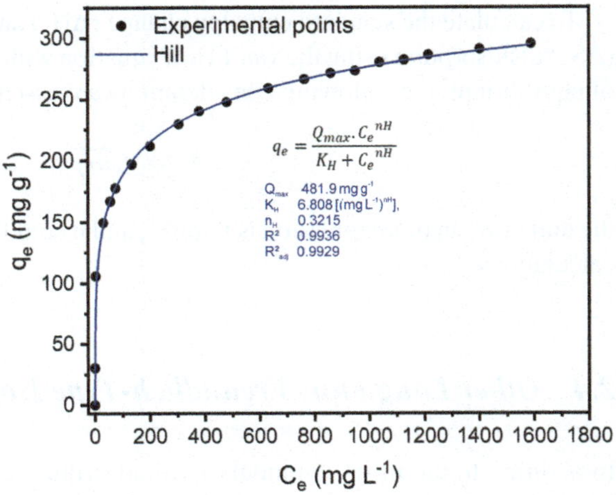


**Fig. 7** Hill adsorption equilibrium isotherm of hydroquinone onto pine activated carbon at 10 °C and pH 7.0

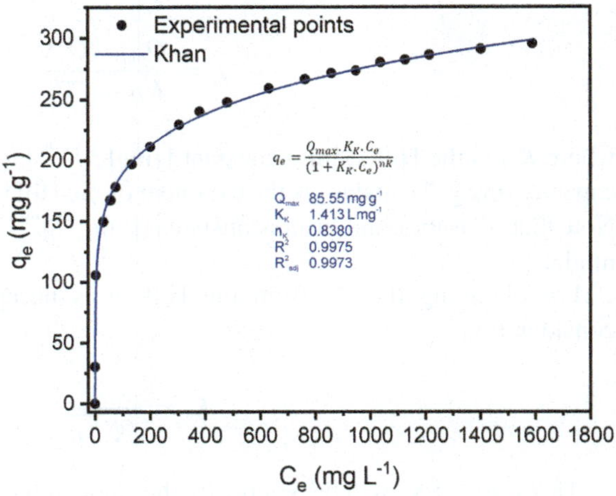


**Fig. 8** Khan adsorption equilibrium isotherm of hydroquinone adsorption onto pine activated carbon at 10 °C and pH 7.0

The Khan isotherm model is a Langmuir–Freundlich type. Consequently, when $n_K = 1$, the isotherm is equal to the Langmuir isotherm, and for low concentrations, it is linear.

$$q_e = \frac{Q_{max} \cdot K_K \cdot C_e}{(1 + K_K \cdot C_e)^1} = \frac{Q_{max} \cdot K_L \cdot C_e}{1 + K_L \cdot C_e} \tag{18}$$

$$q_e = \frac{Q_{max} \cdot K_K \cdot C_e}{(1 + K_K \cdot C_e)^{n_K}} = \frac{Q_{max} \cdot K_K \cdot C_e}{1 + 0} = Q_{max} \cdot K_K \cdot C_e, \ when\ C_e \to 0 \tag{19}$$

To apply the Khan equilibrium constant in thermodynamic calculations:

$$K_e = \mathrm{K}_K \tag{20}$$

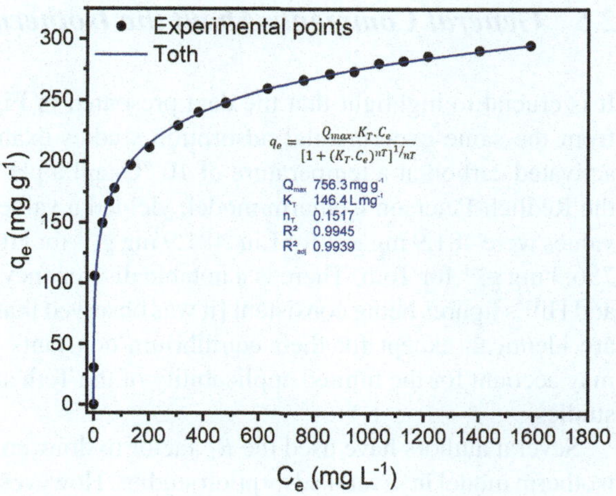


**Fig. 9** The Toth adsorption equilibrium isotherm of adsorption of hydroquinone on-to pine activated carbon at 10 °C and pH 7.0

The Toth isotherm [1, 16, 17] is a Langmuir–Freundlich type isotherm, and it is given by the equation below:

$$q_e = \frac{Q_{max} \cdot K_T \cdot C_e}{[1 + (K_T \cdot C_e)^{n_T}]^{1/n_T}} \tag{21}$$

where, $Q_{max}$ (mg g$^{-1}$) represents the maximum sorption capacity as defined by the Toth model, with $K_T$ denoting the Toth equilibrium constant (L mg$^{-1}$), and $n_T$ being the dimensionless exponent associated with the Toth isotherm model.

The isotherm adsorption of hydroquinone onto pine activated carbon at 10 °C and pH 7.0 is shown in Fig. 9 [15].

The Toth isotherm is a Langmuir–Freundlich type; therefore, when $n_T = 1$, the isotherm is equal to the Langmuir isotherm model, and when the concentration is low, the isotherm tends to be a linear relationship.

$$q_e = \frac{Q_{max} \cdot K_T \cdot C_e}{[1 + (K_T \cdot C_e)^1]^{1/1}} = \frac{Q_{max} \cdot K_L \cdot C_e}{1 + K_L \cdot C_e} \tag{22}$$

$$q_e = \frac{Q_{max} \cdot K_T \cdot C_e}{[1 + (K_T \cdot C_e)^{nT}]^{1/n_T}} = \frac{Q_{max} \cdot K_T \cdot C_e}{1 + 0} = Q_{max} \cdot K_T \cdot C_e, \ when\ C_e \rightarrow 0 \tag{23}$$

For using the Toth equilibrium constant in the van't Hoff equation for the calculation of the thermodynamic parameters of adsorption (ΔH°, ΔS°), it should be considered:

$$K_e = K_T \tag{24}$$

In this equation, $K_e$ presents the unit of L mg$^{-1}$, which could be used in the thermodynamic calculations.

## 2.5 General Comments About the Isotherm Models

It is crucial to highlight that the data presented in Figs. 5, 6, 7, 8 and 9 is derived from the same experimental adsorption studies examining hydroquinone on pine activated carbon at a temperature of 10 °C and a pH of 7.0. All models, excluding the Redlich-Peterson isotherm model, yielded a value for $Q_{max}$. The recorded $Q_{max}$ values were 481.9 mg $g^{-1}$ for Liu, 481.9 mg $g^{-1}$ for Hill, 85.55 mg $g^{-1}$ for Khan, and 756.3 mg $g^{-1}$ for Toth. There is a notable discrepancy among these values, with Liu and Hill's figures being consistent [it was observed that Liu and Hill isotherm models are identical, except for their equilibrium constants ($K_L \neq K_H$)]. This difference may account for the limited applicability of the Toth and Khan models in adsorption studies.

Several authors have used the $R_L$ factor to draw conclusions about the Langmuir isotherm model in several adsorption studies. However, these conclusions are entirely wrong (Fig. 10).

When this equation is used, $R_L$ will always be lower than 1 unless $K_L$ or $C_0$ are negative. However, equilibrium constants ($K_L$) and initial concentrations will always be positive; therefore, this equation is illogical in adsorption equilibrium studies.

Other common errors spread in the adsorption literature are the use of the Dubinin-Radushkevich and Tempkin isotherms. The Dubinin-Radushkevich equation is.

$$q_e = Q_{max} \cdot \exp\left\{-\beta \cdot \left[R \cdot T \cdot Ln\left(1 + \frac{1}{C_e}\right)\right]^2\right\} \tag{25}$$

$$E = \frac{1}{\sqrt{2 \cdot \beta}} \tag{26}$$

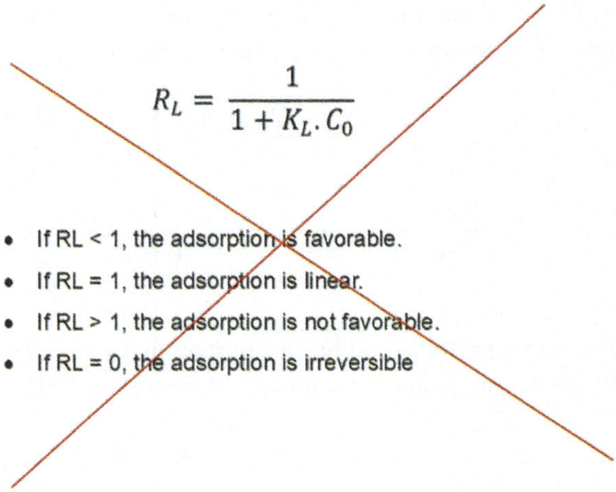


**Fig. 10** $R_L$ factor of the Langmuir Eq. (7)

where $Q_{max}$ (mol kg$^{-1}$) is the maximum sorption capacity, $\beta$ is the Dubinin–Radushkevich constant (mol$^{-2}$ J$^2$), R is 8.314 J K$^{-1}$ mol$^{-1}$, $C_e$ (mol L$^{-1}$) is the equilibrium concentration, and $E$ is the adsorption energy (J). From the $\beta$ Dubinin–Radushkevich equation, it is possible to calculate the adsorption energy.

The problems of using this equation are:

- First of all, the concentration ($C_e$) should be expressed in mol L$^{-1}$, and it cannot be used as mg L$^{-1}$ because the unit of the $\beta$ term is expressed in SI units (mol$^{-2}$ J$^2$). Also, the sorption capacity ($q_e$) should be expressed in mol kg$^{-1}$, which is equivalent to mmol g$^{-1}$.
- When various isotherm models (such as Langmuir, Freundlich, Sips, Liu, Toth, Khan, Redlich–Peterson, Hill, etc.) are displayed as different curves on the same plot, while the user expresses concentration in mg L$^{-1}$ and sorption capacity in mg g$^{-1}$, the Dubinin–Radushkevich plot cannot be included on the same graph.
- Values of about 20 kJ mol$^{-1}$ energy are wrongly attributed as chemisorption. As seen in Fig. 1, chemisorption should be considered for interactions in the same order as a covalent bond (>200 kJ mol$^{-1}$).

Figure 11 depicts the nonlinear Dubinin–Radushkevich isotherm for the uptake of hydroquinone onto pine activated carbon at 10 °C (283 K) and pH 7.0.

Considering that the Temperature is a constant (10 °C = 283 K), and R is 8.314 J K$^{-1}$ mol$^{-1}$, the term RT is 2352.862 J mol$^{-1}$, therefore the Dubinin–Radushkevich equation becomes:

$$q_e = Q_{max} \cdot \exp\left\{ -K \cdot \left[ Ln\left(1 + \frac{1}{C_e}\right)\right]^2 \right\} \tag{27}$$

$$K = \beta \cdot (RT)^2 \tag{28}$$

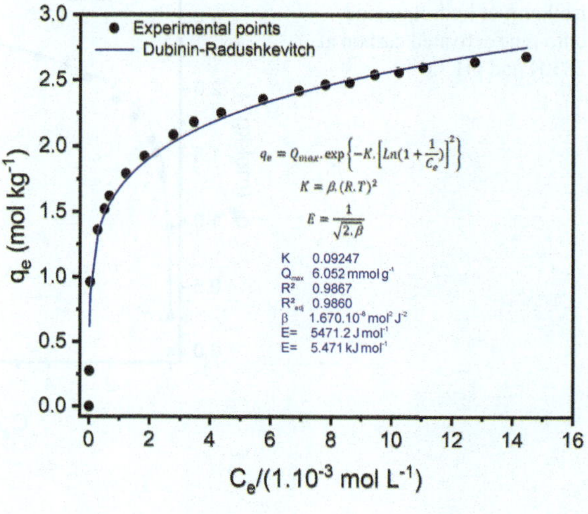


**Fig. 11** The Dubinin–Radushkevich adsorption equilibrium isotherm of hydroquinone onto pine activated carbon at 10 °C (283 K) and pH 7.0

The terms $Q_{max}$ and $K$ are obtained from the nonlinear plot, and the adsorption energy is then calculated. The reader should keep in mind that if the sorption capacity is not expressed in mol kg$^{-1}$ (equivalent to mmol g$^{-1}$) and concentration in mol L$^{-1}$, the values of $K$ and E obtained would be completely different.

The Temkin isotherm is given by the equation [18]:

$$q_e = Q_{max} \cdot \frac{RT}{b} \cdot Ln(K_T \cdot C_e) \tag{29}$$

In this context, $q_e$ (mol kg$^{-1}$) represents the sorption capacity at equilibrium, while $Q_{max}$ (mol kg$^{-1}$) denotes the maximum sorption capacity as specified by the Temkin model. R, which has a value of 8.314 J K$^{-1}$ mol$^{-1}$, signifies the ideal gas constant. $T$ (K) indicates the temperature, $K_T$ (L mol$^{-1}$) is the equilibrium constant, and $b$ (J mol$^{-1}$) refers to the heat of adsorption. Like the Dubinin–Radushkevich model, the Temkin isotherm requires that sorption capacities be stated in mol kg$^{-1}$ and concentrations in mol L$^{-1}$, since the variable b is measured in J mol$^{-1}$, necessitating all units to conform to the SI system [18].

Figure 12 depicts the nonlinear Temkin isotherm for the uptake of hydroquinone onto pine activated carbon at 10 °C (283 K) and pH 7.0.

It is important to emphasize that the β term in the Dubinin–Radushkevich isotherm, valued at 5.471 kJ mol$^{-1}$, differs significantly from the $b$ term in the Temkin isotherm, which is valued at 19.85 kJ mol$^{-1}$, both derived from the same experimental data. We also reaffirm the application of the Langmuir–Freundlich type isotherm model at various temperatures to derive precise equilibrium constants. These constants can then be employed in the van't Hoff equation to estimate ΔH°, ΔS°, and ΔG° of adsorption, which will be elaborated on later in this chapter.

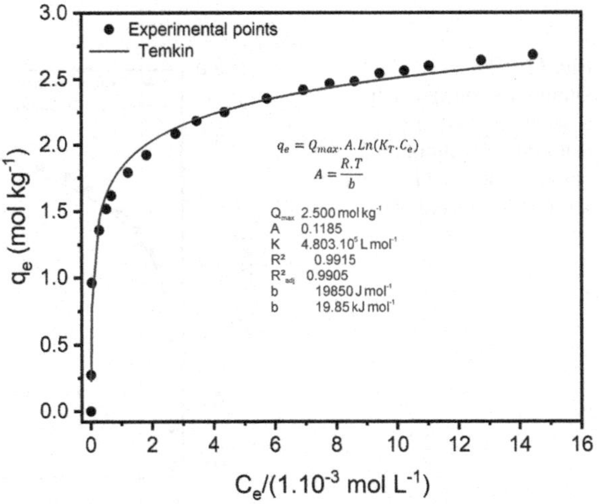


**Fig. 12** The Temkin adsorption equilibrium isotherm of hydroquinone onto pine activated carbon at 283 K and pH 7.0

## 2.6 Statistical Evaluation of Nonlinear Isotherms

The suitability of the kinetic and equilibrium models was evaluated using the residual sum of squares (*RSS*), the determination coefficient ($R^2$), the adjusted determination coefficient ($R^2_{adj}$), the standard deviation of residues (*SD*), and the Bayesian Information Criterion (*BIC*) [1, 19–21]. Equations (30)–(34) are the mathematical expressions for *RSS*, $R^2$, $R^2_{adj}$, *SD*, and *BIC*, respectively.

$$RSS = \sum_{i}^{n} \left(q_{i,exp} - q_{i,model}\right)^2 \tag{30}$$

$$R^2 = \left( \frac{\sum_i^n \left(q_{i,exp} - \overline{q}_{i,exp}\right)^2 - \sum_i^n \left(q_{i,exp} - q_{i,model}\right)^2}{\sum_i^n \left(q_{i,exp} - \overline{q}_{i,exp}\right)^2} \right) \tag{31}$$

$$R^2_{adj} = 1 - (1 - R^2) \cdot \left( \frac{n-1}{n-p-1} \right) \tag{32}$$

$$SD = \sqrt{\left( \frac{1}{n-p} \right) \cdot \sum_{i}^{n} \left(q_{i,exp} - q_{i,model}\right)^2} \tag{33}$$

$$BIC = nLn\left( \frac{RSS}{n} \right) + pLn(n) \tag{34}$$

In the above equations, the $q_{i,\,model}$ is the individual theoretical $q$ value predicted by the model; $q_{i,\,exp}$ is individual experimental $q$ value; $\overline{q}_{i,exp}$ is the average of all experimental $q$ values measured; $n$ is the number of experiments; $p$ is the number of parameters in the fitting model.

The values of $R^2_{adj}$, *SD*, and *BIC* will be presented to compare different models of kinetics and equilibrium presented in this work. The best-fitted model would present $R^2_{adj}$ closer to 1.000, lower values of *SD*, and BIC values. However, the kinetic and equilibrium models could not merely be chosen based on the values of $R^2$ [1, 20, 21] when these models present different numbers of parameters. Therefore, it is necessary to check if the improvements in the $R^2$ values are due to the increase in the parameters [1, 20, 21] or if, physically, the model with more parameters better explains the process taking place [1, 20, 21].

However, the difference in *BIC* values between models could be conclusive if $\Delta BIC$ values $\leq 2.0$ shows no significant difference between the two models [1, 19]. When $2 < \Delta BIC < 6$, there is a positive perspective that the model with the lower BIC is the most suitable [1, 19]. For $6 < \Delta BIC < 10$, there is a strong possibility that the model with a lower BIC value would be the best model to be fitted [1, 19]. However, if $\Delta BIC \geq 10.0$, it can be predicted with accuracy that the model with a lower BIC value is better fitted [1, 19].

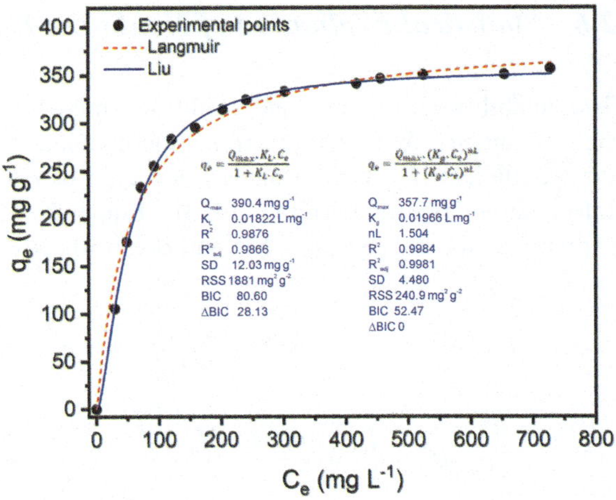


**Fig. 13** Isotherm of Procion Blue MX-R adsorption by a hybrid of wood waste with APTES in the proportion 1.0:0.5 at 30 °C, pH 2.0

To make it clear to the reader, let us evaluate two isotherm models to describe the adsorption of Procion Blue MX-R by a hybrid of wood waste with APTES in the proportion 1.0:0.5 at 30 °C [22] (see Fig. 13).

The fitting data is presented within the plot. According to Eq. (34), as the number of parameters increases, the *BIC* value rises by $pLn(n)$, where $p$ represents the number of parameters and $n$ is the number of experiments. Since $n$ is the same for both the Langmuir and Liu isotherm models, the *RSS* values will determine the *BIC* values. The $\Delta BIC$ for the Liu isotherm is shown to be zero, while the $\Delta BIC$ for the Langmuir isotherm is 28.13. This indicates that the difference between the Langmuir BIC value and the Liu *BIC* value exceeds 10, suggesting that Liu's isotherm is statistically the superior model for characterizing equilibrium isotherms. It is important to mention that the $R^2$ values of Langmuir and Liu cannot be directly compared, as Langmuir has 2 parameters and Liu has 3. However, $R^2_{adj}$ can be compared, as it accounts for the mathematical advantages of having an additional term in the Liu isotherm compared to Langmuir. Hence, if a model's $R^2_{adj}$ is closer to 1.000, even with additional parameters, it reflects improved phenomenological behavior rather than mere mathematical manipulation. Ultimately, the most effective method for model selection is by comparing the *BIC* values of each model.

It should also be noted that $\Delta BIC$ is always a comparison between two models. If there is a third model, compare the first two models beforehand, and finally, the best model between the first and second is compared with the *BIC* value of the third model. Therefore, in Fig. 14, the comparison of the model of Liu and Toth for describing the Isotherm of Procion Blue MX-R adsorption by a hybrid of wood waste with APTES in the proportion 1.0:0.5 at 30 °C, pH 2.0 [22].

Analysing Fig. 14, again, the statistical parameters ($R^2$, $R^2_{adj}$, and *SD*) of Liu's isotherm were better than Khan's isotherm. In addition, the $\Delta BIC$ between Khan and Liu was 21.21, which is higher than 10, indicating clearly that for this set of data, the Liu isotherm model is the best choice.

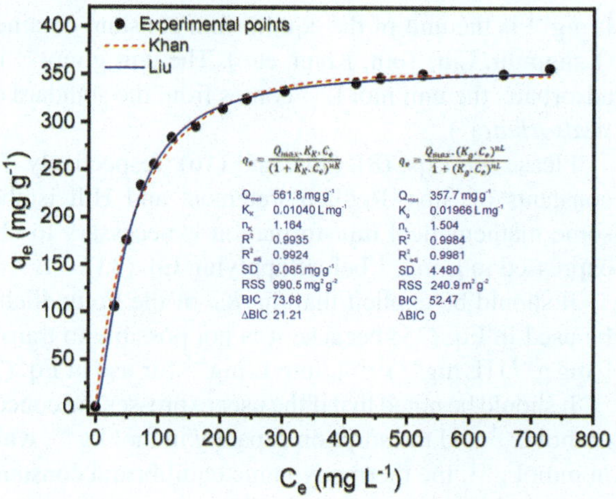


**Fig. 14** Isotherm of Procion Blue MX-R adsorption by a hybrid of wood waste with APTES in the proportion 1.0:0.5 at 30 °C, pH 2.0

## 2.7 *Thermodynamics of Adsorption*

Figure 15 illustrates the equilibrium adsorption of a solute dissolved in aqueous solution and solid adsorbent for a hypothetical system.

Recently Lima et al. [23, 24] deduced the expression of the thermodynamic equilibrium constant of adsorption ($K_e^0$), considering the principles of chemical equilibrium depicted in Eq. (35). The authors arrived at the following expression of the thermodynamic equilibrium constant:

$$K_e^0 = \frac{1000 \cdot K_e \cdot Mw \cdot [adsorbate]^0}{\gamma} \tag{35}$$

where $K_e^0$ is the thermodynamic equilibrium constant (dimensionless), $K_e$ is the equilibrium constant (L mg$^{-1}$) that was obtained from the best isotherm model of isotherms of equilibrium carried out at several temperatures (at least four temperatures are required). The [adsorbate]° is the standard concentration of the adsorbate that, by definition, is 1 mol L$^{-1}$ [1, 25, 26]; $\gamma$ is the activity coefficient (dimensionless).

$$\text{The unit ofis dimensionless}: \frac{1000\,\text{mg}}{\text{g}} \times \frac{\text{L}}{\text{mg}} \times \frac{\text{g}}{\text{mol}} \times \frac{\text{mol}}{\text{L}} = dimensionless$$

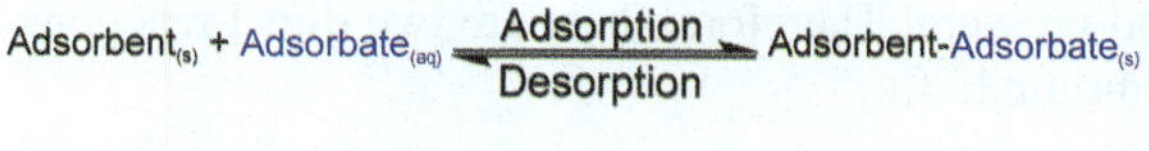


**Fig. 15** Equilibrium adsorption in aqueous/solid phase

L mg$^{-1}$ is the unit of the equilibrium constant obtained by the best isotherm model (Langmuir, Liu, Toth, Khan, etc.). The unit g mol$^{-1}$ is the molecular weight of the adsorbate; the unit mol L$^{-1}$ comes from the standard concentration of the adsorbate ($[adsorbate]^{\circ}$).

Please see Eqs. (8), (14), and (16), respectively, for employing the equilibrium constants of Sips, Redlich–Peterson, and Hill isotherms in Eq. (35). Therefore, some mathematical transformation is necessary to obtain the equilibrium constant expressed in L mg$^{-1}$ before applying Eq. (33).

It should be explicit that the $K_F$ of the Freundlich equilibrium model could not be used in Eq. (35) because it is not possible to transform the unit of $K_F$, which is $[(\text{mg g}^{-1})(\text{L mg}^{-1})^{1/n_F}]$, into L mg$^{-1}$ for use in Eq. (35).

It should be noted that if the user expresses the concentration of the sorbing species in mol L$^{-1}$ and the sorption capacity in mol kg$^{-1}$, which is equivalent to expressing in mmol g$^{-1}$, the thermodynamic equilibrium constant ($K_e^0$) is:

$$K_e^0 = \frac{K_e}{\gamma} = \frac{K_e \cdot [adsorbate]^0}{\gamma} \tag{36}$$

$K_e$ is expressed in L mol$^{-1}$, the standard adsorbate concentration is 1 mol L$^{-1}$, and the thermodynamic equilibrium constant will be dimensionless. Although the conversion of these units seems obvious, the adsorption literature contains many mistakes.

Also, a recurrent error in the literature concerns the arrow for chemical equilibrium and, consequently, for equilibrium isotherms [24]. Many authors use the following symbols for chemical equilibrium [27]:

$$\text{Adsorbent}_{(s)} + \text{Adsorbate}_{(aq)} = \text{Adsorbent} - \text{Adsorbate}_{(s)} \tag{37}$$

$$\text{Adsorbent}_{(s)} + \text{Adsorbate}_{(aq)} \leftrightarrow \text{Adsorbent} - \text{Adsorbate}_{(s)} \tag{38}$$

$$\text{Adsorbent}_{(s)} + \text{Adsorbate}_{(aq)} \rightleftarrows \text{Adsorbent} - \text{Adsorbate}_{(s)} \tag{39}$$

Wrong arrows are utilized throughout the literature to stand for equilibrium. The arrow utilized in Eq. (37) stands for stoichiometry or the sum of several chemical reactions (such as Hess's Law) [27]. The arrow utilized in Eq. (38) stands for resonance [27]. The arrow utilized in Eq. (39) is not a chemical equilibrium. It is used when a reactant A reacts with B, producing C plus D, using one condition of temperature and pressure, and C plus D generate A plus B, using another condition of temperature and pressure. Therefore, there are two direct reactions, and it is not a chemical equilibrium [27].

$$\begin{aligned} \text{A} + \text{B} &\rightarrow \text{C} + \text{D} \text{ (condition 1)} \\ \text{C} + \text{D} &\rightarrow \text{A} + \text{B} \text{ (condition 2)} \end{aligned} \tag{40}$$

The correct symbol for chemical equilibrium and for adsorption equilibrium is

$$\text{Adsorbent}_{(s)} + \text{Adsorbate}_{(aq)} \rightleftharpoons \text{Adsorbent} - \text{Adsorbate}_{(s)} \quad (41)$$

The right arrow that stands for equilibrium (two semi-arrows) is given above.

## 2.8 *Study of Cases for Using Wrong Equilibrium Constant Equations, Frequently Found in the Literature*

Several equations are employed as "*equilibrium constants*" for use in the van't Hoff equation to estimate the thermodynamic parameters of adsorption [1, 23]. Equation (42) (distribution constant) is commonly used as an equilibrium constant:

$$K_d = \frac{q_e}{C_e} \quad (42)$$

By using this equation, $K_d$ presents the unit of L g$^{-1}$.

Frequently, some authors transform the $K_d$ equation into a dimensionless equation.

$$K_d = \frac{q_e}{C_e} \cdot \frac{m}{V} \quad (43)$$

In Eq. (43), $K_d$ has no dimension [1, 23].

Another equation that is frequently confused with "equilibrium constant" is given below:

$$K_c = \frac{C_s}{C_e} = \frac{(C_0 - C_e)}{C_e} \quad (44)$$

In the equation above, $K_c$ has no dimension.

Although at first, seeing it, it seems that Eqs. (43) and (44) are different, they are the same equation. Considering that:

$$q = \frac{(C_0 - C_e)}{m} \cdot V \quad (45)$$

and applying Eq. (45) to (43), it leads to:

$$K_d = \frac{\frac{(C_0 - C_e) \cdot V}{m}}{C_e} \cdot \frac{m}{V} = \frac{(C_0 - C_e) \cdot V}{C_e \cdot m} x \frac{m}{V} = \frac{(C_0 - C_e)}{C_e} = K_d \quad (46)$$

$$\therefore K_d = K_c$$

Equations (42), (43), (44), and (46) are not equilibrium constants [1, 23], since these equations do not have a unique value. Nevertheless, it is frequently used for estimating the thermodynamic parameters of adsorption [17, 28, 29].

The thermodynamic equilibrium constant ($K_e^0$) should be obtained using Eq. (35) [1, 23, 24]. It should not be obtained using only a mathematical strategy to become the equilibrium constant ($K_e$) obtained from the best isotherms model (L mg$^{-1}$) into a dimensionless equation, as given below [17, 30]:

$$K^0 = 1000 \cdot K_e \cdot d \tag{47}$$

In Eq. (47), the equilibrium constant ($K_e$) obtained from fitting the isotherm at different temperatures [1, 23, 24] has the unit of L mg$^{-1}$. When this parameter is multiplied by 1000 mg g$^{-1}$, it results in the unit of L g$^{-1}$. Further, if the parameter is also multiplied by the density of the water (1000 g L$^{-1}$), the $K^\circ$ will become dimensionless.

Another wrong approach that was already reported [31, 32] is to multiply the value of isotherm equilibrium constant (Langmuir, Liu, Toth, Khan) by 1000, to transform the parameter from L mg$^{-1}$ into L g$^{-1}$, and subsequently by the molar concentration of water in an aqueous solution (55.51 mol L$^{-1}$) [31, 32]. Furthermore, the parameter is multiplied by the adsorbate molecular weight (Mw: g mol$^{-1}$) to finalize the mathematical strategy. Performing all these multiplications, the $K$ becomes dimensionless [31, 32].

$$K = 55.51 \cdot K_e \cdot 1000 \cdot Mw \tag{48}$$

Nevertheless, expert users of adsorption should consider the principles of chemical equilibrium described in Eq. (35), and not a simple mathematical strategy to make $K$ dimensionless.

In this section, Eq. (35) ($K_e^0$) will be compared with Eq. (41) ($K_d$) and Eq. (44) ($K_c$). For this purpose, the data of equilibrium isotherms of the dye nipagin adsorbed onto carbon-based material prepared from palm leaves [33] was chosen.

As seen in Fig. 16, the isotherm of adsorption is presented, where the value of $K_g$ (L mg$^{-1}$) is obtained by nonlinear fitting, and after applying Eq. (35), the Thermodynamic Equilibrium Constant ($K_e^0$) is obtained. In Fig. 17 is observed that for each experimental point of the isotherm (from $a$ to $n$), there is one constant value of the supposed $K_d \left(\frac{q_e}{C_e}\right)$or $K_c \left(\frac{(C_o - C_e)}{C_e}\right)$.

Even if some points were deleted, no single value of $K_d$ or $K_c$ is obtained [1, 23, 24]. Therefore, for the plots of Fig. 17a or b, the values of $K_d$ and $K_c$ were obtained for interpolation in the Y-axis, when $C_o \rightarrow 0$. This strategy of obtaining the equilibrium constant for each temperature (283 to 318 K) was carried out, and the result of these constants is depicted in Table 2.

Using the values of equilibrium constants obtained in Table 2 in the linearized Van't Hoff equation depicted below, the van't Hoff curves are presented in Figs. 18 and 19.

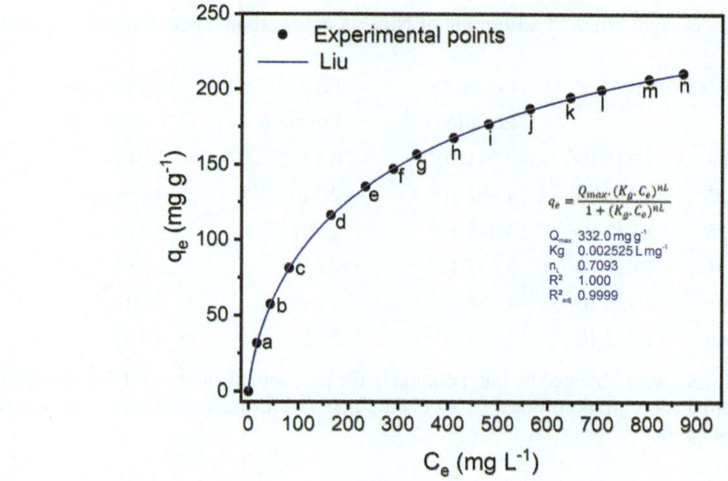


**Fig. 16** Liu's adsorption isotherm of nipagin onto carbon-based material prepared from palm leaves at 298 K

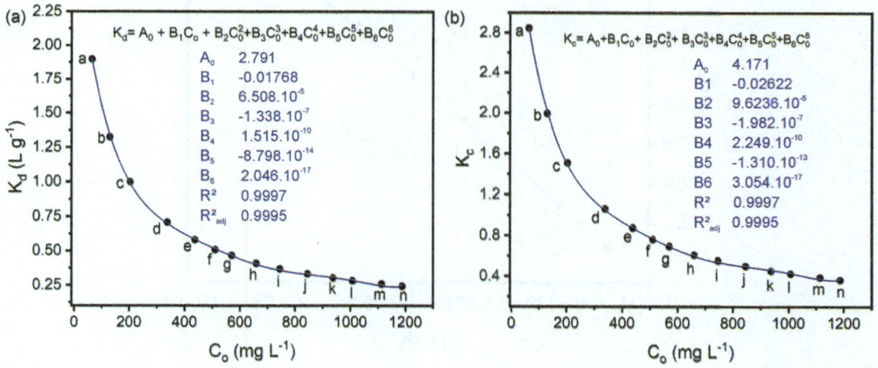


**Fig. 17** **a** Plot of of $K_d\left(\frac{q_e}{C_e}\right)$ versus $C_o$; **b** plot of $K_c\left(\frac{(C_o-C_e)}{C_e}\right)$ versus $C_o$. For the removal of nipagin onto carbon-based material prepared from palm leaves at 298 K

$$\ln K_e^\circ = \frac{\Delta S^\circ}{R} - \left(\frac{\Delta H^\circ}{R}\right) \cdot \frac{1}{T} \tag{49}$$

Observing Figs. 18 and 19, it is possible to verify that the linearized van't Hoff equation could only be obtained from the values of equilibrium constants obtained from the best fitting of nonlinear isotherms of adsorption at several temperatures (at least four temperatures). The Eqs. (41)–(43) did not lead to a regular pattern of the "supposed equilibrium constants" $K_d$ or $K_c$. The only explanation that the authors of this chapter could see for some authors insisting on using Eqs. (41)–(43) to obtain the thermodynamic parameters of adsorption is that they use only one concentration at

**Table 2** Constants of equilibrium of adsorption of Nipagin onto carbon-based material prepared from palm leaves

| Using the strategy shown in | | | Figure 16 | Figure 17a | Figure 17b |
|---|---|---|---|---|---|
| Liu isotherm | | | Equation (35) | Equation (41) | Equation (43) |
| 283 | 0.005088 | $7.741 \cdot 10^2$ | $7.741 \cdot 10^2$ | 0.7362 | 1.044 |
| 293 | 0.003222 | $4.902 \cdot 10^2$ | $4.902 \cdot 10^2$ | 2.746 | 4.086 |
| 298 | 0.002525 | $3.842 \cdot 10^2$ | $3.842 \cdot 10^2$ | 2.791 | 4.171 |
| 303 | 0.001994 | $3.033 \cdot 10^2$ | $3.033 \cdot 10^2$ | 2.656 | 4.012 |
| 313 | 0.001305 | $1.986 \cdot 10^2$ | $1.986 \cdot 10^2$ | 3.696 | 5.501 |
| 318 | 0.001046 | $1.592 \cdot 10^2$ | $1.592 \cdot 10^2$ | 2.755 | 4.106 |

* The values of $K_d$, $K_c$ were obtained by interpolation in the polynomial curves of these constants as a function of initial concentration. When $C_o \rightarrow 0$, the values of these constants were determined
** Nipagin 152.149 g $mol^{-1}$

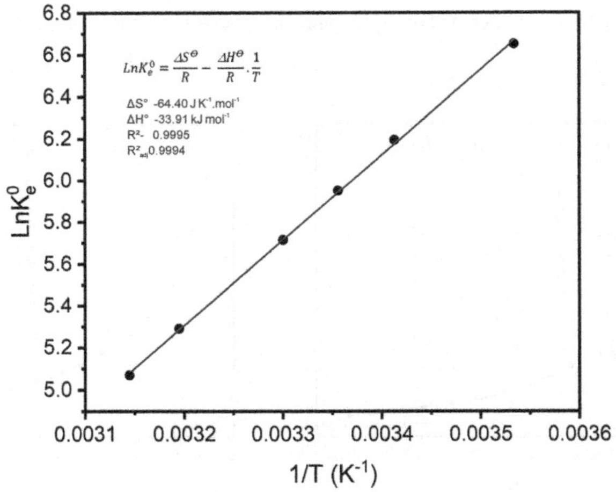


**Fig. 18** Linearized van't Hoff equation for $K_e^0$ obtained from Eq. (35) (Liu isotherm fitting). In Fig. 16, the value of the $K_g$ was obtained at 298 K. In subsequent stages $K_g$ was converted to $K_e^0$ using Eq. (35)

three temperatures. If they use three or more concentrations, they will give up, since, as shown in Fig. 17, $K_d$ and $K_c$ do not have a single value for each concentration of the sorbing species.

As previously reported [1, 23, 24], another recommendation for obtaining accurate thermodynamic parameters is to use at least four different temperatures. Some authors obtain their thermodynamic parameters using only three temperatures. In the plots of Fig. 19, if only three temperatures were chosen, it is possible to obtain lines with better values of $R^2$ and $R^2_{adj}$ [1, 23, 24]. In addition, concerning the values of ΔH° and ΔS°, a significant discrepancy is observed between the plots of Fig. 18, the $K_e^0$ obtained from isotherms at six different temperatures from values of $K_d$ and $K_c$

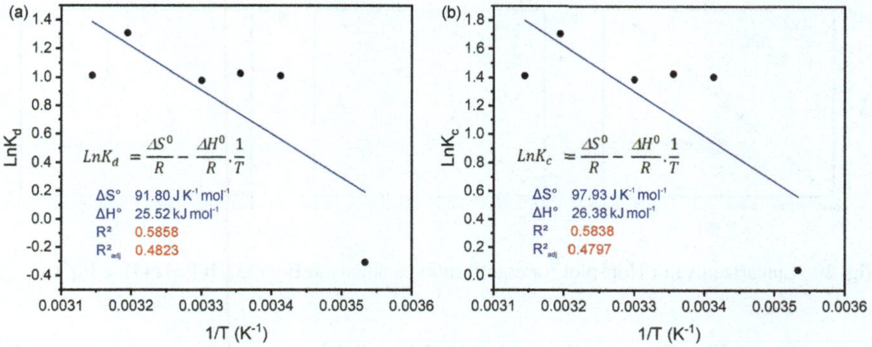


**Fig. 19** Linearized van't Hoff equation for supposed equilibrium constants calculated from data of Table 2

(Fig. 19). Using the Liu isotherms and applying Eq. (35), the nature of adsorption is exothermic. Conversely, both $K_d$ and $K_c$ gave rise to endothermic enthalpies. Also, the signal of ΔS° is inverted. It is negative for Fig. 18 and positive for Fig. 19. Considering several papers in the literature, the use of $K_c$ and $K_d$ leads to various misinterpretations of adsorption thermodynamics that are prevalent throughout the literature.

In addition to the mistake of employing the $K_d$ and $K_c$ equations for accurate thermodynamic outcomes, the challenges in determining the right thermodynamic parameters extend beyond just these equations. Some researchers, even when utilizing the equilibrium constant derived from adsorption isotherms, can encounter mistakes when converting the empirical equilibrium constant from L mg$^{-1}$ to the dimensionless equilibrium constant, using Eqs. (47) and (48) [1, 23]. However, as previously shown [1, 23], this error can impact the ΔS° values.

Table 3 presents the equilibrium constants for the adsorption of Nipagin onto carbon-based material prepared from palm leaves from 283 to 318 K.

**Table 3** The equilibrium constants for the adsorption of Nipagin onto carbon-based material prepared from palm leaves from 283 to 318 K

| | | | Equation (35) | Equation (47) | Equation (48) |
|---|---|---|---|---|---|
| $T$ (K) | $K_g$ (L mg$^{-1}$) | $K_e$ (L mol$^{-1}$) | $K_e^0$ | K° | $K$ |
| 283 | 0.005088 | 7.741·10$^2$ | 7.741·10$^2$ | 5088 | 282.4 |
| 293 | 0.003222 | 4.902·10$^2$ | 4.902·10$^2$ | 3222 | 178.9 |
| 298 | 0.002525 | 3.842·10$^2$ | 3.842·10$^2$ | 2525 | 140.2 |
| 303 | 0.001994 | 3.033·10$^2$ | 3.033·10$^2$ | 1994 | 110.7 |
| 313 | 0.001305 | 1.986·10$^2$ | 1.986·10$^2$ | 1305 | 72.44 |
| 318 | 0.001046 | 1.592·10$^2$ | 1.592·10$^2$ | 1046 | 58.06 |

All the equilibrium constants are dimensionless

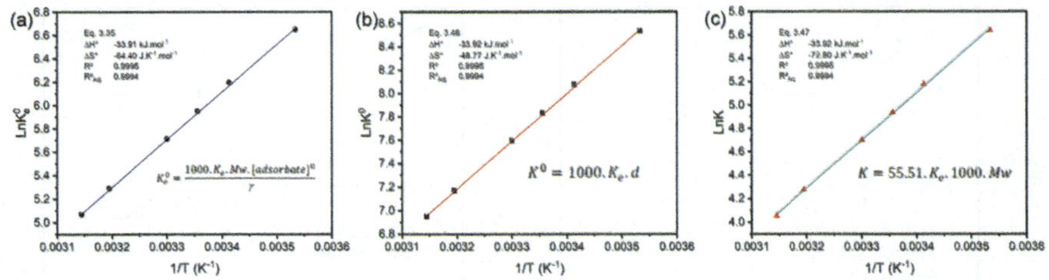


**Fig. 20** Linearized van't Hoff plot for equilibrium constants. **a** Eq. (35); **b** Eq. (47); **c** Eq. (48)

As seen in Fig. 20, the values of ΔH°, $R^2$, and $R^2_{adj}$ were the same for Eqs. (35), (47), and (48). The problems of using Eqs. (47) and (48) will be in the intercept of the linear curves, resulting in changes in ΔS°. Consequently, the values of ΔG° will be different. Calculating these values of ΔG°, the values would be −14.72, −19.39, and −12.23 kJ mol$^{-1}$, for Eqs. (35), (47), and (48), respectively.

Besides the linearized van't Hoff equation for obtaining the values of ΔS° and ΔH° of adsorption, it is also possible to employ the nonlinear van't Hoff equation for the estimation of the thermodynamic parameters of adsorption [34].

$$K_e^\circ = \exp\left[\frac{\Delta S^\circ}{R} - \left(\frac{\Delta H^\circ}{R}\right) \cdot \frac{1}{T}\right] \tag{50}$$

This nonlinearized van't Hoff equation was compared with the linearized form [1, 34] to check if the values of the parameters of adsorption obtained by the linearized van't Hoff equation agree with the values obtained by the nonlinear form.

Figure 21 depicts the nonlinear van't Hoff curve for the adsorption of Nipagin onto carbon-based material prepared from palm leaves. The values of ΔH° and ΔS° obtained from the nonlinear fitting curve are similar to those derived from the linear van't Hoff equation, as previously mentioned [1, 34].

## 3 Kinetic Adsorption Models

### *3.1 Kinetic Models Based on the Order of Reaction*

The study of adsorption kinetics is important in treating aqueous effluents using nanomaterials because it provides valuable information on the reaction pathways and mechanisms of adsorption reactions.

Given that the kinetic equations are well-documented in the literature, this chapter will focus solely on presenting the equations [1].

The first set of kinetic adsorption models, are the one based on the chemical reaction [1].

The pseudo-first-order (PFO) is given by the Eq. (51) [8]:

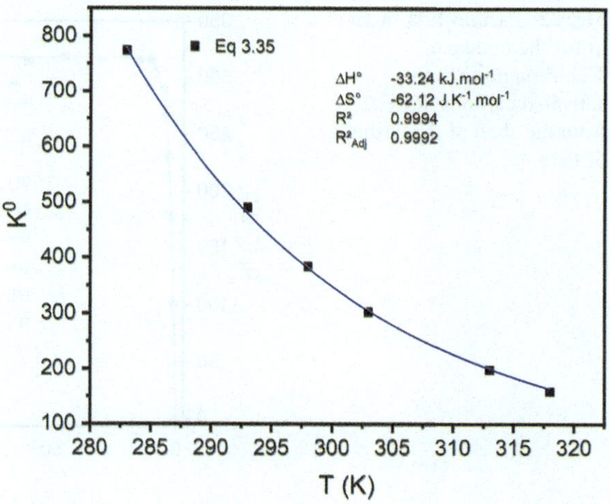


**Fig. 21** Nonlinear van't Hoff plot for the adsorption of Nipagin onto carbon-based material prepared from palm leaves

$$q_t = q_e \cdot \left[1 - \exp(-k_1 \cdot t)\right] \tag{51}$$

where $q_t$ and $q_e$ are the sorption capacities at any time, and at the equilibrium (mg $g^{-1}$), respectively, $k_1$ is the kinetic constant rate ($min^{-1}$ or $h^{-1}$), and $t$ is the time (min, h).

The pseudo-second-order (PSO) is given by the Eq. (52) [9]:

$$q_t = \frac{k_2 \cdot q_e^2 \cdot t}{1 + q_e \cdot k_2 \cdot t} \tag{52}$$

where $q_t$ and $q_e$ are the adsorption capacities (mg $g^{-1}$) at any time and at the equilibrium, respectively, t is the time of contact (min), and $k_2$ is the pseudo-second-order constant rate (g $mg^{-1}$ $min^{-1}$).

The general order kinetic (GO) model [1, 10] complements the adsorption kinetic models based on chemical reaction models [1, 10].

$$q_t = \left( q_e - \frac{q_e}{\left[k_N \cdot (q_e)^{n-1} \cdot t \cdot (n-1) + 1\right]^{1/(1-n)}} \right); \quad n \neq 1 \tag{53}$$

where $q_t$ and $q_e$ are sorption capacities (mg $g^{-1}$) at any time and equilibrium, respectively, t is the contact time (min or h), $k_n$ is the n-order kinetic constant rate $\left[\min^{-1}\left(\text{g mg}^{-1}\right)^{n^{-1}}\right]$, and n is the kinetic order; this value should be different from 1 [1, 10].

Readers should note that kinetic models derived from chemical reactions do not imply that the adsorption process itself is a chemical reaction. This misunderstanding frequently appears in adsorption literature, where many authors mistakenly equate

**Fig. 22** Pseudo-first-order fit for the uptake of 4-chloroaniline onto activated carbons derived from the shell of the cashew of Para

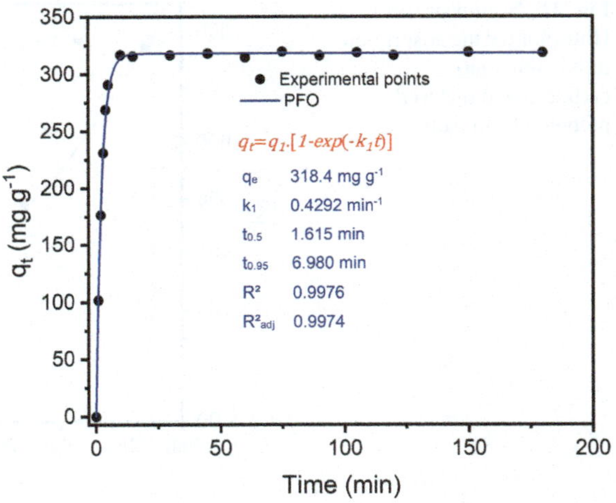


kinetic models based on chemical reactions (PFO, PSO, GO) with actual chemical reactions, similar to those in catalysis [1].

Below are presented plots of PFO, PSO, and GO kinetic models based on chemical reactions.

Figure 22 shows the plot of PFO for the adsorption of 4-chloroaniline onto activated carbons derived from the shell of the cashew of Para [35].

It has been noted that the parameters $t_{0.5}$ and $t_{0.95}$ serve as supplementary metrics within the framework of kinetic models. These parameters denote the duration required to achieve 50% and 95% saturation of the adsorbent, respectively [13, 15, 20–22]. The $t_{0.5}$ and $t_{0.95}$ parameters are essential as they enable the comparison of kinetics behaviors, even when differing kinetic models are employed. The kinetic constant rates ($k_1$, $k_2$, $k_n$, etc.) possess varying units, which complicates the comparison across distinct kinetic models. Conversely, the $t_{0.5}$ and $t_{0.95}$ facilitate this comparison. Figure 23 shows how values of $t_{0.5}$ and $t_{0.95}$ are obtained [36].

Figure 24 depicts the pseudo-second-order kinetic model for Reactive Black 5 dye uptake onto Pinon wastes [37].

Figure 25 depicts the GO kinetic model of adsorption of Reactive Black 5 dye onto Pinon wastes [37].

## 3.2 *Empiric Models*

The Elovich equation is usually employed in adsorption kinetics when chemical reactions occur during the adsorption process [1, 12]. The Elovich kinetic adsorption model has been utilized to describe slow kinetic adsorption rates. Also, for adsorbents with a heterogeneous surface, this model has been utilized [1, 12]. The Elovich chemisorption kinetic model is defined as:

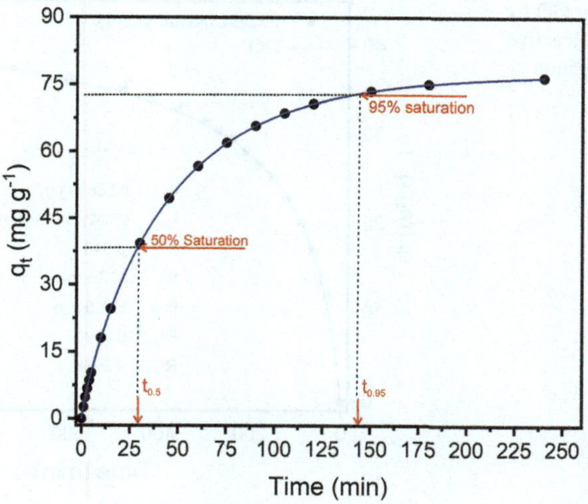


**Fig. 23** The interpolation in the fitted kinetic curve at 50% saturation and 95% saturation allows the determination of the $t_{0.5}$ and $t_{0.95}$, respectively. Kinetic data obtained from [36]

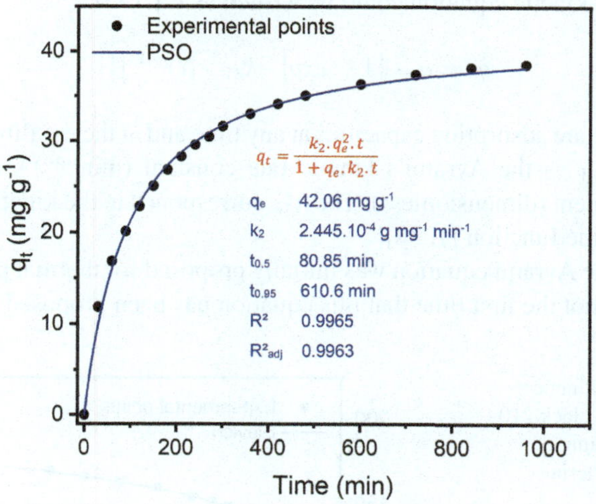


**Fig. 24** Pseudo-second-order fit for the uptake of Reactive Black 5 dye onto Pinon wastes

$$q_t = \frac{1}{\beta}\ln(\alpha \cdot \beta) + \frac{1}{\beta}\ln(t) \quad (54)$$

Equation (53) is defined as the Elovich kinetic model, where $\beta$ is a constant that is related to the activation energy involved in chemisorption and the extent of surface coverage (g mg$^{-1}$), $\alpha$ is the initial sorption rate (mg g$^{-1}$ min$^{-1}$), $q_t$ is the sorption capacity at any time (mg g$^{-1}$) and $t$ is the contact time (min).

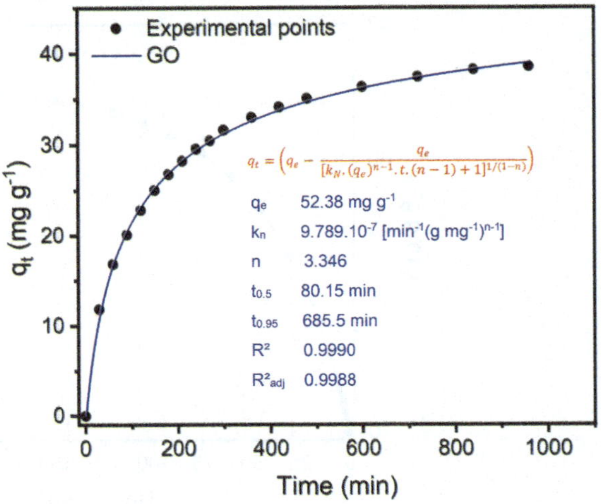


**Fig. 25** Nonlinear GO for the adsorption of Reactive Black 5 dye onto Pinon wastes

Figure 26 shows the nonlinear plot for the kinetics of removing acid black 210 dye from a tannery sludge carbon material using the Elovich kinetic model [41].

The Avrami kinetic equation could be written as Eq. (55).

$$q_t = q_e \cdot \left\{1 - exp\left[-k_{AV} \cdot (t)^{nAV}\right]\right\} \tag{55}$$

where $q_t$ and $q_e$ are adsorption capacities at any time and at the equilibrium (mg g$^{-1}$), respectively, $k_{AV}$ is the Avrami's kinetic rate constant (min$^{-nAV}$), and $n_{AV}$ is the Avrami's exponent (dimensionless). The $n_{AV}$ corresponds to the kinetic constant rate changes as a time function [1, 38].

However, the Avrami equation was initially proposed for thermal phase transition [1, 38]; this is not the first time that this equation has been proposed to describe the

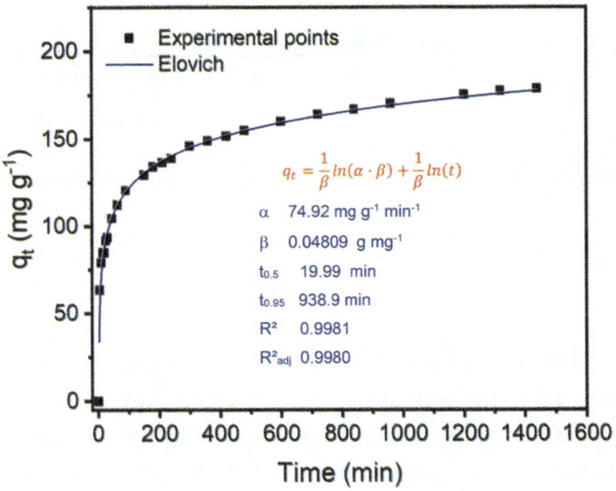


**Fig. 26** Elovich kinetic model fit for acid black 210 adsorption onto tannery sludge carbon material

kinetics of adsorption [39]. The Bangham kinetics equation [39, 40] is referred to as the chemisorption adsorption equation [40]. The nonlinear form of the Bangham equation is given by Eq. (56).

$$q_t = q_e \cdot \{1 - exp[-k_B \cdot (t)^{n_B}]\} \tag{56}$$

In the equation above, $q_t$ is the sorption capacity at any time (mg g$^{-1}$), $q_e$ is the sorption capacity at the equilibrium (mg g$^{-1}$), $n_B$ is the dimensionless Banghan equation, $k_B$ is the Banghan rate constant $\left(\text{min}^{-\text{n}_\text{B}}\right)$, and t is the time of contact between the adsorbent and adsorbate (min).

While some authors treat Avrami and Bangham as separate equations, they are actually identical. This may have gone unverified in the past, as numerous authors typically employ these equations in their linearized forms, making it difficult to tell them apart.

The literature has explored the concept of fractal-like kinetic models, positing that the kinetic constant rate varies as a function of time [11, 42, 43]. This implies that at the onset of the adsorption process, the kinetic constant rate possesses a value that does not remain invariant as time progresses [11, 42, 43]. The main known fractal-like kinetic models are fractal-like pseudo-first-order (FLPFO) and fractal-like pseudo-second-order (FLPSO), whose expressions are given below, respectively:

$$q_t = q_e \cdot \{1 - exp[-k_{1,0} \cdot (t)^n]\} \tag{57}$$

$$q_t = \frac{k_2 \cdot q_e^2 \cdot t^n}{1 + q_e \cdot k_{2,0} \cdot t^n} \tag{58}$$

where $q_t$ and $q_e$ are sorption capacities (mg g$^{-1}$) at any time and at the equilibrium, respectively, $k_{1,0}$ (min$^{-\text{n}}$) is the fractal-like pseudo-first-order constant rate, $k_{2,0}$ (g mg$^{-1}$ min$^{-\text{n}}$) is the fractal-like pseudo-second-order constant rate, and $n$ is the fractional-like exponent ($n > 0$).

It is imperative to emphasize that Eq. (57) is precisely identical to Eqs. (55) and (56). Consequently, the Avrami fractional order, Bangham chemisorption, and fractional-like pseudo-first-order are equivalent to the same kinetic equation. Moreover, concerning the Avrami fractional order concepts, it is indicated that the kinetic constant rate fluctuates over time [1, 38].

Figure 27 depicts the nonlinear Avrami fractional order, or Bangham chemisorption model, or fractal-like pseudo-first-order kinetic model for Reactive Black 5 dye uptake onto Pinon wastes [37].

Figure 28 depicts the fractional-like pseudo-second-order kinetic model for Reactive Black 5 dye uptake onto Pinon wastes [37].

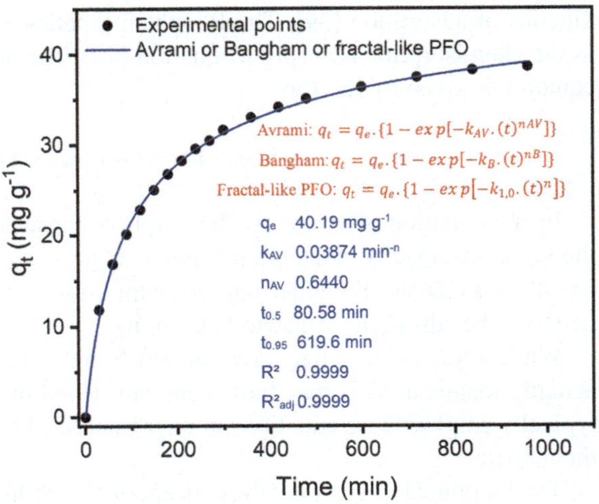


**Fig. 27** Avrami fractional order, or Bangham chemisorption or fractal-like pseudo-first-order fit for Reactive Black 5 dye uptake onto Pinon wastes

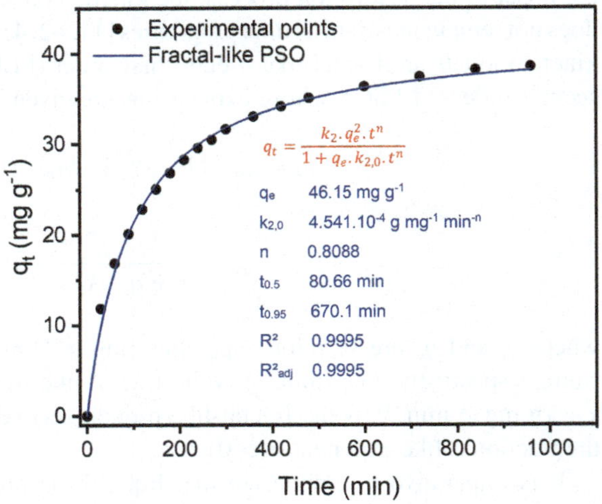


**Fig. 28** Fractal-like pseudo-second-order fit for Reactive Black 5 dye uptake onto Pinon wastes

## 3.3 *Diffusive Mechanisms*

The adsorption process involves five key steps:

(1) the movement of the sorbing species from the bulk solution to the solvent film surrounding the adsorbent. This mass transfer process occurs primarily through convection, which happens rapidly in agitated systems.
(2) transfer of the sorbing species across the film at the solid–liquid interface. This process is referred to as film diffusion.
(3) adsorption occurring on the outer surface of the adsorbent;
(4) diffusion of the sorbing species within the adsorbent's pore structure; and

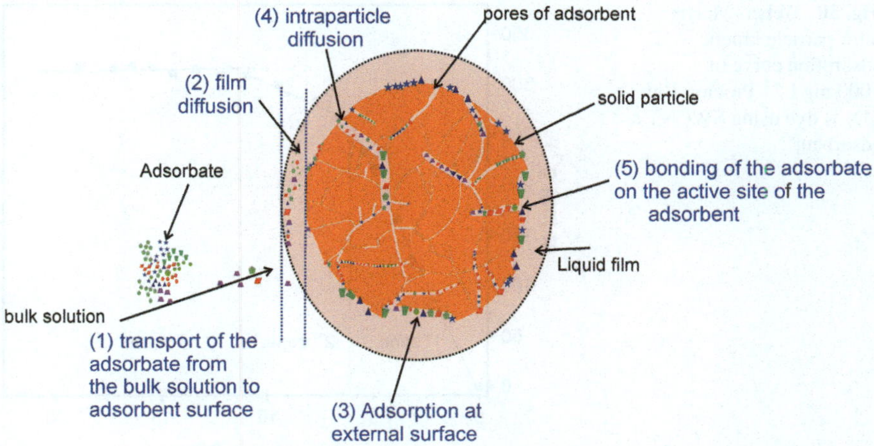


**Fig. 29** Schematic diagram of adsorption of multiple adsorbates in a porous solid adsorbent

(5) adsorption of the adsorbing species into the active sites located within the adsorbent's pores [1, 44].

The fifth step is generally faster than steps 2 and 4. Additionally, step 3 is quick. Therefore, steps 3 and 5 do not limit mass transfer and are not considered in determining the rate constant.

Figure 29 depicts the diagram of steps of adsorption of a solute in a porous adsorbent [1, 44].

The intra-particle diffusion resistance that influences the adsorption process can be studied using the Weber-Morris intra-particle diffusion model depicted in Eq. (59) [44]:

$$q_t = k_{id} \cdot \sqrt{t} + C \tag{59}$$

In this context, $k_{id}$ represents the Weber–Morris intra-particle diffusion constant (mg $g^{-1}$ $min^{-0.5}$), while C denotes the resistance to mass transfer associated with the thickness of the boundary layer (mg $g^{-1}$) [44].

Figure 30 presents the Weber–Morris intra-particle kinetic plot for the Procion Blue MX-R dye adsorption onto single-walled carbon nanotubes (SWCNT) as an adsorbent [10].

The intra-particle diffusion constant ($k_{id}$, mg $g^{-1}$ $min^{-0.5}$) can be obtained from the slope of the plot of $q_t$ versus the square root of time. Figure 30 shows $q_t$ versus $t^{1/2}$, with multi-linearity for the dye's adsorption on the adsorbent.

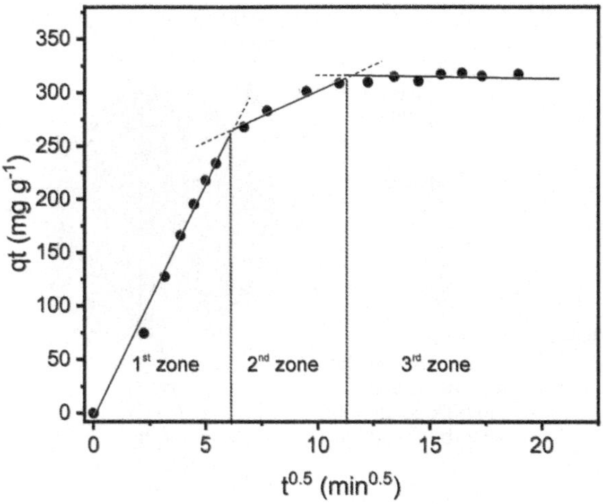


**Fig. 30** Weber–Morris intra-particle kinetic adsorption curve of 1000 mg $L^{-1}$ Procion Blue MX-R dye using SWCNT as adsorbent

The results depicted in Fig. 30 indicate that the adsorption process does not occur in a single kinetic step [1]. The process of adsorption presented three stages that can be assigned to each linear part of Fig. 30. The first stage can be attributed to the diffusion of the dye through the film of solvent covering the adsorbent particles and the adsorption of the adsorbate at the external surface of the adsorbent, which corresponds to steps 2 and 3 of Fig. 29 [1]. This first linear segment corresponds to a fast adsorption step.

The second stage can be assigned as intra-particle diffusion. This stage corresponds to step 4 of Fig. 29. The second linear segment is the kinetic rate-determining step of the adsorption process because this stage is a slow step. The slope of the second linear segment allowed the $k_{id}$ to be determined [1].

The third segment can be assigned to the diffusion of the dye by the adsorbent's smaller pores, followed by the establishment of equilibrium [1], which corresponds to the 5th step of Fig. 29.

It is essential to underscore that the interactions between the solid adsorbent and the sorbing species occurring at the pores and the external surface are fundamentally identical (covalent bonds, electrostatic attraction, complexation, the formation of ion pairs, hydrogen bonding, π–π interaction, n–π donor–acceptor interaction, van der Waals forces, etc.). The process of pore filling by an adsorbate (solute) into the pores of an adsorbent is not an isolated phenomenon. Pore filling occurs when there is a marked interaction between the adsorbent and adsorbate. Even highly porous solids with considerable surface area will show minimal sorption if there is no chemical interaction with the sorbing species.

# 4 Misuse of Linearization for Equilibrium and Kinetics of Adsorption

One of the most prevalent errors in analyzing adsorption studies is the misuse of linearization. This approach was commonly adopted decades ago, back when computers and statistical software were not readily accessible. Nowadays, linearization's use in data analysis is increasing, as many authors opt for simpler tools to evaluate their experimental results. A key reason for favoring linear equations is that they require fewer experimental points to establish a line; this contrasts with nonlinear equations, which need more points to outline a curve. Furthermore, some researchers may discard certain data points to artificially boost $R^2$ values. This raises a significant concern: omitting data points may inflate $R^2$ values, while authors often neglect to adequately address the parameters in the equilibrium and kinetic adsorption models.

There are significant challenges associated with the linearization of an inherently nonlinear equation through various transformations. The primary concern when transforming data to derive a linearized equation lies in understanding the error structure of the data and the way this structure is influenced by the mathematical manipulation of the data [45–47]. Linearization is predicated on the assumption that the variance of all $q$ values (the Y variable of the graph, dependent variable, quantity of adsorbate adsorbed by the adsorbent) remains constant across the entire range of the data. Presuming homoscedasticity of the data (equal variance for all q values throughout the range) when, in fact, the data is heteroscedastic (characterized by unequal variance of all q values across the full range) may lead to an overestimation of the goodness of fit as evaluated by the $R^2$. This misinterpretation results in errors in the $R^2_{adj}$ [45–47]. Consequently, elevated $R^2$ values do not necessarily indicate a superior experimental data fit.

In practice, authors are required to remove specific points from their curves to validate the linearization method. This modification decreases the fitting model's degree of freedom (DOF), thereby restricting the independent ways in which a dynamic system can operate without breaching imposed constraints. The degree of freedom refers to the least number of independent coordinates essential for fully defining a system's position. To better illustrate this for readers, several case studies will be discussed below.

## 4.1 Equilibrium Isotherms

We begin with the equilibrium isotherms. Considering that most published articles utilized the Langmuir model, this equilibrium model will receive thorough examination in this Chapter.

Let us consider the adsorption of paracetamol onto activated biochar prepared from cellulose at 35 °C [48]. According to Kumar [47], there are four linearized equations of the Langmuir as presented in Table 4.

Let us also consider the following experimental data [48] as depicted in Table 5.

It is important to mention that adjusting the parameters led to changes in the parametric units, a factor often overlooked by many authors. Only the nonlinear

**Table 4** Langmuir equilibrium isotherms

| Isotherm model | Linear form | Plot |
|---|---|---|
| $q_e = \frac{Q_{max} \cdot K_L \cdot C_e}{1 + K_L \cdot C_e}$ Langmuir | $\frac{C_e}{q_e} = \frac{1}{K_L \cdot Q_{\max}} + \frac{1}{Q_{max}} C_e$ Langmuir-1 | $\frac{C_e}{q_e}$ vs $C_e$ |
| | $\frac{1}{q_e} = \frac{1}{Q_{\max}} + \left(\frac{1}{K_L \cdot Q_{\max}}\right) \cdot \frac{1}{C_e}$ Langmuir-2 | $\frac{1}{q_e}$ vs $\frac{1}{C_e}$ |
| | $q_e = Q_{max} - \left(\frac{1}{K_L}\right) \frac{q_e}{C_e}$ Langmuir-3 | $q_e$ vs $\frac{q_e}{C_e}$ |
| | $\frac{q_e}{C_e} = K_L \cdot Q_{max} - K_L \cdot q_e$ Langmuir-4 | $\frac{q_e}{C_e}$ vs $q_e$ |

**Table 5** Experimental data of the adsorption of paracetamol onto activated biochar prepared from cellulose at 30 °C [48]

| Experimental values | | Manipulated values | | | |
|---|---|---|---|---|---|
| $C_e$ (mg $L^{-1}$) | $q_e$ (mg $g^{-1}$) | $C_e/q_e$ (g $L^{-1}$) | $1/q_e$ (g $mg^{-1}$) | $1/C_e$ (L $mg^{-1}$) | $q_e/C_e$ (L $g^{-1}$) |
| 0 | 0 | – | – | – | – |
| 8.544408749 | 5.375742995 | 1.5894377 | 0.1860208 | 0.117035599 | 0.629153304 |
| 16.91199627 | 10.3099349 | 1.6403592 | 0.0969938 | 0.059129625 | 0.609622586 |
| 22.693064 | 13.1425134 | 1.7266913 | 0.0760889 | 0.044066328 | 0.579142305 |
| 31.9207109 | 16.16402987 | 1.9747991 | 0.0618658 | 0.031327623 | 0.506380636 |
| 48.86269841 | 20.19388622 | 2.4196778 | 0.0495199 | 0.020465509 | 0.413278163 |
| 66.91552941 | 22.53750354 | 2.9690746 | 0.0443705 | 0.014944214 | 0.336805279 |
| 89.3156367 | 24.18058307 | 3.6936924 | 0.0413555 | 0.011196248 | 0.2707318 |
| 115.3281991 | 25.36725163 | 4.5463419 | 0.0394209 | 0.008670906 | 0.21995706 |
| 146.958305 | 26.2786584 | 5.5923062 | 0.0380537 | 0.006804651 | 0.178817103 |
| 185.0389135 | 26.93178602 | 6.8706514 | 0.0371308 | 0.005404269 | 0.145546607 |
| 215.2422932 | 27.26294933 | 7.8950480 | 0.0366798 | 0.004645927 | 0.126661675 |
| 260.6932256 | 27.59411265 | 9.4474219 | 0.0362396 | 0.003835926 | 0.105848982 |
| 312.7178933 | 27.83611661 | 11.2342500 | 0.0359246 | 0.00319777 | 0.089013508 |
| 348.5314044 | 28.05906376 | 12.4213483 | 0.0356391 | 0.002869182 | 0.080506558 |

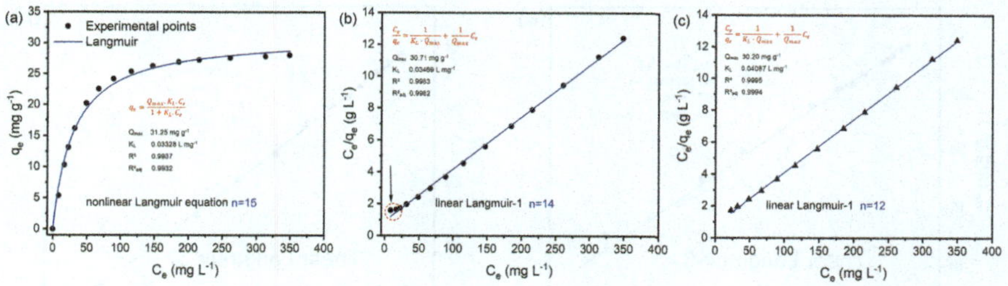


**Fig. 31** **a** Nonlinearized Langmuir equation ($np = 15$); **b** Linearized Langmuir-1 equation ($np = 14$); **c** Linearized Langmuir-1 equation ($np = 12$). The arrow indicates that some points were deleted to improve the values of $R^2$

isotherm incorporates experimental data directly; this includes the concentrations of paracetamol remaining in the supernatant after the adsorption process (x-axis) and the sorption capacity (y-axis), which were derived using Eq. (60):

$$q_e = \frac{(C_0 - C_e)}{m} \cdot V \tag{60}$$

where $q_e$ is the sorption capacity in the equilibrium (mg g$^{-1}$), $C_o$ and $C_e$ (mg L$^{-1}$) are the initial and equilibrium concentration of the adsorbate in contact with the adsorbent, $m$ is the mass of adsorbent (g), and $V$ is the volume of adsorbate solution (L).

The nonlinear, linear Langmuir-1, (see Fig. 31), and linear Langmuir-2, linear Langmuir-3, and linear Langmuir-4 (see Fig. 32) isotherms are presented.

Table 6 presents the statistical analysis of all isotherms shown in Figs. 31 and 32. For the linear Langmuir-1 plot, without deleting any experimental points, the values of $R^2$ and $R^2_{adj}$ were greater than 0.99. However, for linearized Langmuir-2, linearized Langmuir-3, and linearized Langmuir-4, it was necessary to delete some data points to enhance the values of $R^2$ and $R^2_{adj}$. Typically, authors focus only on the value of $R^2$ while neglecting other statistical parameters. When points on an isotherm are deleted, the slope and intercept of the plot are modified, which invariably influences the values of $Q_{max}$ and $K_L$, in addition to reducing the values of DOF. The determination coefficient of any equation is given by Eq. (61).

$$R^2 = \left( \frac{\sum_i^n \left(Y_{i,exp} - \overline{Y}_{i,exp}\right)^2 - \sum_i^n \left(Y_{i,exp} - Y_{i,model}\right)^2}{\sum_i^n \left(Y_{i,exp} - \overline{Y}_{i,exp}\right)^2} \right) \tag{61}$$

In nonlinear isotherm fitting, the Y-axis shows the sorption capacity at equilibrium ($q_e$), whereas in kinetic studies, it reflects $q_t$. However, as seen in Table 4, the linearized Langmuir-1, Langmuir-2, and Langmuir graphs exhibit distinct Y-axis placements: $\frac{C_e}{q_e}$, $\frac{1}{q_e}$, and $\frac{q_e}{C_e}$. Since these Y-axes differ, the $R^2$ and $R^2_{adj}$ values cannot be compared. This is a prevalent error observed in the literature pertaining to adsorption, whereby authors compare various isotherm models, such as Langmuir

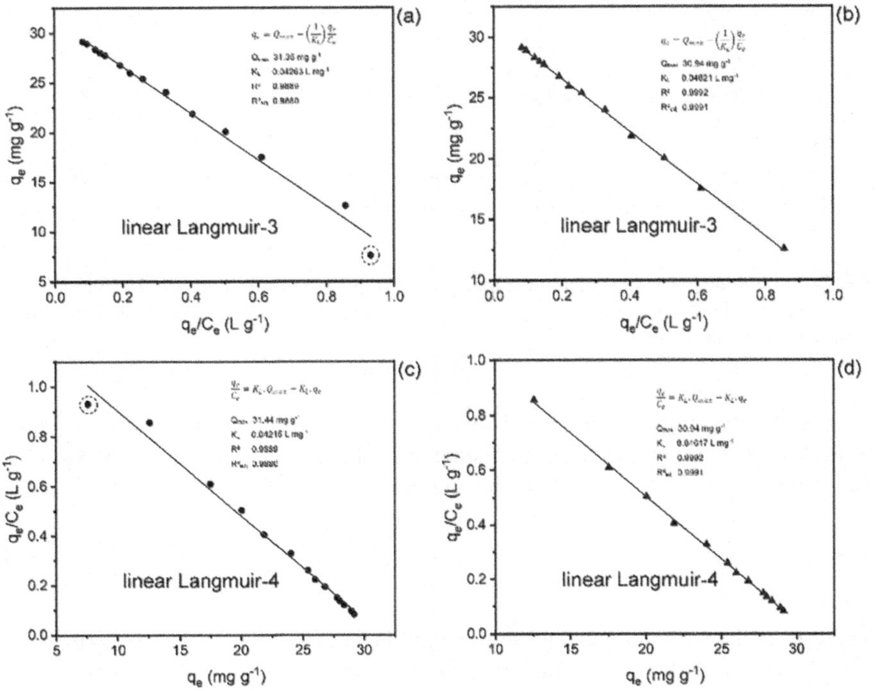


**Fig. 32** **a** Linearized Langmuir-2 equation ($np = 14$); **b** linearized Langmuir-2 equation ($np = 12$), **c** linearized Langmuir-3 equation ($np = 14$); **d** linearized Langmuir-3 equation after ($np = 9$); **e** linearized Langmuir-4 equation ($np = 14$); **f** linearized Langmuir-4 equation ($np = 9$). The arrow indicates that some points were deleted to improve the values of $R^2$

and Freundlich, or kinetic models, including pseudo-first-order and pseudo-second-order. Although $R^2$ and $R^2_{adj}$ are dimensionless, it is crucial to underscore that the standard deviation of the residues (*SD*) modifies the units of the dependent variable (Y-axis). The linearized Langmuir-3 equation is the only one that maintains consistent units for the statistical parameters in comparison to the nonlinear Langmuir equation. Consequently, it cannot be concluded that the linearized Langmuir-1 equation yields more precise results than its nonlinear counterpart, as the *SD* value presented in Table 6 illustrates varying scenarios (Y-axis of each graph) for both the nonlinear and linearized Langmuir-1 equations (see Table 6).

Considering that the nonlinear fit represents the most effective method for modeling equilibrium isotherms [1, 46, 47], the values of $Q_{max}$ and $K_L$ derived from both nonlinear and linearized versions of Langmuir-1, Langmuir-2, Langmuir-3, and Langmuir-4 were subjected to comparison. Table 7 delineates the relative differences among the parameters of the isotherm plots, as illustrated in Figs. 31 and 32. It is noted that, except for Langmuir-2, after deleting experimental points to improve the values of $R^2$ and $R^2_{adj}$, the relative error of all $Q_{max}$ and $K_L$ values has increased in relation to the values of these parameters obtained by nonlinear fitting.

**Table 6** Statistical analysis of data presented in Figs. 31 and 32

| Parameter | Value | Parameter | Value |
|---|---|---|---|
| Nonlinear | | | |
| $Q_{max}$ (mg g$^{-1}$) | 31.25 | | |
| $K_L$ (L mg$^{-1}$) | 0.03328 | | |
| Number of points | 15 | | |
| Degrees of freedom | 13 | | |
| $R^2$ (COD) | 0.9937 | | |
| $R^2_{adj}$ | 0.9932 | | |
| $SD$ (mg g$^{-1}$) | 0.7480 | | |
| Langmuir-1 | | Langmuir-1 | |
| $Q_{max}$ (mg g$^{-1}$) | 30.71 | $Q_{max}$ (mg g$^{-1}$) | 30.20 |
| $K_L$ (L mg$^{-1}$) | 0.03469 | $K_L$ (L mg$^{-1}$) | 0.04087 |
| Number of points | 14 | Number of points | 12 |
| Degrees of freedom | 12 | Degrees of freedom | 10 |
| $R^2$ (COD) | 0.9983 | $R^2$ (COD) | 0.9995 |
| $R^2_{adj}$ | 0.9982 | $R^2_{adj}$ | 0.9994 |
| $SD$ (g L$^{-1}$) | 0.1590 | $SD$ (g L$^{-1}$) | 0.08804 |
| Langmuir-2 | | Langmuir-2 | |
| $Q_{max}$ (mg g$^{-1}$) | 36.10 | $Q_{max}$ (mg g$^{-1}$) | 33.71 |
| $K_L$ (L mg$^{-1}$) | 0.02162 | $K_L$ (L mg$^{-1}$) | 0.02726 |
| Number of points | 14 | Number of points | 10 |
| Degrees of freedom | 12 | Degrees of freedom | 8 |
| $R^2$ (COD) | 0.9860 | $R^2$ (COD) | 0.9917 |
| $R^2_{adj}$ | 0.9849 | $R^2_{adj}$ | 0.9906 |
| $SD$ (g mg$^{-1}$) | 0.005046 | $SD$ (g mg$^{-1}$) | 0.001960 |
| Langmuir-3 | | Langmuir-3 | |
| $Q_{max}$ (mg g$^{-1}$) | 32.26 | $Q_{max}$ (mg g$^{-1}$) | 29.89 |
| $K_L$ (L mg$^{-1}$) | 0.02854 | $K_L$ (L mg$^{-1}$) | 0.04712 |
| Number of points | 14 | Number of points | 9 |
| Degrees of freedom | 12 | Degrees of freedom | 7 |
| $R^2$ (COD) | 0.9361 | $R^2$ (COD) | 0.9942 |
| $R^2_{adj}$ | 0.9308 | $R^2_{adj}$ | 0.9933 |
| $SD$ (mg g$^{-1}$) | 1.950 | $SD$ (mg g$^{-1}$) | 0.1528 |
| Langmuir-4 | | Langmuir-4 | |
| $Q_{max}$ (mg g$^{-1}$) | 32.99 | $Q_{max}$ (mg g$^{-1}$) | 29.91 |
| $K_L$ (L mg$^{-1}$) | 0.02671 | $K_L$ (L mg$^{-1}$) | 0.04684 |
| Number of points | 14 | Number of points | 9 |

(continued)

**Table 6** (continued)

| Parameter | Value | Parameter | Value |
|---|---|---|---|
| Degrees of freedom | 12 | Degrees of freedom | 7 |
| $R^2$ (COD) | 0.9361 | $R^2$ (COD) | 0.9942 |
| $R^2_{adj}$ | 0.9308 | $R^2_{adj}$ | 0.9933 |
| $SD$ (L $g^{-1}$) | 0.05384 | $SD$ (L $g^{-1}$) | 0.007177 |

Authors should note that these values were derived from only one specific system: paracetamol adsorption onto activated biochar made from cellulose at 30 °C [48]. Variations in temperature or different adsorbent/adsorbate systems could result in different data.

Considering that $K_L$ values are very important equilibrium parameters for estimating equilibrium constants (determination of thermodynamic parameters) [1, 23, 24], the use of Langmuir-1 linearized isotherms could lead to erroneous estimation of Gibbs' free energy change ($\Delta G°$), enthalpy change ($\Delta H°$), and entropy change ($\Delta S°$) of adsorption.

**Table 7** Relative differences of $Q_{max}$ and $K_L$ isotherm parameters of linearized Langmuir-1, Langmuir-2, Langmuir-3, and Langmuir-4 in relation to the nonlinear Langmuir equation

| Isotherm Langmuir-1 | % Difference | Isotherm Langmuir-1 | % Difference |
|---|---|---|---|
| $n = 14$ | | $n = 12$ | |
| $Q_{max}$ $(mg\ g^{-1})$ | 1.73% | $Q_{max}$ $(mg\ g^{-1})$ | 3.36% |
| $K_L$ $(L\ mg^{-1})$ | −4.24% | $K_L$ $(L\ mg^{-1})$ | −22.81% |
| Isotherm Langmuir-2 | % Difference | Isotherm Langmuir-2 | % Difference |
| $n = 14$ | | $n = 10$ | |
| $Q_{max}$ $(mg\ g^{-1})$ | −15.52% | $Q_{max}$ $(mg\ g^{-1})$ | −7.87% |
| $K_L$ $(L\ mg^{-1})$ | 35.04% | $K_L$ $(L\ mg^{-1})$ | 18.09% |
| Isotherm Langmuir-3 | % Difference | Isotherm Langmuir-3 | % Difference |
| $n = 14$ | | $n = 9$ | |
| $Q_{max}$ $(mg\ g^{-1})$ | −3.23% | $Q_{max}$ $(mg\ g^{-1})$ | 4.35% |
| $K_L$ $(L\ mg^{-1})$ | 14.24% | $K_L$ $(L\ mg^{-1})$ | −41.59% |
| Isotherm Langmuir-4 | % Difference | Isotherm Langmuir-4 | % Difference |
| $n = 14$ | | $n = 9$ | |
| $Q_{max}$ $(mg\ g^{-1})$ | −5.57% | $Q_{max}$ $(mg\ g^{-1})$ | 4.29% |
| $K_L$ $(L\ mg^{-1})$ | 19.74% | $K_L$ $(L\ mg^{-1})$ | −40.75% |

## 4.2 Kinetics of Adsorption

The most employed pseudo-first-order and pseudo-second-order kinetic equations are given in Table 8 in nonlinear and linearized forms [1, 49]. Let us consider the kinetics of 4-chloroaniline onto activated carbons derived from the shell of the cashew of Para [35, 49]. The experimental kinetic data of adsorption of 4-chloroaniline onto activated carbons derived from the shell of the cashew of Para [35, 49] are in Table 9. The nonlinear pseudo-first order and linearized pseudo-first order plots are shown in Fig. 33 and Table 10. In contrast, nonlinear pseudo-second order, and linearized pseudo-second order-1, pseudo-second order-2 are presented in Fig. 34 and Table 11, and in Fig. 35 and Table 12 are presented the linearized pseudo-second order-3, and pseudo-second order-4 plots. It is observed from Table 9 that the manipulation of the data alters the units of the variables. The analysis of linear pseudo-first-order kinetics necessitated the deletion of eight experimental points to achieve a suitable value of $R^2$; however, the degree of freedom (DOF) decreases from 13 to 5.

The examination of pseudo-second order linearized models shows that the linearized pseudo-second order-1 equation offers the highest $R^2$ values among the four models. However, it is important to note that this analysis did not account for the units of *SD*, which differ from those in the nonlinearized model. Thus, different scenarios are being assessed; nonetheless, some authors use the $R^2$ value to suggest that a model is a suitable fit.

**Table 8** Kinetics of adsorption

| Kinetic model | Linear form | Plot |
|---|---|---|
| $q_t = q_1 \cdot [1 - \exp(-k_1 t)]$<br>Pseudo-first order | $ln(q_e - q_t) = ln q_e - k_1 t$ | $\ln(q_e - q_t)$ vs $t$ |
| $q_t = \frac{k_2 \cdot q_e^2 \cdot t}{1 + q_e \cdot k_2 \cdot t}$<br>Pseudo second-order | $\frac{t}{q_t} = \frac{1}{k_2 q_e^2} + \frac{1}{q_e^2} \cdot t$<br>Pseudo-second-1 | $\frac{t}{q_t}$ vs $t$ |
| 1 | $\frac{1}{q_t} = \frac{1}{q_e} + \left(\frac{1}{k_2 \cdot q_e^2}\right) \times \frac{1}{t}$<br>Pseudo-second-2 | $\frac{1}{q_t}$ =vs $\frac{1}{t}$ |
| | $q_t = q_e - \left(\frac{1}{k_2 \cdot q_e}\right) \frac{q_t}{t}$<br>Pseudo-second-3 | $q_t$ vs $\frac{q_t}{t}$ |
| | $\frac{q_t}{t} = k_2 \cdot q_e^2 - k_2 \cdot q_e \cdot q_t$<br>Pseudo-second-4 | $\frac{q_t}{t}$ vs $q_t$ |

**Table 9** Experimental data of the adsorption of 4-chloroaniline onto activated carbons derived from the shell of the cashew of Para [35, 49]

| Experimental data | | | Manipulated data | | | |
|---|---|---|---|---|---|---|
| $t$ (min) | $q_t$ (mg g$^{-1}$) | Ln($q_e - q_t$) | $t/q_t$ (min g mg$^{-1}$) | $1/q_t$ (g mg$^{-1}$) | $1/t$ (min$^{-1}$) | $q_t/t$ (mg g$^{-1}$ min$^{-1}$) |
| 0 | 0.000000 | 6.1687192 | | | | |
| 1 | 144.895413 | 5.8071769 | 0.0069015 | 0.0069015 | 1.00000 | 144.895413 |
| 2 | 267.845124 | 5.3458157 | 0.0074670 | 0.0037335 | 0.50000 | 133.922562 |
| 3 | 350.541226 | 4.8444453 | 0.0085582 | 0.0028527 | 0.33333 | 116.847075 |
| 4 | 403.254178 | 4.3083780 | 0.0099193 | 0.0024798 | 0.25000 | 100.813545 |
| 5 | 434.214580 | 3.7695245 | 0.0115150 | 0.002303 | 0.20000 | 86.842916 |
| 10 | 473.021580 | 1.5156642 | 0.0211407 | 0.0021141 | 0.10000 | 47.302158 |
| 15 | 474.125890 | 1.2378332 | 0.0316372 | 0.0021091 | 0.06667 | 31.608393 |
| 30 | 475.021580 | 0.9370513 | 0.0631550 | 0.0021052 | 0.03333 | 15.834053 |
| 45 | 477.574024 | #NÚM! | 0.0942262 | 0.0020939 | 0.02222 | 10.612756 |
| 60 | 474.012590 | 1.2701633 | 0.1265789 | 0.0021096 | 0.01667 | 7.900210 |
| 75 | 476.251458 | 0.2795738 | 0.1574798 | 0.0020997 | 0.01333 | 6.350019 |
| 90 | 475.214560 | 0.8584345 | 0.1893881 | 0.0021043 | 0.01111 | 5.280162 |
| 105 | 476.210257 | 0.3102508 | 0.2204908 | 0.0020999 | 0.00952 | 4.535336 |
| 120 | 475.201257 | 0.8640568 | 0.2525246 | 0.0021044 | 0.00833 | 3.960010 |
| 150 | 474.698745 | 1.0561496 | 0.3159899 | 0.0021066 | 0.00667 | 3.164658 |
| 180 | 476.021458 | 0.4399090 | 0.3781342 | 0.0021007 | 0.00556 | 2.644564 |

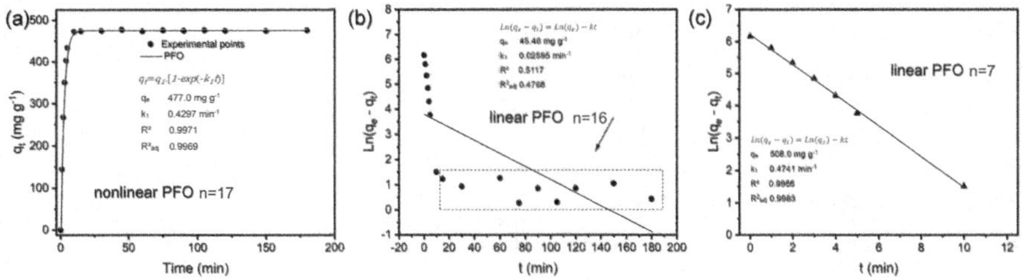


**Fig. 33** **a** Nonlinearized pseudo-first order equation ($n = 17$); **b** linearized pseudo-first-order ($n = 16$); **c** linearized pseudo-first order equation ($n = 7$). The dashed line shows the region to delete points just to improve the values of $R^2$

Only linearized pseudo-second order-3 presents the same units as the nonlinearized kinetic equation (Table 9). However, the *SD* is worse than the nonlinearized equation, even after deleting 12 experimental points.

**Table 10** Statistical analysis of data presented in Fig. 33

| Parameter | Value | Parameter | Value |
|---|---|---|---|
| Nonlinear pseudo-first order | Figure 33a | Pseudo-first order-1 | Figure 33b |
| $k_1$ (min$^{-1}$) | 0.4297 | $k_1$ (min$^{-1}$) | 0.02595 |
| $q_e$ (mg g$^{-1}$) | 477.0 | $q_e$ (mg g$^{-1}$) | 45.46 |
| Number of points | 17 | Number of points | 16 |
| Degrees of freedom | 15 | Degrees of freedom | 14 |
| $R^2$ (COD) | 0.9971 | $R^2$ (COD) | 0.5117 |
| $R^2_{adj}$ | 0.9969 | $R^2_{adj}$ | 0.4768 |
| $SD$ (mg g$^{-1}$) | 7.702 | $SD$ (min g mg$^{-1}$) | 1.571 |
| Linear pseudo-first order-2 | Figure 33c | | |
| $k_1$ (min$^{-1}$) | 0.4741 | | |
| $q_e$ (mg g$^{-1}$) | 508.0 | | |
| Number of points | 7 | | |
| Degrees of freedom | 5 | | |
| $R^2$ (COD) | 0.9986 | | |
| $R^2_{adj}$ | 0.9983 | | |
| $SD$ (min g mg$^{-1}$) | 0.06524 | | |

Considering that the kinetics of adsorption is a nonlinear equation, the parameter values ($q_e$ and $k_1$, $k_2$) of this model were taken as a reference for computing the relative differences of the linearized models in relation to the nonlinear equations; these values are depicted in Table 13.

It is observed that the strategy of deleting some points to improve the values of $R^2$ causes a serious error in the values of $q_e$ and $k_1$ or $k_2$. The activation energy is calculated based on the values of kinetic rate constants of the Arrhenius equation [1] (Eq. 35):

$$\ln k_{kinetic} = \ln A - \frac{E_a}{RT} \tag{62}$$

where $k_{kinetic}$ is the kinetic rate constant of adsorption, $A$ is the Arrhenius constant, $E_a$ is the activation energy (kJ mol$^{-1}$), $R$ is the universal gas constant (8.314 J K$^{-1}$ mol$^{-1}$), and $T$ is the absolute temperature (K). If the values of $k_{kinetic}$ are biased, the values of the calculated activation energy will also be compromised.

Using the examples shown (Figs. 31, 32, 33, 34 and 35; Tables 7 and 11), we illustrate the challenges of applying linear equations to Langmuir isotherm, pseudo-first order, and pseudo-second order adsorption kinetics. Notably, while the linearized Langmuir-1 and linearized pseudo-second order-1 equations produce $R^2$ values close to one, indicating a good fit, the parameters (equilibrium and kinetics) obtained from these models differ from those derived from nonlinear equations. Additionally, it is important to note that the manipulation of data to transform a nonlinear equation into

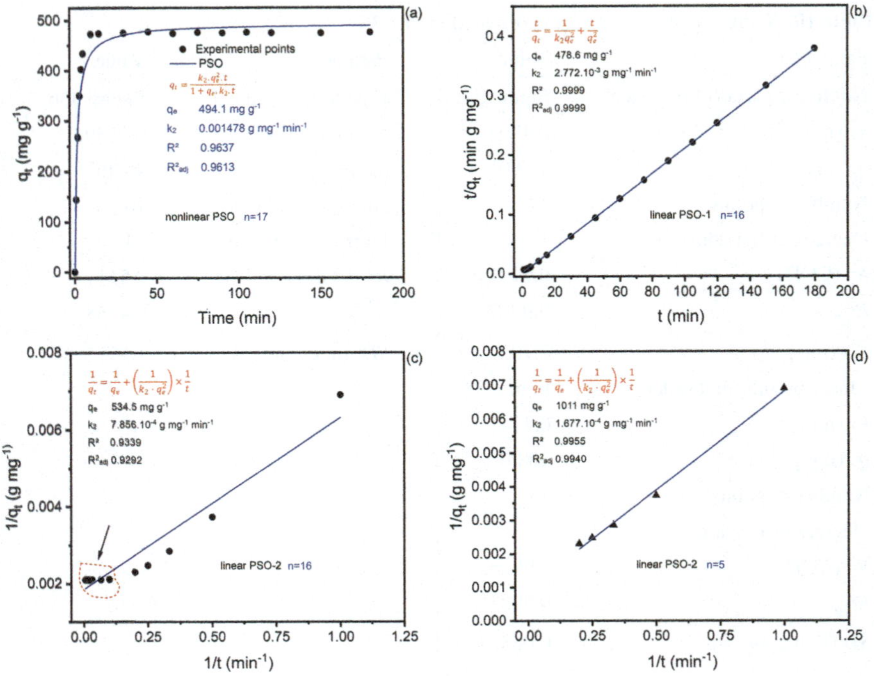


**Fig. 34** **a** Nonlinearized pseudo-second order equation ($n = 17$); **b** linearized pseudo-second-order 1 ($n = 16$); **c** linearized pseudo-second-order 2 equation ($n = 16$); **d** linearized pseudo-second-order 2 ($n = 5$). The dashed line shows the region to delete points just to improve the values of $R^2$

a linear one leads to significant changes in the units of the Y and X axes. Thus, the standard deviation should be taken into account when assessing a model's fit. Only the linearized model that maintains the same Y-axis ($q_e$ and $q_t$ for equilibrium and kinetic equations, respectively) can be directly compared to the nonlinear equations. Although the linearized equations show different units and may seem to offer slightly better values in comparison to the nonlinear equations, these numerical values do not align with the *SD* values of the nonlinear models, rendering them incomparable (see Eq. (61), regarding the Y-axis of the plots). We trust that reviewing the figures and tables in this section will clarify for readers that the nonlinear approach to equilibrium and kinetics in adsorption models provides values that are reliable and statistically significant for modeling an isotherm and kinetic adsorption curve. We advise against using linearized equilibrium and kinetics adsorption models, even in the instances of the linearized Langmuir-1 and linearized pseudo-second order-1 equations.

**Table 11** Statistical analysis of data presented in Fig. 34

| Parameter | Value | Parameter | Value |
|---|---|---|---|
| Nonlinear pseudo-second-order | Figure 34a | Linear pseudo-second-order-1 | Figure 34b |
| $k_2$ (g mg$^{-1}$ min$^{-1}$) | 1.478·10$^{-3}$ | $k_2$ (g mg$^{-1}$ min$^{-1}$) | 2.772·10$^{-3}$ |
| $q_e$ (mg g$^{-1}$) | 494.1 | $q_e$ (mg g$^{-1}$) | 478.6 |
| Number of points | 17 | Number of points | 16 |
| Degrees of freedom | 15 | Degrees of freedom | 14 |
| $R^2$ (COD) | 0.9637 | $R^2$ (COD) | 0.9999 |
| $R^2_{adj}$ | 0.9613 | $R^2_{adj}$ | 0.9999 |
| $SD$ (mg g$^{-1}$) | 27.31 | $SD$ (min g mg$^{-1}$) | 0.001282 |
| Linear pseudo-first order-2 | Figure 34c | Linear pseudo-first order-2 | Figure 34d |
| $k_2$ (g mg$^{-1}$ min$^{-1}$) | 7.856·10$^{-4}$ | $k_2$ (g mg$^{-1}$ min$^{-1}$) | 1.677·10$^{-4}$ |
| $q_e$ (mg g$^{-1}$) | 534.5 | $q_e$ (mg g$^{-1}$) | 1011.0 |
| Number of points | 16 | Number of points | 5 |
| Degrees of freedom | 14 | Degrees of freedom | 3 |
| $R^2$ (COD) | 0.9339 | $R^2$ (COD) | 0.9955 |
| $R^2_{adj}$ | 0.9292 | $R^2_{adj}$ | 0.9940 |
| $SD$ (g mg$^{-1}$) | 3.271·10$^{-4}$ | $SD$ (g mg$^{-1}$) | 1.474·10$^{-4}$ |

Table 13 shows the relative difference in the kinetic parameters of pseudo-first order and pseudo-second order. In some cases, where experimental data points were deleted to increase the values of $R^2$ (see Table 13), the relative difference in the kinetic parameters worsened compared to the parameters before the deletion of experimental data points. Therefore, we reiterate that linearization of equilibrium and kinetic adsorption models can render the models' parameter values meaningless.

Another point that was not considered in this analysis is that for linearization, nearly all manuscripts indicate that the linearized pseudo-second order model is followed. Reviewing Table 9, one could erroneously conclude this observation. However, the linearized pseudo-first-order model is not a suitable fit in any situation as already reported [49]. When comparing the $R^2_{adj}$ and $SD$ values for nonlinearized kinetic models, in the example provided in this chapter (Table 8 and Fig. 33a), it becomes apparent that the kinetic data fit better to the nonlinear pseudo-first order ($R^2_{adj}$ 0.9969; and $SD$ 7.702 mg g$^{-1}$) than to the nonlinear pseudo-second order ($R^2_{\mathrm{adj}}$ 0.9613; $SD$ 27.31 mg g$^{-1}$). It is crucial to emphasize that the standard deviation of the nonlinearized pseudo-second order is 6.54 times higher than that of the nonlinearized pseudo-first order. Therefore, it is clear that the kinetics should preferably follow pseudo-first order. Nevertheless, the use of distorted linearized equations yields the opposite result. This might explain why the majority of papers published in the literature, almost all of which suggest that the kinetics should follow pseudo-second order, to the detriment of pseudo-first order [49]. However, if nonlinear equations were utilized, the opposite result could be achieved. This is one of the reasons the

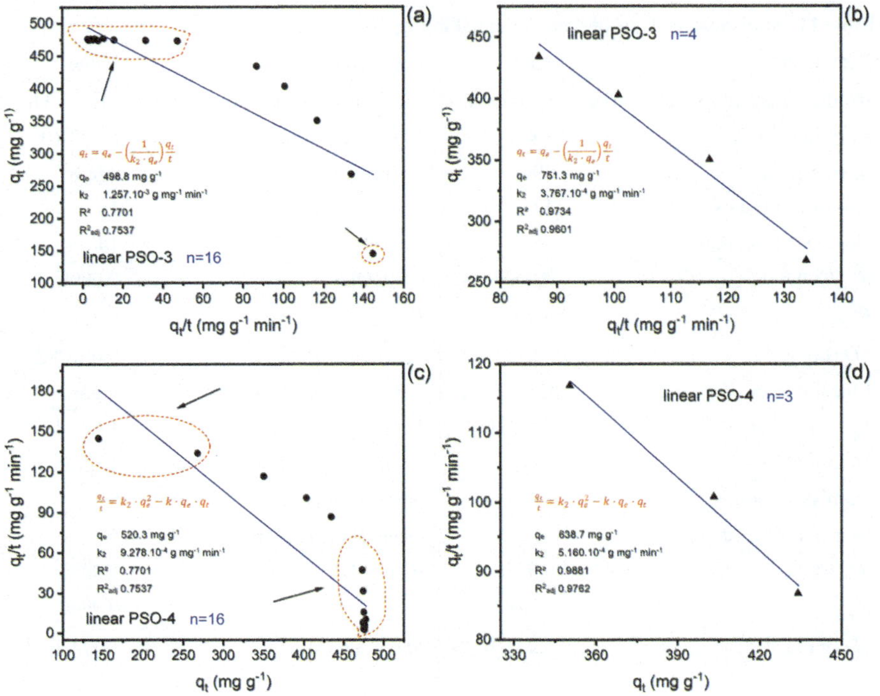


**Fig. 35** **a** Linearized pseudo-second-order 3 ($n = 16$); **b** linearized pseudo-second-order 3 ($n = 4$); **c** linearized pseudo-second-order 4 equation ($n = 16$); **d** linearized pseudo-second-order 4 ($n = 3$). The dashed line shows the region to delete points just to improve the values of $R^2$

authors of this chapter do not recommend using linearized equations for the equilibrium and kinetics of adsorption, and we also reiterate that numerous errors are found in the published literature on adsorption research.

## 5 Concluding Remarks

In this chapter, the commonly employed and less favored adsorption equilibrium isotherm models, such as Langmuir, Freundlich, Sips, Liu, Redlich–Peterson, Hill, Khan, and Toth, are discussed with suitable illustrations. Adsorption kinetic models are adequately described, including pseudo-first order, pseudo-second order, general order, and empirical models (Avrami fractional, Elovich chemisorption, fractal-like pseudo-first-order, fractal-like pseudo-second-order). The usage of different isotherm models at various temperature values is suggested in this chapter. The importance of statistical parameters such as $R^2_{adj}$ and $SD$ is emphasized, besides the use of ΔBIC to

**Table 12** Statistical analysis of data presented in Fig. 35

| Parameter | Value | Parameter | Value |
|---|---|---|---|
| Linear pseudo-first order-3 | Figure 35a | Linear pseudo-first order-3 | Figure 35b |
| $k_2$ (g mg$^{-1}$ min$^{-1}$) | $1.257 \cdot 10^{-3}$ | $k_2$ (g mg$^{-1}$ min$^{-1}$) | $3.767 \cdot 10^{-4}$ |
| $q_e$ (mg g$^{-1}$) | 498.8 | $q_e$ (mg g$^{-1}$) | 751.3 |
| Number of points | 16 | Number of points | 4 |
| Degrees of freedom | 14 | Degrees of freedom | 2 |
| $R^2$ (COD) | 0.7701 | $R^2$ (COD) | 0.9734 |
| $R^2_{adj}$ | 0.7537 | $R^2_{adj}$ | 0.9601 |
| $SD$ (mg g$^{-1}$) | 47.42 | $SD$ (mg g$^{-1}$) | 14.55 |
| Linear pseudo-first order-4 | Figure 35c | Linear pseudo-first order-4 | Figure 35d |
| $k_2$ (g mg$^{-1}$ min$^{-1}$) | $9.278 \cdot 10^{-4}$ | $k_2$ (g mg$^{-1}$ min$^{-1}$) | $5.160 \cdot 10^{-4}$ |
| $q_e$ (mg g$^{-1}$) | 520.3 | $q_e$ (mg g$^{-1}$) | 683.7 |
| Number of points | 16 | Number of points | 3 |
| Degrees of freedom | 14 | Degrees of freedom | 1 |
| $R^2$ (COD) | 0.7701 | $R^2$ (COD) | 0.9881 |
| $R^2_{adj}$ | 0.7537 | $R^2_{adj}$ | 0.9762 |
| $SD$ (mg g$^{-1}$ min$^{-1}$) | 26.09 | $SD$ (mg g$^{-1}$ min$^{-1}$) | 2.316 |

compare different kinetic and equilibrium models. Statistical analysis guides adsorption researchers to select the most appropriate adsorption isotherm and kinetic model for their work, which is coupled with the obtainability of useful constants or meaningful parameters. The model with the best value of $R^2_{adj}$ and the lowest value of $SD$ should be regarded as the optimal isotherm and kinetic model for describing the adsorption process. Based on a detailed analysis of linearized and nonlinearized models in this chapter, we do not recommend using linearized equations to analyze adsorption equilibrium and kinetic data. Analytical techniques, together with thermodynamic data on enthalpy and entropy changes, are necessary to determine whether an adsorption process is chemical or physical. It is advisable to evaluate the value of the adsorption equilibrium constant ($K$) from the best-fit nonlinear isotherm equilibrium model instead of using the calculated equilibrium constant ($K_d$) from the initial and equilibrium concentrations of an adsorbate. Overall, this chapter offers solutions to the common errors arising from the use of linearized equilibrium and kinetic adsorption models.

**Table 13** Relative differences of $q_e$ and $k_1$ or $k_2$ kinetic parameters of linearized pseudo-first order, linearized pseudo-second order-1, pseudo-second order-2, pseudo-second order-3 and pseudo-second order-4 in relation to nonlinear pseudo-first order and nonlinear pseudo-second order equations

| Pseudo-first order 1 | % Difference | | % Difference |
|---|---|---|---|
| $k_1$ | 93.96 | $k_1$ | −10.33 |
| $q_e$ | 90.47 | $q_e$ | −6.50 |
| | Figure 33b | | Figure 33c |
| Pseudo second order-1 | % Difference | | % Difference |
| $k_2$ | −87.55 | | – |
| $q_e$ | 3.14 | | – |
| | Figure 34b | | |
| Pseudo second order-2 | % Difference | | % Difference |
| $k_2$ | −46.85 | $k_2$ | 88.65 |
| $q_e$ | −8.18 | $q_e$ | −104.6 |
| | Figure 34c | | Figure 34d |
| Pseudo second order-3 | % difference | | % difference |
| $k_2$ | 14.95 | $k_2$ | 74.51 |
| $q_e$ | −0.95 | $q_e$ | −52.05 |
| | Figure 35a | | Figure 35b |
| Pseudo second order-4 | % Difference | | % Difference |
| $k_2$ | 37.23 | $k_2$ | 65.09 |
| $q_e$ | −5.30 | $q_e$ | −38.37 |
| | Figure 35c | | Figure 35d |

Values are expressed as a percentage

**Acknowledgements** The authors gratefully acknowledge financial support from the Brazilian funding agencies: Conselho Nacional de Desenvolvimento Científico e Tecnológico (CNPq), Coordenação de Aperfeiçoamento de Pessoal de Nível Superior (CAPES), and Fundação de Amparo à Pesquisa do Estado do Rio Grande do Sul (FAPERGS). Prof. Edér C. Lima acknowledges support from CNPq (Process 303612/2021-5), CAPES-PROEX (Processes 88881.844968/2023-01 and 001), and FAPERGS (Process 24/2551-0001291-5). Prof. Fernando M. Machado thanks CNPq for research grant support (Processes 308203/2022-4, 408413/2022-0, 443031/2023-1, and 403522/2025-0) and FAPERGS (Process 23/2551-0001863-2).

## Abbreviations

| | |
|---|---|
| APTES | (3-Aminopropyl)Triethoxysilane |
| BIC | Bayesian Information Criterion |
| DOF | Degrees of Freedom |
| DR | Dubinin-Radushkevich |
| FLPFO | Fractal-Like Pseudo-First-Order |
| FLPSO | Fractal-Like Pseudo-Second-Order |
| FTIR | Fourier Transform Infrared Spectroscopy |

| | |
|---|---|
| GO | General Order Kinetic Model |
| PFO | Pseudo-First-Order |
| PSO | Pseudo-Second-Order |
| RSS | Residual Sum of Squares |
| SD | Standard Deviation |
| SEM–EDS | Scanning Electron Microscopy with Energy Dispersive X-Ray Spectroscopy |
| SWCNT | Single-Walled Carbon Nanotubes |
| XPS | X-Ray Photoelectron Spectroscopy |

## References


1. Lima, E.C., Dehghani, M.H., Guleria, A., et al.: Adsorption: fundamental aspects and applications of adsorption for effluent treatment. In: Dehghani, M.H., Karri, R., Lima, E.C. (eds.) Green Technologies for the Defluoridation of Water, pp. 41–88. Elsevier (2021). https://doi.org/10.1016/B978-0-323-85768-0.00004-X
2. Chang, R., Thoman, Jr J.W.: Chapter 17: Intermolecular forces. In: Physical Chemistry for the Chemical Sciences, p. 792. University Science Books (2014)
3. Langmuir, I.: The adsorption of gases on plane surfaces of glass, mica and platinum. J. Am. Chem. Soc. **40**, 1361–1403 (1918). https://doi.org/10.1021/ja02242a004
4. Freundlich, H.: Adsorption in solution. Phys. Chem. Soc. **40**, 1361–1368 (1906)
5. Sips, R.: On the structure of a catalyst surface. J. Chem. Phys. **16**, 490–495 (1948). https://doi.org/10.1063/1.1746922
6. Liu, Y., Xu, H., Tay, J.H.: Errata for "Derivation of a general adsorption isotherm model." J. Environ. Eng. **132**, 147 (2006). https://doi.org/10.1061/(ASCE)0733-9372(2006)132:1(147)
7. Redlich, O., Peterson, D.L.: A useful adsorption isotherm. J. Chem. Phys. **63**, 1024–1027 (1959). https://doi.org/10.1021/j150576a611
8. Lagergren, S.: About the theory of so-called adsorption of soluble substances. Kungliga Svenska Vetenskapsakademiens Handlingar **24**, 1–39 (1898)
9. Blanchard, G., Maunaye, M., Martin, G.: Removal of heavy metals from waters by means of natural zeolites. Water Res. **18**, 1501–1507 (1984). https://doi.org/10.1016/0043-1354(84)90124-6
10. Machado, F.M., Bergmann, C.P., Lima, E.C., et al.: Adsorption of Reactive Blue 4 dye from water solutions by carbon nanotubes: experiment and theory. Phys. Chem. Chem. Phys. **14**, 11139–11153 (2012). https://doi.org/10.1039/C2CP41475A
11. Haerifar, M., Azizian, S.: Fractal-like kinetics for adsorption on heterogeneous solid surfaces. J. Phys. Chem. C **118**, 1129–1134 (2014). https://doi.org/10.1021/jp4110882
12. Vaghetti, J.C.P., Lima, E.C., Royer, B., et al.: Application of Brazilian-pine fruit coat as a biosorbent to the removal of Cr(VI) from aqueous solution: kinetics and equilibrium study. Biochem. Eng. J. **42**, 67–76 (2008). https://doi.org/10.1016/j.bej.2008.05.021
13. Cunha, M.R., Lima, E.C., Lima, D.R., et al.: Removal of captopril pharmaceutical from synthetic pharmaceutical-industry wastewaters: use of activated carbon derived from Butia catarinensis. J. Environ. Chem. Eng. **8**, 104506 (2020). https://doi.org/10.1016/j.jece.2020.104506
14. da Silva, L.G., Ruggiero, R., Gontijo, P.M., et al.: Adsorption of Brilliant Red 2BE dye from water solutions by a chemically modified sugarcane bagasse lignin. Chem. Eng. J. **168**, 620–628 (2011). https://doi.org/10.1016/j.cej.2011.01.040

15. Cimirro, N.F.G.M., Lima, E.C., Cunha, M.R., et al.: Removal of diphenols using pine biochar: kinetics, equilibrium, thermodynamics, and mechanism of uptake. J. Mol. Liq. **364**, 119979 (2022). https://doi.org/10.1016/j.molliq.2022.119979
16. Kumar, K.V., Porkodi, K.: Letter to the Editor: relation between some two- and three-parameter isotherm models for the sorption of methylene blue onto the lemon peel. J. Hazard. Mater. **138**, 633–635 (2006). https://doi.org/10.1016/j.jhazmat.2006.06.078
17. Anastopoulos, I., Kyzas, G.Z.: Are the thermodynamic parameters correctly estimated in liquid-phase adsorption phenomena? J. Mol. Liq. **218**, 174–185 (2016). https://doi.org/10.1016/j.molliq.2016.02.059
18. Chu, K.H.: Revisiting the Temkin isotherm: dimensional inconsistency and approximate forms. Ind. Eng. Chem. Res. **60**, 13140–13147 (2021). https://doi.org/10.1021/acs.iecr.1c01788
19. Schwarz, G.E.: Estimating the dimension of a model. Ann. Stat. **6**, 461–464 (1978). http://www.jstor.org/stable/2958889
20. Mello, B.L., Thue, P.S., da Silva, P.V., et al.: Grafting 3-trimethoxysilylpropyl)diethylenetriamine on microcrystalline cellulose for the adsorption of dyes: experimental and modeling studies. React. Funct. Polym. **196**, 105836 (2024). https://doi.org/10.1016/j.reactfunctpolym.2024.105836
21. Thue, P.S., Teixeira, R.A., Hounfodji, J.W., et al.: Intercalation of organosilane in clay mineral for the removal of Procion red MX-5B: investigational and theoretical studies. Sep. Purif. Technol. **347**, 127491 (2024). https://doi.org/10.1016/j.seppur.2024.127491
22. Teixeira, R.A., Lima, E.C., Benetti, A.D., et al.: Preparation of hybrids of wood sawdust with 3-aminopropyltriethoxysilane: application as an adsorbent to remove Reactive Blue 4 dye from wastewater effluents. J. Taiwan Inst. Chem. Eng. **125**, 141–152 (2021). https://doi.org/10.1016/j.jtice.2021.06.007
23. Lima, E.C., Hosseini-Bandegharaei, A., Moreno-Piraján, J.C., Anastopoulos, I.: A critical review of the estimation of the thermodynamic parameters on adsorption equilibria: wrong use of equilibrium constant in the Van't Hoff equation for calculation of thermodynamic parameters of adsorption. J. Mol. Liq. **273**, 425–434 (2019). https://doi.org/10.1016/j.molliq.2018.10.048
24. Lima, E.C., Hosseini-Bandegharaei, A., Anastopoulos, I.: Response to "Some remarks on a critical review of the estimation of the thermodynamic parameters on adsorption equilibria: wrong use of equilibrium constant in the van't Hoff equation for calculation of thermodynamic parameters of adsorption." J. Mol. Liq. **280**, 298–300 (2019). https://doi.org/10.1016/j.molliq.2019.01.160
25. Atkins, P., de Paula, J.: Physical Chemistry. 9th edn. W. H. Freeman and Company, New York (2010)
26. Levine, I.N.: Physical Chemistry, 6th edn. McGraw-Hill, New York (2009)
27. Renner, T.: Quantities, Units, and Symbols in Physical Chemistry. Royal Society of Chemistry (2007)
28. Abe, I., Hayashi, K., Kitagawa, M., Urahata, T.: Adsorptive mechanism on activated carbon in the liquid phase. I. Free energy change for adsorption of organic compounds from aqueous solution on activated carbon. Bull. Chem. Soc. Jpn. **52**, 1899–1904 (1979). https://doi.org/10.1246/bcsj.52.1899
29. Tran, H.N., You, S.J., Bandegharaei, A.H., Chao, H.P.: Mistakes and inconsistencies regarding adsorption of contaminants from aqueous solutions: a critical review. Water Res. **120**, 88–116 (2017). https://doi.org/10.1016/j.watres.2017.04.014
30. Dotto, G.L., Santos, J.M.N., Rodrigues, I.L., et al.: Adsorption of Methylene Blue by ultrasonic surface-modified chitin. J. Colloid Interface Sci. **446**, 133–140 (2015). https://doi.org/10.1016/j.jcis.2015.01.046
31. Tran, H.N., You, S.J., Chao, H.P.: Thermodynamic parameters of cadmium adsorption onto orange peel calculated from various methods: a comparison study. J. Environ. Chem. Eng. **4**, 2671–2682 (2016). https://doi.org/10.1016/j.jece.2016.05.009
32. Milonjić, S.K.: A consideration of the correct calculation of thermodynamic parameters of adsorption. J. Serb. Chem. Soc. **72**, 1363–1367 (2007). https://doi.org/10.2298/JSC0712363M

33. Lima, E.C., Ponce-Vargas, M., Naushad, M., et al.: Removal of methylparaben from aqueous effluents using biobased carbon material: experimental and DFT calculations. J. Mol. Liq. **397**, 124194 (2024). https://doi.org/10.1016/j.molliq.2024.124194
34. Lima, E.C., Gomes, A.A., Tran, H.N.: Comparison of the nonlinear and linear forms of the van't Hoff equation for calculation of adsorption thermodynamic parameters (ΔS° and ΔH°). J. Mol. Liq. **311**, 113315 (2020). https://doi.org/10.1016/j.molliq.2020.113315
35. Thue, P.S., Lima, D.R., Naushad, M., et al.: High removal of emerging contaminants from wastewater by activated carbons derived from the shell of cashew of Para. Carbon Lett. **31**, 13–28 (2021). https://doi.org/10.1007/s42823-020-00145-x
36. Mello, B.L., Thue, P.S., da Silva, P.V., et al.: Hybrid nanocellulose material as an adsorbent to remove reactive yellow 2 dye. Sci. Rep. **14**, 20074 (2024). https://doi.org/10.1038/s41598-024-70906-5
37. Cardoso, N.F., Pinto, R.B., Lima, E.C., et al.: Removal of remazol black B textile dye from aqueous solution by adsorption. Desalination **269**, 92–103 (2011). https://doi.org/10.1016/j.desal.2010.10.047
38. Lopes, E.C.N., dos Anjos, F.S.C., Vieira, E.F.S., Cestari, A.R.: An alternative Avrami equation to evaluate kinetic parameters of the interaction of Hg(II) with thin chitosan membranes. J. Colloid Interface Sci. **263**, 542–547 (2003). https://doi.org/10.1016/S0021-9797(03)00326-6
39. Černý, S., Ponec, V., Hládek, L.: Analogue simulation of adsorption calorimetry. J. Chem. Thermodyn. **2**, 391–397 (1970). https://doi.org/10.1016/0021-9614(70)90009-1
40. Hashem, A., Badawy, S.M., Farag, S., Mohamed, L.A., Fletcher, A.J., et al.: Nonlinear adsorption characteristics of modified pinewood sawdust optimized for adsorption of Cd(II) from aqueous systems. J. Environ. Chem. Eng. **8**, 103966 (2020). https://doi.org/10.1016/j.jece.2020.103966
41. Puchana-Rosero, M.J., Adebayo, M.A., Lima, E.C., Machado, F.M., Thue, P.S., et al.: Microwave-assisted activated carbon obtained from the sludge of the tannery treatment effluent plant for the removal of leather dyes. Colloids Surf. A Physicochem. Eng. Asp. **504**, 105–115 (2016). https://doi.org/10.1016/j.colsurfa.2016.05.059
42. Haerifar, M., Azizian, S.: Fractal-like adsorption kinetics at the solid/solution interface. J. Phys. Chem. C **116**, 13111–13119 (2012). https://doi.org/10.1021/jp301261h
43. Nayak, A.K., Pal, A.: Development and validation of an adsorption kinetic model at solid-liquid interface using normalized Gudermannian function. J. Mol. Liq. **276**, 67–77 (2019). https://doi.org/10.1016/j.molliq.2018.11.089
44. Weber, W.J., Jr., Morris, J.C.: Kinetics of adsorption on carbon from solution. J. Sanit. Eng. Div. Am. Soc. Civil Eng. **89**, 31–59 (1963). https://doi.org/10.1061/JSEDAI.0000430
45. El-Khaiary, M.I., Malash, G.F.: Common data analysis errors in batch adsorption studies. Hydrometallurgy **105**, 314–320 (2011). https://doi.org/10.1016/j.hydromet.2010.11.005
46. El-Khaiary, M.I., Malash, G.F., Ho, Y.S.: On the use of linearized pseudo-second-order kinetic equations for modelling adsorption systems. Desalination **257**, 93–101 (2010). https://doi.org/10.1016/j.desal.2010.02.041
47. Kumar, K.V.: Optimum sorption isotherm by linear and non-linear methods for malachite green onto lemon peel. Dyes Pigments **74**, 595–597 (2007). https://doi.org/10.1016/j.dyepig.2006.03.014
48. Thue, P.S., Wamba, A.G.N., Mello, B.L., Machado, F.M., Petroman, K.F., et al.: Magnetic composite carbon from microcrystalline cellulose to tackle paracetamol contamination: kinetics, mass transfer, equilibrium, and thermodynamic studies. Polymers **16**, 3538 (2024). https://doi.org/10.3390/polym16243538
49. Lima, E.C., Sher, F., Guleria, A., Saeb, M.R., Anastopoulos, I., et al.: Is one performing the treatment data of adsorption kinetics correctly? J. Environ. Chem. Eng. **9**, 104813 (2021). https://doi.org/10.1016/j.jece.2020.104813

# Experimental Approaches to Adsorption Processes

**Fernando Machado Machado and Eder Claudio Lima**

**Abstract** Various factors can influence adsorptive systems, making it essential to understand the phenomena involved in adsorption processes to optimize the interaction between the adsorbent and the adsorbate. This chapter outlines the different components and experimental conditions that can impact the batch adsorption method. The discussion covers the role of adsorbents' morphology and textural properties, including pore size, total pore volume, surface area, and surface characteristics. Furthermore, it examines the effects of initial adsorbate concentration and adsorbent dosage on adsorption. Key experimental conditions are also assessed, such as the pH of the adsorbate solution, contact time between the adsorbent and adsorbate, coexisting ions in the solution, agitation speed, and temperature. Finally, the significance of optimizing experimental conditions is emphasized, and insights are offered on maximizing removal efficiency and enhancing the overall performance of the adsorption process.

**Keywords** Adsorption · Carbon-based nanomaterials · Morphology and textural properties · pH of solution · Adsorbate concentrations · Adsorbent dosage · Temperature

## Abbreviations

| | |
|---|---|
| BA | Benzoic Acid |
| BET | Brunauer–Emmett–Teller |
| BJH | Barrett–Joyner–Halenda |

F. Machado Machado (✉)
Technology Development Center, Federal University of Pelotas (UFPel), Pelotas, RS, Brazil
e-mail: fernando.mahado@ufpel.edu.br

E. C. Lima (✉)
Institute of Chemistry, Federal University of Rio Grande Do Sul (UFRGS), Porto Alegre, RS, Brazil
e-mail: profederlima@gmail.com

© The Author(s), under exclusive license to Springer Nature Switzerland AG 2026 
C. P. Bergmann and F. Machado Machado (eds.), *Carbon Nanomaterials as Adsorbents for Environmental and Biological Applications*, Carbon Nanostructures,
https://doi.org/10.1007/978-3-032-08669-3_7

| | |
|---|---|
| CAHB | Charge-Assisted Hydrogen Bond |
| CQDs | Carbon quantum dots |
| CNTs | Carbon nanotubes |
| FTIR | Fourier Transform Infrared Spectroscopy |
| GFL | Graphene Few-Layers |
| GO | Graphene Oxide |
| HK | Horváth–Kawazoe |
| IEP | Isoelectric point |
| IUPAC | International Union of Pure and Applied Chemistry |
| MWCNT | Multi-walled carbon nanotubes |
| NOM | Natural Organic Matter |
| OC | Organic Contaminant |
| PD | Potassium diclofenac |
| $pH_{PZC}$ | PH of the point of zero charge |
| PNT | Phenanthrene |
| PW | Pore width |
| *Qmax* | Maximum adsorption capacity |
| SEM | Scanning electron microscopy |
| SMR | Sulfamerazine |
| SWCNT | Single-walled carbon nanotubes |
| TCE | Trichloroethylene |
| XPS | X-ray Photoelectron Spectroscopy |

## 1 Introduction

The increasing contamination of water resources by industrial, agro-industrial, and domestic waste has become a global concern, particularly due to the direct and indirect impacts on human health and aquatic ecosystems. As a result, developing and implementing effective wastewater treatment methods have emerged as an environmental priority. Among the various available technologies, adsorption is a promising alternative due to its operational simplicity, low cost, reusability of adsorbents, and high efficiency in removing contaminants from aqueous media [1, 2].

In this context, using adsorptive materials, especially carbon-based nanomaterials, has led to significant advances in the efficiency of treatment processes, substantially reducing pollutant concentrations and, consequently, environmental risks [3, 4]. However, for an adsorption process to reach its full potential, it is essential to understand the fundamental mechanisms and to optimize the operational parameters that govern the technique.

The effectiveness of the adsorption process is directly influenced by several experimental factors, including the morphology, surface characteristics, and textural properties of the adsorbent; the pH of the adsorbate solution; contact time; initial concentration of the adsorbate; adsorbent dosage; the presence of competing ions and natural organic matter in the solution, agitation speed and temperature [5]. Accurate evaluation and adjustment of these variables are essential to maximizing pollutant removal capacity, enabling adsorbent reuse, and ensuring a more sustainable and effective water treatment process.

Therefore, the study and optimization of the conditions that influence adsorption are crucial for the practical application of this technique in real wastewater treatment systems.

The following section will present a detailed discussion on how these factors affect the adsorption process.

## 2 Morphology, Textural, and Surface Properties of Carbon-Based Nanomaterials

The successful application of adsorption in wastewater treatment largely depends on the appropriate selection of adsorbent material. Key factors to consider include the adsorbent's sorption capacity as well as its mechanical and chemical stability under the intended operating conditions [5]. Additionally, the potential for regenerating and reusing the spent adsorbent should be evaluated as part of the selection process [3, 5].

Among the most critical adsorbent characteristics influencing adsorption performance are the adsorbent's textural properties (such as specific surface area, pore size distribution, and total pore volume) and surface characteristics (such as the presence of functional groups). Since adsorption is a surface-based phenomenon, these properties directly affect how efficiently pollutants can be captured from aqueous environments.

Regarding specific surface area, carbon-based nanomaterials such as carbon nanotubes (CNTs) [6–9], graphene family materials [10–12], and carbon quantum dots (CQDs) [13, 14] are well known for their exceptionally high specific surface areas, which make them highly attractive for adsorption applications [5]. A substantial surface area provides more active sites for interacting with pollutants, enhancing adsorption capacity. These features, combined with their tunable surface chemistry and structural versatility, make carbon-based nanomaterials highly effective for removing a wide range of contaminants from water.

Adsorbates may either adhere to the external surface of the material or diffuse into its internal pores; this latter process is known as intraparticle diffusion [3, 15]. In such cases, the pore size directly impacts the rate at which the adsorbate can move into the interior of the adsorbent, thereby affecting both the kinetics and the overall sorption capacity.

According to the classification established by the International Union of Pure and Applied Chemistry (IUPAC), pores are divided into three main types based on their width (pore width, PW): micropores ($PW < 2$ nm), mesopores ($2 \text{ nm} < PW < 50$ nm), and macropores ($PW > 50$ nm) [14]. Some authors use the term "nanopore"; however, it is generally accepted that nanopores refer to pores with widths smaller than 100 nm [5]. Despite this, the use of the term is discouraged, as all material pores exist at the nanoscale to some extent, making "nanopores" a conceptually imprecise term. Unfortunately, some researchers, relying on images obtained by scanning electron microscopy (SEM), describe visible holes and channels in the solid sample as pores. This is a common mistake found in several adsorption studies. Most micrographs are captured at the micrometer (μm) scale. Therefore, holes sawn observed in adsorbents at this scale using SEM cannot be accurately classified as pores.

Carbon-based nanoadsorbents are typically characterized by a high proportion of micropores, contributing to a large specific surface area and enhanced adsorption capacity. These micropores are particularly effective for the adsorption of small molecules, whereas larger molecules are more likely to be retained within mesopores. Additionally, an increased total pore volume generally correlates with a higher overall sorption capacity [2, 15]. PW is commonly represented by the diameter of the opening, modeled either as a slit width or assuming a cylindrical (or elliptical) shape, as shown in Fig. 7.1. These models are applied to various carbon-based nanomaterials, such as stacked graphene layers (Fig. 7.1a, c) and CNTs (Fig. 7.1b, d).

Accurate determination of the pore size distribution, particularly within the microporous range, is a critical aspect of sorbent characterization. Several instrumental techniques have been developed for this purpose, with nitrogen adsorption–desorption analysis being the most widely applied [16]. This method is commonly coupled with mathematical models such as the Kelvin equation (or Barrett–Joyner–Halenda method—BJH) [16], the Brunauer–Emmett–Teller (BET) equation [17], the Density Functional Theory (DFT) method [18], and the Horváth–Kawazoe (HK) approach [19]. Each technique presents specific advantages depending on the pore size range and the nature of the adsorbent material under investigation.

When using nanomaterials as adsorbents, the adsorption process typically involves interactions between the functional groups of the pollutant (adsorbate) and the active sites located on the surface of the nanomaterial [7, 20]. These interactions are strongly influenced by the structural and textural properties of the adsorbent, as well as by the presence of functional groups on its surface [7, 20]. Carbon-based nanomaterials usually possess a high concentration of surface functional groups, particularly oxygen-containing groups [5]. The amount and type of these functional groups depend on various factors, including the synthesis route, purification methods, and any functionalization procedures applied to the nanostructures [21, 22].

Instrumental techniques such as Fourier Transform Infrared Spectroscopy (FTIR), X-ray Photoelectron Spectroscopy (XPS), and Raman Spectroscopy [8, 9] can be used to identify the functional groups present on the surface of carbon-based nanomaterials and to determine which of these groups are involved in the adsorption of a specific adsorbate. When combined with experimental data from adsorption studies, these spectroscopic analyses provide valuable insights into the interactions between adsorbent and adsorbate.

Typically, shifts in the FTIR vibration bands observed after the adsorption process indicate interactions between particular functional groups on the adsorbent's surface

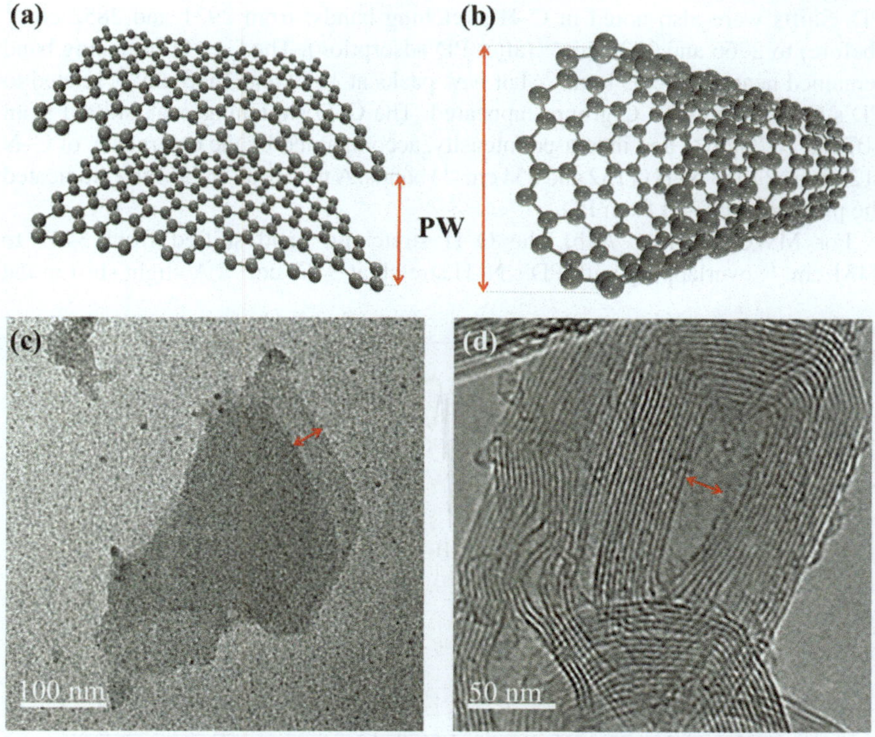


**Fig. 7.1** The schematic representation includes: **a** a slit-shaped pore, **b** a cylindrical pore, **c** a slit pore formed between graphene few-layers (GFL) sheets, and **d** a cylindrical pore CNT. (Images C and D were obtained via transmission electron microscopy)

and the adsorbate molecules [9]. However, the user should consider the resolution of the FTIR equipment before considering the band shift. The default resolution of the FTIR spectrophotometers is 4 $cm^{-1}$. Therefore, any band shift lower than 12 $cm^{-1}$ should not be considered a band shift. To use it, it is necessary to increase the FTIR resolution to 1 $cm^{-1}$ and increase the number of scans (usually higher than 150–200). However, the signal/noise ratio could be worsened. A shift to lower wavenumbers suggests that the adsorbate has withdrawn electrons from the surface group, reducing its electron density. In contrast, a shift to higher wavenumbers indicates that the adsorbate has donated electrons to the adsorbent. An example of this application is shown in Fig. 7.2, which presents FTIR analysis revealing significant spectral changes after potassium diclofenac (PD) adsorption onto both single-walled carbon nanotubes (SWCNT) and multi-walled carbon nanotubes (MWCNT) (resolution of 1 $cm^{-1}$, 200 scans, KBr pellets), indicating interactions between the medicine molecules and specific functional groups on the surfaces of carbon-based nanomaterials. For SWCNT (Fig. 7.2a), a shift of the O–H stretching band from 3445 to 3425 $cm^{-1}$ was observed, along with reduced intensity and broadening of this band. A secondary peak at 3401 $cm^{-1}$ appeared, attributed to N–H stretching of aromatic amines from

PD. Shifts were also noted in C–H stretching bands: from 2921 and 2852 cm$^{-1}$ (before) to 2966 and 2920 cm$^{-1}$ (after PD adsorption). The C = O stretching band remained near 1635–1654 cm$^{-1}$, but new peaks at 1629 and 1565 cm$^{-1}$, related to PD's C = O and C = C groups, appeared. The C–O stretching peak shifted from 1072 to 1079 cm$^{-1}$ and increased intensity, accompanied by the emergence of C–N (1290 cm$^{-1}$) and N–H (1152 and 854 cm$^{-1}$) bands. A new peak at 756 cm$^{-1}$ indicated the presence of C–Cl from PD.

For MWCNT (Fig. 7.2b), the O–H stretching band shifted from 3437 to 3431 cm$^{-1}$, overlapping with PD's N–H stretch at 3467 cm$^{-1}$. A slight shift in the

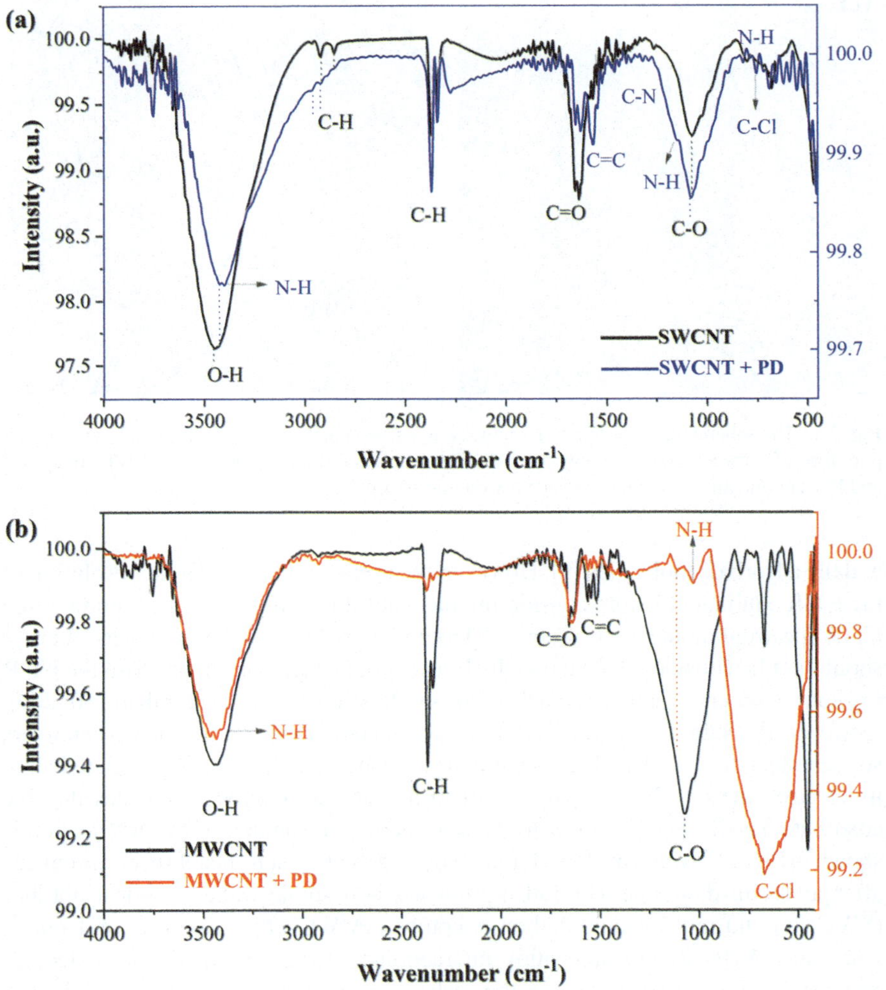


**Fig. 7.2** FTIR spectra of **a** pure SWCNT and SWCNT loaded with PD, and **b** pure MWCNT and MWCNT loaded with PD. Analysis conditions: resolution of 1 cm$^{-1}$, KBr pellets, and 200 scans

$C = O$ band was observed (from 1654 to 1649 $cm^{-1}$), with a broad absorption band from 1721 to 1587 $cm^{-1}$ indicating $C = O$ overlap. The C–O peak showed a notable blueshift from 1072 to 1107 $cm^{-1}$, suggesting strong involvement in the adsorption. Additional peaks at 1027 $cm^{-1}$ (C–N) and a broad band between 944–442 $cm^{-1}$ (C–Cl) confirmed PD's presence on MWCNT.

Overall, FTIR results indicate that the adsorption of PD onto CNTs involves mainly O–H, $C = O$, C–O, and aromatic ring interactions.

## 3 pH of the Solution

One of the most influential factors in adsorption studies is the effect of the working solution's pH, particularly when cationic or anionic dyes are used as adsorbates. Different adsorbents show specific optimal pH ranges based on their characteristics [4–6, 15]. The adsorption of contaminant molecules onto carbon-based nanostructures is influenced by the adsorbent's surface charge and the chemical structure of the pollutant in solution, which are both affected by the solution's pH [21, 23]. The pH not only impacts the adsorption capacity but also the overall performance of the adsorption process [21, 23]. Furthermore, the pH of the adsorbate solution also affects the solubility of the adsorbate in solution.

Electrochemical methods have been widely used to evaluate the effect of pH on the surface modification of carbon-based adsorbents, particularly in determining the isoelectric point (IEP) and the pH of the point of zero charge ($pH_{PZC}$). The IEP corresponds to the pH at which the particle of the adsorbents exhibits no electrophoretic mobility, reflecting only the charge of the external surface of the material [24]. In contrast, the $pH_{PZC}$ represents the pH at which the total surface charge of the material (internal and external) is zero (that is, when the charges are in balance) [25]. The difference between the $pH_{PZC}$ and the IEP ($pH_{PZC}$—IEP) provides insight into the radial charge distribution within the porous structure of the adsorbents. For instance, a positive value indicates a greater concentration of charged groups on the external surface relative to the internal structure. In contrast, values close to zero suggest a more homogeneous distribution [24]. This distinction is particularly useful in analyzing surface modifications of carbon-based materials, allowing for controlling properties such as acidity, basicity, and affinity toward specific adsorbates [21].

Figure 7.3 presents the IEP and $pH_{PZC}$ values for SWCNT and MWCNT nanoadsorbents, as well as the influence of working solution pH on removing the nonsteroidal anti-inflammatory PD. Figure 7.3a, b show that the $pH_{PZC}$ values were 5.0 for SWCNT and 9.0 for MWCNT, while the IEP values were 5.5 and 5.29, respectively. Based on the $pH_{PZC}$ values, the CNTs display a negative surface charge at $pH > pH_{PZC}$ and a positive charge at $pH < pH_{PZC}$. Accordingly, these nanoadsorbents can preferentially interact (via electrostatic interactions) with negatively charged adsorbates at $pH < pH_{PZC}$ and positively charged adsorbates at $pH > pH_{PZC}$ [25, 26].

Figure 7.3c shows the effect of the initial solution pH on the percentage of PD removal by CNTs across a pH range of 4–10. At $pH = 4$, PD removal reached

**Fig. 7.3** **a** IEP and the **b** $pH_{PZC}$ of CNTs. **c** The influence of the initial pH of a solution on the percentage of PD removal by CNTs varies across a pH range of 4–10

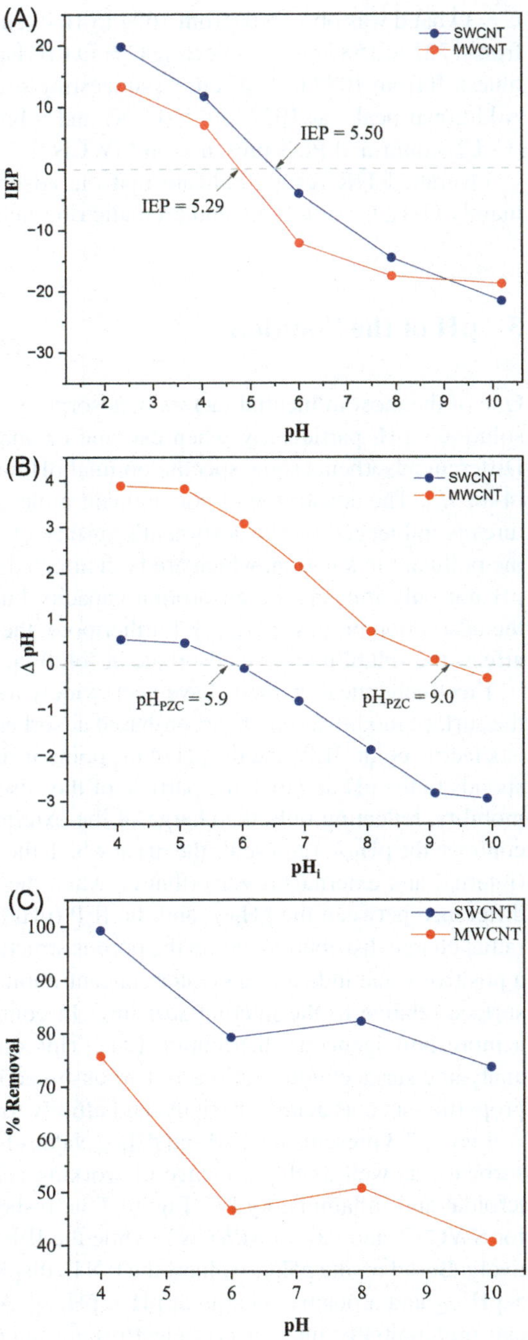

99% for SWCNT and 75% for MWCNT. For both nanomaterials, removal efficiency decreased between pH 4 and 6 and remained nearly constant at higher pH values (between 73% for SWCNT and 40% for MWCNT). The higher removal efficiency under acidic conditions can be attributed to the properties of both the nanoadsorbents and the pharmaceutical compound. The $pK_a$ of PD is 4.10, meaning that at $pH < pK_a$, the molecule exists in its protonated (neutral) form, and at $pH > pK_a$, it becomes deprotonated (anionic) [27]. Additionally, it is worth noting that at pH values below 4, PD solubility may decrease [27]. The highest removal percentages were observed at acidic pH values, under which the adsorbents are positively charged, and PD is in its deprotonated form, promoting electrostatic interaction ($CNT^+$—$PD^-$). However, at pH values close to the adsorbents' $pH_{PZC}$, a slight reduction in removal efficiency is observed, possibly due to repulsive interactions between adsorbate and adsorbent molecules.

The $pH_{PZC}$—IEP difference for SWCNT is −0.5, whereas for MWCNT it is 3.71. The low and close values of $pH_{PZC}$ and IEP for SWCNT suggest a relatively homogeneous and acidic surface, both internally and externally. On the other hand, the positive value observed for MWCNT indicates a higher concentration of charged groups on its external surface than on the internal one [24].

Finally, it is important to emphasize that carbon-based materials' $pH_{PZC}$ and IEP depend on the surface functional groups [21, 28, 29], which can significantly influence the adsorbent–adsorbate interaction process.

## 4 Effect of Adsorbent Dosage

The effect of adsorbent dosage should be optimized to balance removal efficiency, adsorption capacity, and economic feasibility [30].

The study of adsorbent dosage variation in removing adsorbates from aqueous media is commonly performed using doses between 1.0 and 10.0 g $L^{-1}$, with a constant adsorbate concentration [30–33]. At first, increasing the amount of adsorbent improves the removal efficiency until it stabilizes, forming a plateau on the curve (see Fig. 7.4a). The removal percentage remains nearly unchanged above a specific dosage, referred to as the threshold. This initial improvement is mainly due to the greater surface area made available, which increases the number of active sites for adsorption, as reported in previous studies [31, 32].

In contrast, increasing the adsorbent dose beyond this point significantly reduces the dye molecules uptake per gram of adsorbent ($q$), as shown in Fig. 7.4b. This behavior can be quantitatively described by Eq. 7.1 [31, 32], which shows that the uptake capacity ($q$) is inversely related to the adsorbent dose ($X$). Because the solution volume ($V$) and the initial adsorbate concentration ($C_0$) are held constant, increasing the mass of adsorbent results in a lower $q$ value.

$$q = \frac{\%\,\mathbf{Remoção}.C_0}{100.X} \tag{7.1}$$

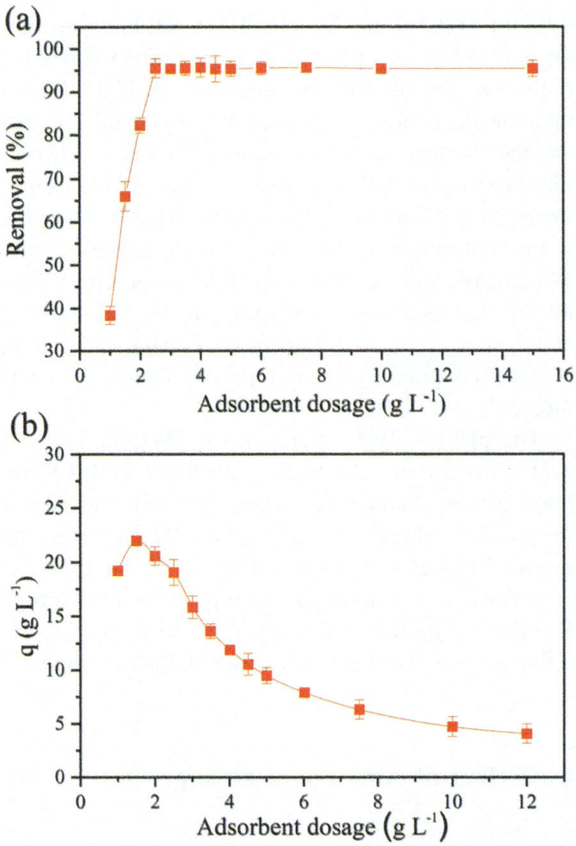


**Fig. 7.4** **a** Effect of adsorbent dosage on adsorbate removal efficiency and **b** effect of adsorbent dosage on sorption capacity

Consequently, the most efficient dosage corresponds to the threshold value, representing the minimum mass of adsorbent needed to maintain constant removal efficiency [31]. The reduction in $q$ with increasing dosage can be attributed to two main effects [31]:

(i) the underutilization of adsorption sites, since excess adsorbent is present about the fixed amount of adsorbate, and
(ii) the tendency for particles to aggregate at higher dosages, which decreases the effective surface area and increases the diffusion path length.

# 5 Role of Contact Time in Adsorbate–Adsorbent Interactions

Contact time is another important parameter influencing the adsorption process, including its economic aspects. It reflects the interaction dynamics between the adsorbent and the adsorbate. Once equilibrium is achieved, the necessary contact time is considered complete [7–9, 33].

Kinetic studies allow the determination of the minimum time required to reach this equilibrium. After this point, the system is regarded as stable. However, it is important to note that the initial concentration of the adsorbate can affect the adsorption rate. For this reason, even after identifying the equilibrium time through kinetic experiments, it is advisable to extend the contact time to ensure sufficient interaction, especially at higher initial concentrations, to achieve accurate equilibrium conditions [7–9].

Figure 7.5 illustrates the influence of contact time on the adsorption of Direct Blue 53 (DB-53) dye onto MWCNT [7].

Kinetic data provide other relevant parameters in addition to identifying the equilibrium time. For example, it is possible to determine the time required for the carbon-based nanoadsorbent to remove 50% ($t_{0.5}$) and 95% ($t_{0.95}$) of the adsorbate from the solution (see Chap. 6), which contributes to a deeper understanding of the adsorption dynamics and helps optimize the process [2, 15].

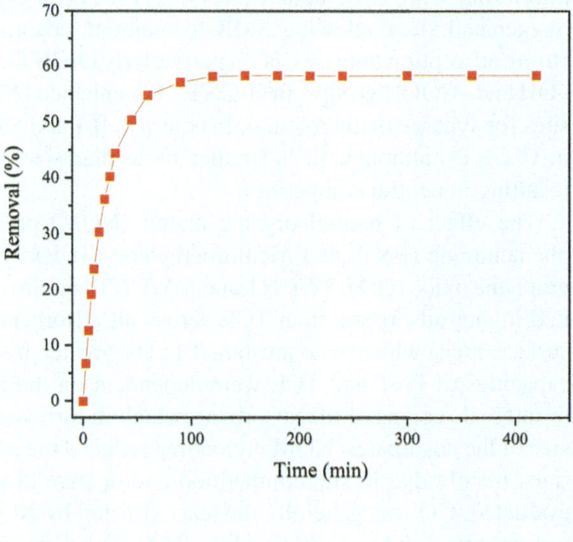


**Fig. 7.5** Effect of contact time on the interaction between MWCNT and DB-53 synthetic dye. Conditions: temperature = 25 °C and $C_0$ = 300.0 mg $L^{-1}$

## 6 Coexisting Ions in the Solution

Coexisting ions play a fundamental role in adsorption processes, directly influencing the interaction between adsorbents and contaminants in aqueous systems. Studies have shown that the presence of coexisting ions in solutions leads to competitive adsorption on the adsorbent surface, significantly affecting the removal efficiency of the adsorbate, particularly when carbon-based materials such as graphene family materials [22] and CNTs [22, 34] are involved.

Several factors may explain this behavior, including competition for adsorption sites and/or pore or interstitial blockage [22], as well as variations in ionic strength [34–36].

Zhang et al. [37] investigated the coadsorption behavior of Pb(II), sulfamerazine (SMR), and benzoic acid (BA) on oxidized (O-CNTs) and graphitized multi-walled carbon nanotubes (G-CNTs) [37]. Their findings revealed that charge-assisted hydrogen bonding (CAHB) played a key role in the adsorption of SMR and BA onto oxygen-rich nanomaterials like O-CNTs. Pb(II) exhibited greater affinity for CNT surfaces than SMR, primarily due to cation–π interactions (involving strong interactions between $Pb^{2+}$ and the π-electrons of the graphitic structure) and surface complexation with oxygen-containing groups (–COOH and –OH). In G-CNTs, Pb(II) did not directly compete with SMR, which adsorbed via π–π and hydrophobic interactions, while Pb(II) adsorbed through the cation-π mechanism without significantly interfering with SMR uptake. In O-CNTs, Pb(II) exhibited weak competition for oxygenated sites, allowing SMR to maintain adsorption via CAHB. BA showed strong adsorption onto all CNTs, particularly O-CNTs, where it formed CAHBs with –OH and –COOH groups. In G-CNTs, BA enhanced SMR adsorption by exposing π-sites for synergistic interaction. In contrast, BA and SMR competed for CAHB sites in O-CNTs, although BA's smaller molecular size prevented complete blockage, resulting in neutral competition.

The effect of natural organic matter (NOM) preloading on the adsorption of phenanthrene (PNT) and trichloroethylene (TCE) by graphene few-layers (GFL), graphene oxide (GO), SWCNT and MWCNT was also investigated [22]. PNT exhibited higher adsorption than TCE across all adsorbents, both in terms of mass and surface area, which was attributed to its greater hydrophobicity. The adsorption capacities of PNT and TCE were dependent on the accessibility of the molecules to internal regions of the adsorbent, which in turn was influenced by the molecular size of the adsorbates. NOM preloading reduced the adsorption capacity of all materials, mainly due to site competition and/or pore blockage. Among all adsorbents evaluated, GO was generally the least affected by NOM for PNT, while no specific competitive trend was observed for TCE. This behavior is likely due to the surface characteristics of GO [38]. The polarity of surface oxides improves GO dispersion in water, reducing NOM interference. Additionally, the presence of oxygen functional groups can limit NOM accumulation on GO due to electrostatic repulsion between negatively charged NOM and the acidic GO surface.

Ionic strength also significantly affects the adsorption of organic contaminants (OC) onto carbon-based materials [36, 39]. Ionic salts influence various mechanisms, including interactions with OC molecules in solution and on the adsorbent surface, as well as changes in the surface charge of the adsorbent [38–40]. Two main mechanisms have been proposed to describe the effect of ionic strength on OC adsorption: the salting-out effect and the squeezing-out effect [35, 36].

High ionic strength can reduce the solubility of hydrophobic OCs in water (salting-out effect), promoting their partitioning onto the hydrophobic surfaces of graphene-based adsorbents and enhancing adsorption [35]. For instance, the adsorption of pharmaceuticals such as ketoprofen increases with higher ionic strength [39]. On the other hand, elevated ionic strength can also cause a squeezing-out effect, in which ions compress the diffuse double layer surrounding the adsorbent, leading to aggregation of adsorbent particles, reduced surface area, and hindered access to adsorption sites [35, 36]. Jiang et al. [36] demonstrated this behavior during the removal of 17β-estradiol using GFL [36]. The net effect of ionic strength on adsorption depends on the balance between these two mechanisms: at low NaCl concentrations (<0.001 M), the salting-out effect tends to dominate, while at higher concentrations, the squeezing-out effect becomes more pronounced.

## 7 Influence of Initial Adsorbate Concentration on Adsorption Performance

The initial adsorbate concentration ($C_0$) is a fundamental parameter in evaluating adsorption behavior, as it directly influences the interaction dynamics between the adsorbate and the adsorbent [41]. Figure 7.6 presents an example in which a fixed amount of adsorbent was used, while the initial concentration of the adsorbate was varied. At lower concentrations, the abundance of available active sites and high surface area on the adsorbent favor rapid uptake of the adsorbate [42]. As $C_0$ increases, the adsorption capacity also rises due to the higher probability of collisions between adsorbate molecules and the adsorbent surface, as well as the greater concentration gradient [43], which enhances the driving force for mass transfer from the liquid phase to the solid surface [44].

This effect is particularly evident in carbon-based nanomaterials adsorbents, such as those incorporating MWCNT, which demonstrate improved adsorption capacities at higher initial concentrations [7]. The increased adsorbate concentration promotes the occupation of adsorption sites more efficiently, intensifying the overall adsorption process [43]

However, while low initial concentrations often lead to faster initial adsorption rates, systems with higher $C_0$ values typically require longer contact times to reach equilibrium [45]. For this reason, it is common in isotherm studies to employ agitation times that exceed the equilibrium time when working with more concentrated solutions [1, 3, 7, 15].

**Fig. 7.6** The impact of the initial adsorbate concentration on the adsorption capacity of an adsorbent

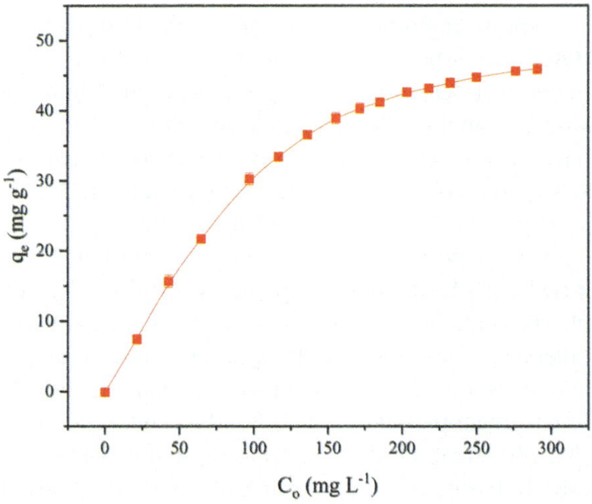


A common issue with using carbon-based materials to absorb emerging contaminants (such as pharmaceuticals and pesticides) is the concentration range of these micropollutants. Typically, the environmental concentration of these contaminants in samples (like water and soil) is found at parts per billion or parts per trillion. Conversely, for adsorbents with high sorption capacity, such as carbon-based adsorbents (activated biochars, graphene family materials, CNT, etc.), usually, concentrations of up to 1–2 g L$^{-1}$ (1000–2000 mg L$^{-1}$) were employed to achieve the saturation of the adsorbent. Furthermore, the working concentrations in the adsorption studies range from 100 to 500 mg L$^{-1}$. However, many researchers dealing with chromatographic analysis state that these concentrations have no practical significance, as the concentration level is $10^6$–$10^9$ times higher than the actual environmental sample. At this point, we need to clarify that when using an adsorbent with a high sorption capacity (>100 mg g$^{-1}$) and an adsorbate at a concentration of 1–10 µg L$^{-1}$, it will be totally adsorbed after a specific contact time. For adsorbents with a high surface area and high total pore volume presenting high sorption capacities, the complete removal of the adsorbate will occur in a few minutes. Therefore, it is impossible to carry out adsorption experiments using 0.2–2 mg L$^{-1}$ of adsorbate with an adsorbent presenting high sorption capacity because the removal rate will be 100% a few minutes, and no remaining adsorbate at the final solution will be determined by the analytical technique (chromatographic, spectrophotometric, fluorometric, etc.). The plot below (Fig. 7.7) corresponds to the uptake of methylparaben, an emerging contaminant (initial concentrations ranging from 1 to 350,000 µg L$^{-1}$) by Butia Palm activated biochar [46].

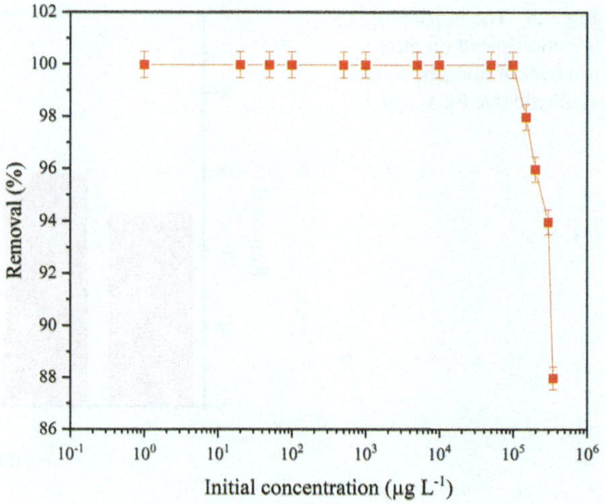


**Fig. 7.7** Effect of the initial concentration of methylparaben on its removal using butia palm activated biochar

## 8 Influence of Agitation Speed

Agitation speed is another parameter that can influence the performance of an adsorption system [47]. Although it is not always thoroughly investigated, agitation plays a critical role in facilitating effective contact between adsorbate molecules and the active sites of the adsorbent. Proper agitation enhances mass transfer, allowing the adsorbate to move closer to the surface of the adsorbent, thereby improving adsorption efficiency.

For instance, Fig. 7.8 illustrates the effect of agitation speed in the removal of Malachite Green synthetic dye using *S. asper* as adsorbent. The results indicate that an agitation speed of 150 rpm yielded the highest removal efficiency. Studies have reported optimal agitation speeds for adsorption processes to fall within the range of 100–200 rpm, depending on the specific adsorbent-adsorbate system.

In many cases, the uptake of pollutants by nanostructured adsorbents increases with agitation speed up to a certain threshold. Beyond this point, further speed increases may result in a decline or plateau in adsorption performance. This decrease is often attributed to the potential desorption of adsorbate molecules at excessively high speeds, which can disrupt the interaction between the adsorbent and adsorbate [47].

It is important to note that the optimal agitation speed may vary depending on the combination of adsorbent and adsorbate. Therefore, this parameter should not be overlooked or generalized but carefully evaluated for each specific system to ensure accurate results and optimal performance.

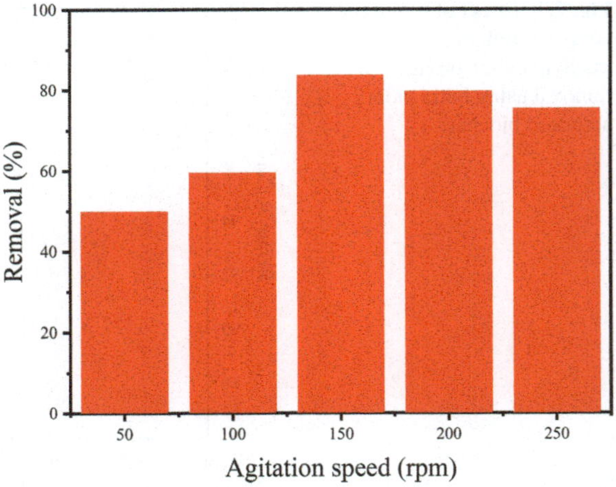


**Fig. 7.8** The impact of agitation speed on the removal of malachite green synthetic dye by *S. asper*

## 9 The Impact of Temperature on Adsorption Behavior

Temperature is a critical factor influencing adsorption behavior, as it affects both the carbon-based nanoadsorbents' structural properties and the adsorbate's kinetics. Temperature variations can alter the porosity, thermal expansion, and surface chemistry of carbon nanomaterials, modify ion mobility in solution, change the solution's viscosity, and impact interactions at the solid–liquid interface [48–50].

Conversely, it is impossible to predict the thermal behavior of an adsorbate/adsorbent system. The sorption capacity could increase, decrease, or remain at the same level when the temperature is increased. The adsorbent user should be concerned that it is incorrect to state that when the sorption capacity increases, the adsorption is endothermic, or it is exothermic when the sorption capacity decreases with the temperature increase. To state whether an adsorption system is endothermic or exothermic, it is necessary to evaluate the equilibrium constant obtained from the adsorption isotherms carried out at different temperatures. Suppose the equilibrium constant of a given isotherm model (Langmuir, Liu, Sips, Redlich-Peterson, etc., see Chap. 6) increases with the increase of temperature. In that case, the adsorption process is endothermic; if the value of $K$ decreases with the temperature, the adsorption process is exothermic.

Below is presented Table 7.1 for the uptake of resorcinol using pine-activated biochar using the Liu equilibrium isotherm [51].

If the user only examines the $Q_{max}$ values, they may draw an incorrect conclusion; it has been noted that an increase in temperature raises the $Q_{max}$ value. On the other hand, the value of $K_g$ (Liu equilibrium constant) decreases with the increase in temperature. Applying the van´t Hoff equation (see Chap. 6), the calculated adsorption enthalpy change was $-22.7$ kJ mol$^{-1}$ [50], indicating that the adsorption process is exothermic instead of endothermic.

**Table 7.1** Liu's equilibrium isotherm for the uptake of resorcinol onto pine-activated biochar

| Liu | 10 °C | 20 °C | 25 °C | 30 °C | 40 °C | 45 °C |
|---|---|---|---|---|---|---|
| $Q_{max}$ (mg g$^{-1}$) | 179.6 | 184.3 | 249.8 | 182.3 | 263.8 | 249.4 |
| $K_g$ (L mg$^{-1}$) | 0.01058 | 0.007635 | 0.006558 | 0.005727 | 0.004070 | 0.003659 |
| $n_L$ | 2.267 | 1.052 | 0.2882 | 0.9470 | 0.1571 | 0.1631 |
| $R^2_{adj}$ | 0.9999 | 0.9999 | 0.9999 | 0.9998 | 0.9999 | 0.9999 |
| $SD$ (mg g$^{-1}$) | 0.6991 | 0.3978 | 0.4612 | 0.7694 | 0.3220 | 0.3498 |
| BIC | −6.091 | −29.77 | −23.57 | −2.071 | −38.66 | −35.17 |

## 9.1 Conclusions

This chapter discusses and supports examples from the literature of the main experimental factors and conditions that influence the batch adsorption process. Understanding these variables is essential for optimizing the interaction between the adsorbent and the adsorbate, which is fundamental to improving water treatment and environmental remediation strategies.

Among the most significant factors are the morphological and textural properties of the carbon-based nanomaterials adsorbents, including surface area, pore size, and total pore volume. Small molecules typically adsorb in micropores, which are associated with higher surface areas, while larger molecules are generally retained in mesopores. A greater total pore volume is often correlated with increased sorption capacity.

The pH of the solution also plays a critical role in adsorption efficiency, as it can affect the charge and speciation of both the adsorbate and the adsorbent. Different systems may exhibit distinct optimal pH ranges. Additionally, the adsorbent dosage should be optimized to a threshold level that ensures efficient removal without excessive material use. The initial concentration of the adsorbate is another important factor; increasing this concentration typically leads to greater adsorption until the adsorbent becomes saturated.

Other key parameters include contact time, which should exceed the equilibrium time to ensure complete adsorption at higher concentration levels, and agitation speed, which can influence mass transfer and, consequently, adsorption capacity. Furthermore, the presence of competing ions and natural organic matter in the solution can modulate the adsorption process.

Finally, depending on the system's specific characteristics, temperature can enhance, reduce, or have minimal impact on adsorption. Therefore, integrated control of these parameters is essential to achieving high-performance adsorption systems.

**Acknowledgements** The authors gratefully acknowledge financial support from the Brazilian funding agencies: Conselho Nacional de Desenvolvimento Científico e Tecnológico (CNPq), Coordenação de Aperfeiçoamento de Pessoal de Nível Superior (CAPES), and Fundação de Amparo à Pesquisa do Estado do Rio Grande do Sul (FAPERGS). Prof. Fernando M. Machado thanks CNPq for research grant support (Processes 308203/2022-4, 408413/2022-0, 443031/2023-1, and 403522/2025-0) and FAPERGS (Process 23/2551-0001863-2). Prof. Edér C. Lima acknowledges support from CNPq (Process 303612/2021-5), CAPES-PROEX (Processes 88881.844968/2023-01 and 001), and FAPERGS (Process 24/2551-0001291-5).

## References

1. Rodrigues, D.L.C., Fuhr, A.C.F.P., Guido, J.A., et al.: Olive biomass-derived magnetic activated biochar for ciprofloxacin removal: integrated kinetic, isotherm, thermodynamic, and spectroscopic analysis. Sep. Purif. Technol. **360**, 131014 (2025). https://doi.org/10.1016/j.seppur.2024.131014
2. Mello, B.L., Thue, P.S., Silva, P.V., et al.: Grafting 3-(trimethoxysilylpropyl)diethylenetriamine on microcrystalline cellulose for the adsorption of dyes: experimental and modeling studies. React. Funct. Polym. **196**, 105836 (2024). https://doi.org/10.1016/j.reactfunctpolym.2024.105836
3. Fuhr, A.C.F., Rodrigues, D.L.C., Guido, J.A., et al.: Modeling the adsorption of ciprofloxacin on magnetic biochar: a comparative study of traditional and advanced approaches. J Water Process Eng **74**, 107858 (2025). https://doi.org/10.1016/j.jwpe.2025.107858
4. Thue, P.S., Wamba, A.G.N., Mello, B.L., et al.: Magnetic composite carbon from microcrystalline cellulose to tackle paracetamol contamination: kinetics, mass transfer, equilibrium, and thermodynamic studies. Polymers **16**, 3538 (2024). https://doi.org/10.3390/polym16243538
5. Lima, É.C., Adebayo, M.A., Machado, F.M.: Experimental adsorption. In: Bergmann C, Machado F (eds) Carbon Nanomaterials as Adsorbents for Environmental and Biological Applications. Carbon Nanostructures. Springer (2015). https://doi.org/10.1007/978-3-319-18875-1_4
6. Machado, F.M., Carmalin, S.A., Lima, E.C., et al.: Adsorption of Alizarin Red S dye by carbon nanotubes: an experimental and theoretical investigation. J. Phys. Chem. C **120**, 18296–18306 (2016). https://doi.org/10.1021/acs.jpcc.6b03884
7. Prola, L.D.T., Machado, F.M., Bergmann, C.P., et al.: Adsorption of direct blue 53 dye from aqueous solutions by multi-walled carbon nanotubes and activated carbon. J. Environ. Manage. **130**, 166–175 (2013). https://doi.org/10.1016/j.jenvman.2013.09.003
8. Machado, F.M., Bergmann, C.P., Lima, E.C., et al.: Adsorption of reactive blue 4 dye from water solutions by carbon nanotubes: experiment and theory. Phys. Chem. Chem. Phys. **14**, 11139–11153 (2012). https://doi.org/10.1039/C2CP41475A
9. Azevedo, C.F., Machado, F.M., Souza, N.F., et al.: Comprehensive adsorption and spectroscopic studies on the interaction of carbon nanotubes with diclofenac anti-inflammatory. Chem. Eng. J. **454**, 140102 (2023). https://doi.org/10.1016/j.cej.2022.140102
10. Liao, Y., Zhu, R., Zhang, W., et al.: Ultrafast synthesis of 3D porous flash graphene and its adsorption properties. Colloids Surf A Physicochem Eng Asp **676**, 132178 (2023). https://doi.org/10.1016/j.colsurfa.2023.132178
11. Luong, D.X., Bets, K.V., Algozeeb, W.A., et al.: Gram-scale bottom-up flash graphene synthesis. Nature **577**, 647–651 (2020). https://doi.org/10.1038/s41586-020-1938-0
12. Jauris, I.M., Matos, C.F., Zarbin, A.J.G., et al.: Adsorption of anti-inflammatory nimesulide by graphene materials: a combined theoretical and experimental study. Phys. Chem. Chem. Phys. **19**, 22099–22110 (2017). https://doi.org/10.1039/C7CP04272H

13. Zhao, J., Zhu, J., Li, Y., et al.: Quantum dot reinforced electrospun carbon nanofiber fabrics with high surface area for ultrahigh rate supercapacitors. ACS Appl. Mater. Interfaces **12**, 11669–11678 (2020). https://doi.org/10.1021/acsami.9b22408
14. Thommes, M., Kaneko, K., Neimark, A.V., et al.: Physisorption of gases, with special reference to the evaluation of surface area and pore size distribution (IUPAC Technical Report). Pure Appl. Chem. **87**(9–10), 1051–1069 (2015). https://doi.org/10.1515/pac-2014-1117
15. Azevedo, C.F., Rodrigues, D.L.C., Silveira, L.L., et al.: Comprehensive adsorption and spectroscopic studies on the interaction of magnetic biochar from black wattle sawdust with betablocker metoprolol. Bioresour. Technol. **388**, 129708 (2023). https://doi.org/10.1016/j.biortech.2023.129708
16. Barrett, E.P., Joyner, L.G., Halenda, P.P.: The determination of pore volume and area distributions in porous substances. I. Computations from nitrogen isotherms. J. Am. Chem. Soc. **73**(1), 373–380 (1951). https://doi.org/10.1021/ja01145a126
17. Brunauer, S., Emmett, P.H., Teller, E.: Adsorption of gases in multimolecular layers. J. Am. Chem. Soc. **60**(2), 309–319 (1938). https://doi.org/10.1021/ja01269a023
18. Lastoskie, C., Gubbins, K.E., Quirke, N.: Pore size distribution analysis of microporous carbons: a density functional theory approach. J. Phys. Chem. **97**(18), 4786–4796 (1993). https://doi.org/10.1021/j100120a035
19. Horvath, G., Kawazoe, K.: Method for the calculation of effective pore size distribution in molecular sieve carbon. J. Chem. Eng. Jpn. **16**, 470–475 (1983). https://doi.org/10.1252/jcej.16.470
20. Jauris, I.M., Matos, C.F., Saucier, C., et al.: Adsorption of sodium diclofenac on graphene: a combined experimental and theoretical study. Phys. Chem. Chem. Phys. **18**, 1526–1536 (2016). https://doi.org/10.1039/C5CP06167A
21. Sajid, M., Asif, M., Baig, N., et al.: Carbon nanotubes-based adsorbents: properties, functionalization, interaction mechanisms, and applications in water purification. J. Water Process Eng. **47**, 102815 (2022). https://doi.org/10.1016/j.jwpe.2022.102815
22. Ersan, G., Apul, O.G., Perreault, F., Karanfil, T.: Adsorption of organic contaminants by graphene nanosheets: a review. Water Res. **126**, 385–398 (2017). https://doi.org/10.1016/j.watres.2017.08.010
23. Zhang, X.R., Guo, J.G., Zhou, L.J.: Experimental and theoretical investigation on the adsorption properties of benzene on graphene surface: influence of pH and edge effects. Chem. Eng. J. **440**, 135794 (2022). https://doi.org/10.1016/j.cej.2022.135794
24. Menéndez, J.A., Illán-Gómez, M.J., León, C.A., Radovic, L.R.: On the difference between the isoelectric point and the point of zero charge of carbons. Carbon **33**(11), 1655–1657 (1995). https://doi.org/10.1016/0008-6223(95)96817-R
25. Escobar, O.S., Azevedo, C.F., Swarowsky, A., et al.: Utilization of different parts of Moringa oleifera Lam. seeds as biosorbents to remove acid blue 9 synthetic dye. J. Environ. Chem. Eng. **9**(4), 105553 (2021). https://doi.org/10.1016/j.jece.2021.105553
26. Thue, P.S., Teixeira, R.A., Hounfodji, J.W., et al.: Intercalation of organosilane in clay mineral for the removal of procion red MX-5B: investigational and theoretical studies. Sep. Purif. Technol. **347**, 127491 (2024). https://doi.org/10.1016/j.seppur.2024.127491
27. Alessandretti, I., Rigueto, C.V.T., Nazari, M.T., et al.: Removal of diclofenac from wastewater: a comprehensive review of detection, characteristics and tertiary treatment techniques. J. Environ. Chem. Eng. **9**, 106743 (2021). https://doi.org/10.1016/j.jece.2021.106743
28. Liao, Q., Sun, J., Gao, L.: Adsorption of chlorophenols by multi-walled carbon nanotubes treated with $HNO_3$ and $NH_3$. Carbon **46**, 544–561 (2008). https://doi.org/10.1016/j.carbon.2007.12.009
29. Pillay, K., Cukrowska, E.M., Coville, N.J.: Multi-walled carbon nanotubes as adsorbents for the removal of parts per billion levels of hexavalent chromium from aqueous solution. J. Hazard. Mater. **166**, 1067–1075 (2009). https://doi.org/10.1016/j.jhazmat.2008.12.011
30. Ahmed, S., Fatema-Tuj-Zohra, C.T.R., et al.: Adsorption of Cu(II) and Cd(II) with graphene based adsorbent: adsorption kinetics, isotherm and thermodynamic studies. Desalination Water Treat. **285**, 167–179 (2023). https://doi.org/10.5004/dwt.2023.29290

31. Oladoja, N.A., Akinlabi, A.K.: Congo Red biosorption on palm kernel seed coat. Ind. Eng. Chem. Res. **48**, 6188–6196 (2009). https://doi.org/10.1021/ie801003v
32. Calvete, T., Lima, E.C., Cardoso, N.F., et al.: Application of carbon adsorbents prepared from the Brazilian-pine fruit shell for removal of procion red MX 3B from aqueous solution—kinetic, equilibrium, and thermodynamic studies. Chem. Eng. J. **155**, 627–636 (2009). https://doi.org/10.1016/j.cej.2009.08.019
33. Nascimento, V.X., Pinto, D., Lütke, S.F., et al.: Brilliant blue FCF dye adsorption using magnetic activated carbon from Sapelli wood sawdust. Environ. Sci. Pollut. Res. **30**, 58684–58696 (2023). https://doi.org/10.1007/s11356-023-26646-6
34. Ghosh, D., Chakraborty, K., Bharti, et al.: The effects of pH, ionic strength, and natural organics on the transport properties of carbon nanotubes in saturated porous medium. Colloids Surf. A Physicochem. Eng. Asp. **647**, 129025 (2022). https://doi.org/10.1016/j.colsurfa.2022.129025
35. Zhou, Y., Apul, O.G., Karanfil, T.: Adsorption of halogenated aliphatic contaminants by graphene nanomaterials. Water Res. **79**, 57–67 (2015). https://doi.org/10.1016/j.watres.2015.04.017
36. Jiang, L.H., Liu, Y.G., Zeng, G.M., et al.: Removal of 17β-estradiol by few-layered graphene oxide nanosheets from aqueous solutions: external influence and adsorption mechanism. Chem. Eng. J. **284**, 93–102 (2016). https://doi.org/10.1016/j.cej.2015.08.139
37. Zhang, J., Zhai, J., Zheng, H., et al.: Adsorption, desorption and coadsorption behaviors of sulfamerazine, Pb(II) and benzoic acid on carbon nanotubes and nanosilica. Sci. Total. Environ. **738**, 139685 (2020). https://doi.org/10.1016/j.scitotenv.2020.139685
38. Apul, O.G., Karanfil, T.: Adsorption of synthetic organic contaminants by carbon nanomaterials: a critical review. Water Res. **68**, 34–55 (2015). https://doi.org/10.1016/j.watres.2014.09.037
39. Liu, F.F., Zhao, J., Wang, S., et al.: Effects of solution chemistry on adsorption of selected pharmaceuticals and personal care products (PPCPs) by graphenes and carbon nanotubes. Environ. Sci. Technol. **48**(22), 13197–13206 (2014). https://doi.org/10.1021/es5034684
40. Zhang, S., Shao, T., Kose, H.S., et al.: Adsorption of aromatic compounds by carbonaceous adsorbents: a comparative study on granular activated carbon, activated carbon fiber, and carbon nanotubes. Environ. Sci. Technol. **44**(16), 6377–6383 (2010). https://doi.org/10.1021/es100874y
41. Nascimento, V.X., Schnorr, C., Lütke, S.F., et al.: Adsorptive features of magnetic activated carbons prepared by a one-step process towards brilliant blue dye. Molecules **28**(4), 1821 (2023). https://doi.org/10.3390/molecules28041821
42. Pal, S., Ahamed, Z., Pal, P.: Removal of antibiotics and pharmaceutically active compounds from water environment: experiments towards industrial scale up. Sep. Purif. Technol. **295**, 121249 (2022). https://doi.org/10.1016/j.seppur.2022.121249
43. Jiang, L., Yu, H., Pei, L., et al.: Initial concentrations of adsorbate influence on synergistic effect between a magnetic field and graphene oxide-carbon nanotube composite for $Pb^{2+}$ and phenol adsorption. Desalination Water Treat. **133**, 92–102 (2018). https://doi.org/10.5004/dwt.2018.22997
44. Dehmani, Y., Dridi, D., Lamhasni, T., et al.: Review of phenol adsorption on transition metal oxides and other adsorbents. J. Water Process. Eng. **49**, 102965 (2022). https://doi.org/10.1016/j.jwpe.2022.102965
45. Li, Y., Du, Q., Liu, T., et al.: Comparative study of methylene blue dye adsorption onto activated carbon, graphene oxide, and carbon nanotubes. Chem. Eng. Res. Des. **91**, 361–368 (2013). https://doi.org/10.1016/j.cherd.2012.07.007
46. Lima, E.C., Ponce-Vargas, M., Naushad, M., et al.: Removal of methylparaben from aqueous effluents using biobased carbon material. Experimental and DFT calculations. J. Mol. Liq. **397**, 124194 (2024). https://doi.org/10.1016/j.molliq.2024.124194.
47. Omri, A., Benzina, M., Ammar, B.: Preparation, modification and industrial application of activated carbon from almond shell. J. Ind. Eng. Chem. **19**, 2092–2099 (2013). https://doi.org/10.1016/j.jiec.2013.03.025

48. Leroy, M., Flahaut, E., Larue, C.: Carbon based nanomaterial interactions with metals and metalloids in terrestrial environment: a review. Carbon **206**, 325–339 (2023). https://doi.org/10.1016/j.carbon.2023.02.047
49. Thue, P.S., Lima, D.R., Lima, E.C., et al.: Comparative studies of physicochemical and adsorptive properties of biochar materials from biomass using different zinc salts as activating agents. J. Environ. Chem. Eng. **10**(3), 107632 (2022). https://doi.org/10.1016/j.jece.2022.107632
50. Cimirro, N.F.G.M., Lima, E.C., Cunha, M.R., et al.: Removal of diphenols using pine biochar. Kinetics, equilibrium, thermodynamics, and mechanism of uptake. J. Mol. Liq. **364**, 119979 (2022). https://doi.org/10.1016/j.molliq.2022.119979
51. Machado, F.M., Bergmann, C.P., Fernandes, T.H.M., et al.: Adsorption of reactive red M-2BE dye from water solutions by multi-walled carbon nanotubes and activated carbon. J. Hazard. Mater. **192**(3), 1122–1131 (2011). https://doi.org/10.1016/j.jhazmat.2011.06.020

# Molecules with Biological Interest Adsorbed on Graphene and Derivatives

**Júlia Vaz Schultz, Mariana Zancan Tonel, Mirkos Ortiz Martins, Ivana Zanella, Telma Domingues, Mafalda Abrantes, Jérôme Borme, Pedro Alpuim, and Solange Binotto Fagan**

**Abstract** Carbon nanomaterials are a class of nanostructures that have garnered significant attention from the scientific community due to their diverse applications in nanotechnology. The two-dimensional structure of graphene endows it with a range of physicochemical properties that make it a key material in several nanotechnology applications. In addition, graphene oxide is one of the most popular derivatives of graphene, gaining special attention due to its biological interest. This chapter highlights recent studies that investigate the interactions between graphene and graphene oxide, as well as the adsorption of molecules for biological research, particularly in the context of controlled release systems and biosensor development. The analyses consider both pristine and functionalized forms of graphene, showing how the presence of specific chemical groups can enhance the selectivity, stability, and efficiency of adsorption processes. By compiling and analysing these studies, this work seeks to demonstrate how graphene can serve as a versatile and practical platform for biomedical applications, contributing to significant advancements in diagnostics, monitoring, and therapeutic strategies through electrolyte-gated graphene field-effect transistors. Understanding the molecular interactions involved represents a crucial step toward developing innovative approaches that integrate nanotechnology

J. V. Schultz · M. Z. Tonel · M. O. Martins · I. Zanella · S. B. Fagan (✉)
Área de Ciências Tecnológicas, Universidade Franciscana, Santa Maria, RS, Brazil
e-mail: solange.fagan@gmail.com

M. O. Martins
e-mail: mirkos@ufn.edu.br

T. Domingues · M. Abrantes · J. Borme · P. Alpuim
International Iberian Nanotechnology Laboratory, Av. Mestre José Veiga, Braga, Portugal
e-mail: telma.domingues@udc.es

J. Borme
e-mail: jerome.borme@inl.int

P. Alpuim
e-mail: pedro.alpuim.us@inl.int

T. Domingues · P. Alpuim
Centro de Física das, Universidades Do Minho E Porto, Universidade Do Minho, Braga, Portugal

© The Author(s), under exclusive license to Springer Nature Switzerland AG 2026 
C. P. Bergmann and F. Machado Machado (eds.), *Carbon Nanomaterials as Adsorbents for Environmental and Biological Applications*, Carbon Nanostructures,
https://doi.org/10.1007/978-3-032-08669-3_8

and biomedicine, aiming to meet the growing demand for intelligent and functional materials in the healthcare context.

**Keywords** Biointerface engineering · Controlled release systems · Functional surface interactions · Nanobiotechnology · Sensor platforms

## Abbreviations

| | |
|---|---|
| 0D | Zero-dimensional |
| 2D | Two-dimensional |
| 1D | One-dimensional |
| AO | Acridine Orange |
| ASP | Aspartic Acid |
| BSSE | Basis Set Superposition Error |
| CAT | Catechin |
| CN | Carbon Nanomaterial |
| CNS | Central Nervous System |
| ctDNA | Circulating Tumor DNA |
| DFT | Density Functional Theory |
| DNA | Deoxyribonucleic Acid |
| DOS | Density of States |
| DOX | Doxorubicin |
| DZP | Double-Zeta Basis Set with Polarization Functions |
| EF | Fermi Energy |
| EGFET | Electrolyte-Gated Field-Effect Transistor |
| EGFRvIII | Epidermal Growth Factor Receptor variant III |
| FET | Field-Effect Transistor |
| FSC | Fluorenyl Methyl Succinimidyl Carbonate |
| GLN | Glutamine |
| GLU | Glutamic Acid |
| GO | Graphene Oxide |
| HB | Hydrogen Bond |
| HOMO | Highest Occupied Molecular Orbital |
| IDH1 | Isocitrate Dehydrogenase 1 |
| IDS | Drain-Source Current |
| ILE | Isoleucine |
| LDA | Local Density Approximation |
| LDOS | Local Density of States |
| LEU | Leucine |
| LUMO | Lowest Unoccupied Molecular Orbital |
| LYS | Lysine |
| NGS | Next-Generation Sequencing |
| NHS | N-Hydroxysuccinimide (used for blocking non-reacted linkers) |

| | |
|---|---|
| NLRP3 | NOD-, LRR-, and pyrin domain-containing protein 3 |
| PBSE | 1-Pyrene-Butyric Acid Succinimidyl Ester |
| pDNA | Probe DNA |
| PHE | Phenylalanine |
| PRO | Proline |
| PYD | Pyrin Domain |
| RMSD | Root Mean Square Deviation |
| RNA | Ribonucleic Acid |
| SARS | Severe Acute Respiratory Syndrome |
| SARS-CoV-2 | Severe Acute Respiratory Syndrome Coronavirus 2 |
| SER | Serine |
| SNP | Single-Nucleotide Polymorphism |
| SPR | Surface Plasmon Resonance |
| ssDNA | Single-Stranded Deoxyribonucleic Acid |
| TERT | Telomerase Reverse Transcriptase |
| TP53 | Tumour Protein p53 |
| VDirac | Dirac Voltage |
| VDS | Drain-Source Voltage |
| VGS | Gate-Source Voltage |

## 1 Introduction

Nanoscience and nanotechnology have attracted significant interest from the scientific community due to the new technological possibilities offered by the properties of nanomaterials [1]. A wide range of applications has been suggested in fields spanning from electronics to materials science. A significant amount of research has also been conducted with a biological focus, aiming to discover new methods for disease detection and therapy [1–4].

Carbon nanomaterials (CN) are promising systems that can be used for various applications in nanotechnology. Among these, graphene, carbon nanotubes, and fullerenes are considered two-dimensional (2D), one-dimensional (1D), and zero-dimensional (0D) materials, respectively, based on their electronic confinement. These nanomaterials have attracted considerable interest due to their structural and electronic properties [5], particularly those that offer the possibility of performing biofunctionalizations [6] through applications such as drug delivery systems or biological sensors.

Novoselov and collaborators first isolated graphene through the mechanical exfoliation of graphite [4]. This material is defined as a one-carbon layer with atoms arranged in a honeycomb lattice [4, 7, 8]. A 2D material in which the electrons are free to move in two directions and confined in one, graphene is the thinnest material known (a thickness of one atom) and one of the strongest ever measured.

Graphene also exhibits interesting features that offer possibilities for various applications, including as a charge carrier due to its high electron mobility and zero effective mass [5, 6]. Further promising applications in biomedicine include drug delivery [9], cancer therapy [10], biological imaging [11], and disease diagnosis [11, 12].

One of the most important derivatives of graphene is graphene oxide (GO), whose molecular structure is based on graphene with functional groups such as carboxyl and hydroxyl [13]. The functionalization of graphene involves surface modifications through covalent and non-covalent interactions, which are crucial for tailoring its properties to specific applications [5, 12, 14–16]. This process enhances graphene's biocompatibility and optical properties [15]. The ability of graphene to interact with the biomolecule through $\pi$-$\pi$ stacking and its high drug-loading capacity make it a promising material for drug delivery systems, allowing for controlled release and targeted delivery of therapeutic agents [17].

Finally, graphene-based nanomaterials exhibit stimuli-responsive behaviour, allowing for intelligent drug release in response to various triggers, such as pH changes, which is particularly useful in targeting the acidic environment of tumour cells [18].

Moreover, the integration of CNs with other compounds and biomolecules has led to the development of advanced biosensors with high sensitivity and specificity, which are essential for early disease detection and personalized treatment management [19–24]. These advancements highlight the transformative potential of surface-functionalized CNs in the biomedical field, offering new opportunities for targeted drug delivery and efficient biosensing systems [25]. Overall, the strategic functionalization of CNs is a critical step in enhancing their performance and expanding their applications in biological nanotechnologies. The adsorption method of functionalization on the CN surface can be a non-destructive and reversible method to improve its properties. It happens as a spontaneous binding, generally caused by the $\pi$ orbitals of the carbon aromatic rings, and allows the specific interaction of CN with biological molecules [13].

In the literature, several research publications on various types of CNs related to bioapplications can be found. However, graphene leads the number of publications for biosensing and drug delivery applications. Figure 8.1 illustrates the exponential increase in the number of publications associated with the subject's 'biosensor' and 'COVID-19', 'graphene' or 'graphene oxide' that has taken place during the last decade (2014–2024), indicating the relevance of the study of biomolecules interacting with each nanomaterial.

The COVID-19 pandemic was used to understand its impact on biosensing research, which is justified due to the emergency for fast, reliable, and secure diagnostic methods. In 2024, COVID-19 accounted for approximately 6, 7% and 11% of publications related to biomolecule adsorption and biosensors, respectively, to both graphene and GO nanomaterials.

The following sections will discuss some of the key issues related to graphene and GO, including the adsorption of biomolecules and the application of graphene nanosensors.

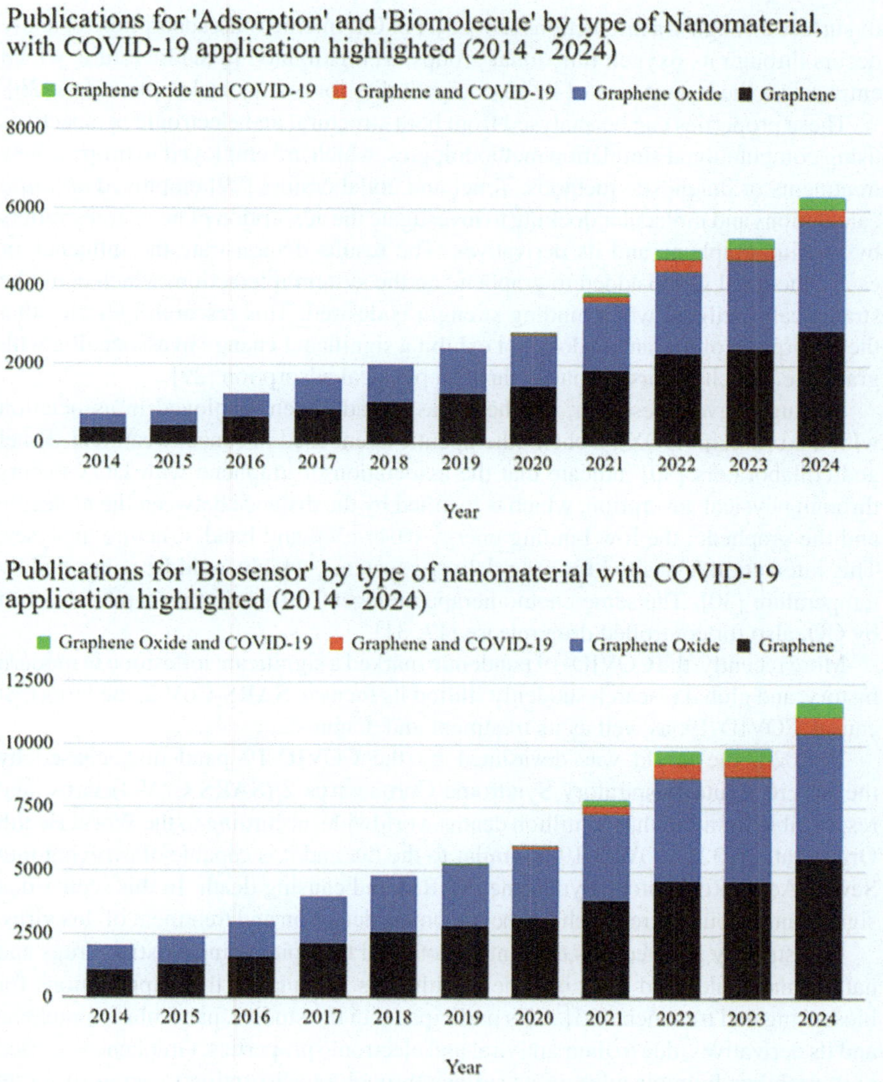


**Fig. 8.1** Articles published since 2014 regarding the application of the nanomaterials for the descriptors **a** 'biomolecule adsorption' and **b** 'biosensor' associated with the descriptors 'graphene' (black), 'graphene oxide' (blue), 'graphene' and 'COVID-19' (red), and 'graphene oxide' and 'COVID-19' (green) (www.sciencedirect.com, last accessed 02nd June 2025)

## 2 Biomolecule Adsorption on Graphene and Derivatives

Several experimental studies have been conducted to develop graphene for use in biomedical applications. These studies have revealed graphene to be efficient for both drug delivery and biosensing purposes due to its ability to anchor molecules on its surface through $\pi$-$\pi$ interactions [26–28]. For GO interactions, adsorption typically occurs through its oxygen functional groups, as highlighted in the literature, which emphasizes the importance of its biocompatibility for biomedical applications [26].

These properties can be analysed from both structural and electronic perspectives using computational simulation methodologies, which are employed to propose new treatments or diagnostic methods. Tonel and collaborators [29] employed *ab initio* calculations and molecular docking to investigate the adsorption of neurotransmitters by pristine graphene and its derivatives. The results demonstrate the influence of each functional group added to graphene on the system interaction, which could be strategically utilized when binding strength is desired. This research indicates that the adsorption of glutamate does not exhibit a significant change in association with graphene, which occurs primarily through physical adsorption [29].

In drug delivery research, graphene has already been employed in association with doxorubicin (DOX), a chemotherapeutic agent used in cancer treatment. Tonel and collaborators [30] indicate that the association of graphene with DOX occurs through physical adsorption, which is justified by the distance between the molecule and the graphene, the low binding energy (0.49 eV), and band structure analyses. This interaction is desired for drug delivery systems, which could be controlled by temperature [30]. The same chemotherapeutic agent is found in research adsorbed by GO, also for controlled drug release [13, 31].

More recently, the COVID-19 pandemic marked a significant milestone in modern history, and global research suddenly shifted its focus to SARS-CoV-2, the virus that causes COVID-19, as well as its treatment and diagnosis.

In 2020, the world was devastated by the COVID-19 pandemic, caused by the Severe Acute Respiratory Syndrome Coronavirus 2 (SARS-CoV-2) virus, and responsible for more than 7 million deaths worldwide, according to the World Health Organization [32]. COVID-19 is similar to the flu, and it is capable of evolving into Severe Acute Respiratory Syndrome (SARS) and causing death. In this scenario, a significant amount of research has begun on the detection and treatment of this virus.

One strategy adopted was to identify potential therapies using existing drugs and natural molecules and associate them with CNs to enhance their applications for biosensing and treatment [33]. CNs participated in the studies, particularly graphene and its derivatives, due to their antiviral and electronic properties. Graphene was used to immobilize biomolecules on its surface through physisorption to act as different biosensors [34], and GO is a graphene derivative that is highlighted in research for its ease of functionalization.

The adsorption of drugs in GO was studied by Dos Santos and co-workers [35], who investigated through *ab initio* simulations, molecular docking, and *in vitro* assays the interaction between GO and heparin as a system in the face of the spike protein, which is present in the SARS-CoV-2 surface, and is key for cell internalization. Their results indicate that the system formed by GO and heparin enhances the binding affinity with the protein compared to the isolated molecules.

Similarly, the GO was also proposed in association with natural compounds, such as the flavonoids orientin and apigenin, which are found in the extract of *açaí*, a fruit native to the Amazon biodiversity [36]. Schultz and collaborators [37] investigate the interaction between GO and these flavonoids, and the results indicate a stable interaction when the molecules are disposed of in parallel, with binding energies indicating physical adsorption resulting from a weak $\pi$-$\pi$ stacking interaction. In the same study, GO was proposed as a ligand to the spike protein of SARS-CoV-2, as this protein is involved in the process of cell internalization. The molecular docking results indicate that it is a potential inhibitor of the spike protein for both the omicron and delta variants of the virus.

To advance the understanding of graphene-based nanomaterials interacting with biomolecules, we present a systematic investigation that combines density functional theory (DFT) and molecular docking. It is essential to highlight the complementary roles of these two computational methodologies. DFT simulations enable an in-depth analysis of the electronic structure and interaction mechanisms at the atomic scale, particularly for systems modelled within a periodic unit cell. This framework is suitable for representing extended nanomaterials, such as graphene. Molecular docking is valuable for predicting binding orientations and affinities in complex biological environments; however, it is limited by the finite size of the receptor model and cannot accurately represent an infinite graphene sheet due to computational constraints.

In this context, we performed a detailed study of the adsorption behaviour of catechin (CAT) on graphene and GO surfaces. The DFT calculations were performed using the SIESTA code [38], which self-consistently solves the spin-polarized Kohn–Sham equations [39] employing atomic numerical basis sets. A double-zeta basis set with polarization functions (DZP) was adopted to describe the pseudo-orbitals in all simulations. The real-space electronic density was modelled using a mesh cut-off of 200 Ry. Exchange and correlation effects were treated within the Local Density Approximation (LDA), employing the Perdew and Zunger parametrization [38]. Geometry optimizations were performed until all residual atomic forces were below 0.05 eV/Å. A cubic supercell with dimensions of $40 \times 40 \times 40$ Å$^3$ was used to eliminate interactions between periodic images. The binding energies were computed with the counterpoise correction method for the Basis Set Superposition Error (BSSE) [39], as described in Eq. (8.1).

$$E = -\left[E(a+b) - E\left(a_{ghost} + b\right) - E\left(a + b_{ghost}\right)\right] \tag{8.1}$$

System A corresponds to the adsorbate molecule (CAT), whereas system B represents the graphene or GO surface. All simulations were performed using the respective atomic basis sets, in which the basis set of system A {B} is centered on the atomic coordinates of system B {A}. The term ghost denotes basis functions located at the positions associated with graphene or GO without the inclusion of corresponding atomic potentials—i.e., no physical atoms are present at these sites. Positive binding energy values indicate attractive interactions between the components. This computational strategy has been previously employed by the research group in similar studies [35, 37, 40].

The selected biomolecule is CAT, a bioactive compound present in the chemical matrix of Euterpe oleracea Mart., commonly known as *açaí*. Davidson and colleagues

[41] have demonstrated that catechin exhibits one of the highest binding affinities to the NLRP3 (NOD-, LRR- and pyrin domain-containing protein 3) inflammasome, which is associated with various inflammatory diseases, including COVID-19. This study aims to propose and systematically analyse the interaction between CAT and graphene or GO by combining DFT simulations and molecular docking, to elucidate the adsorption mechanisms and the underlying electronic properties. Three distinct interaction configurations were evaluated, and the corresponding results are presented in Table 8.1, including the binding energy values, the difference between the Highest Occupied Molecular Orbital (HOMO) and the Lowest Unoccupied Molecular Orbital (LUMO), and the electronic charge transfer.

The interactions between graphene and the biomolecule were modeled using three distinct configurations: a parallel alignment (configuration A), a T-shaped orientation (configuration B), and a parallel-displaced arrangement (configuration C), as illustrated in Fig. 8.2a–c.

The DFT calculations reveal that the interaction between the carbon-based nanomaterials and the catechin molecule follows a pattern consistent with other results presented in this chapter. The most favorable interaction for the Gr_CAT_C system, characterized by a $\pi$–$\pi$ stacking energy of 0.62 eV, occurs when the aromatic rings of both structures are aligned in a parallel-displaced configuration. The results exhibit a consistent trend across the evaluated configurations, with binding energies following the order: configuration C > A > B (Table 8.1).

The molecular systems representing the interaction between GO and CAT were constructed by positioning distinct functional groups of the biomolecule in proximity to the oxygen-containing groups of the GO surface. The modelled configurations are shown in Fig. 8.2d–f. Notably, in these representations, the GO surface exhibits a curved morphology, a feature attributed to the presence of functional groups that induce structural distortion. In contrast, the pristine graphene surface remains planar [16].

**Table 8.1** Minimum interaction distances, binding energies, HOMO–LUMO difference, and electronic charge transfer values were calculated for the most stable configurations of catechin (CAT) interacting with graphene (Gr) and graphene oxide (GO). Positive binding energy values indicate exothermic interactions, and positive charge transfer values correspond to electron transfer from the carbon nanomaterial to the biomolecule

| Molecule or system | Minimum interaction distance (Å) | Binding energy (eV) | HOMO–LUMO difference (eV) | Electronic charge transfer ($e^-$) |
|---|---|---|---|---|
| Gr_CAT_A | 2.42 | 0.58 | 1.26 | +0.208 |
| Gr_CAT_B | 2.32 | 0.10 | 1.32 | +0.050 |
| **Gr_CAT_C** | **2.32** | **0.62** | **1.24** | **+0.214** |
| **GO_CAT_A** | **1.04** | **1.15** | **0.64** | **+0.03** |
| GO_CAT_B | 1.70 | 0.86 | 0.59 | +0.176 |
| GO_CAT_C | 1.53 | 0.64 | 0.40 | +0.106 |

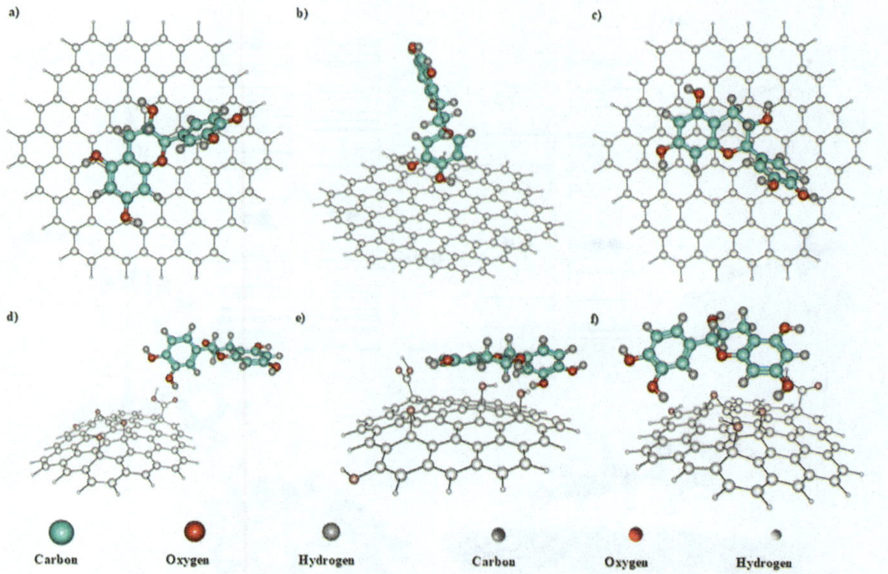


**Fig. 8.2** Configurations of CAT interacting with Gr (**a**, **b**, and **c**) and GO (**d**, **e**, and **f**). The systems are named Gr_CAT_A (**a**), Gr_CAT_B (**b**), Gr_CAT_C (**c**), GO_CAT_A (**d**), GO_CAT_B (**e**), and GO_CAT_C (**f**)

The results for the most stable GO_CAT_A system, exhibiting a binding energy of 1.04 eV, demonstrate the presence of hydrogen bonding interactions. Additionally, this system shows proximity between the biomolecule and the GO surface, with initial contact distances of approximately 2.50 Å. After optimization, the shortest binding distance decreases to 1.04 Å, accompanied by a charge transfer of 0.03 $e^{-}$.

Overall, the interactions characterized in this study are classified as physisorption due to the relatively low binding energies involved. Notably, the HOMO–LUMO difference for the systems interacting with graphene (1.24 eV) and GO (0.64 eV) remains comparable to those of pristine graphene (1.33 eV) and GO (0.59 eV), respectively, indicating that the electronic properties of the nanomaterials are essentially preserved. To further elucidate these interactions, charge density plots of the HOMO and LUMO orbitals, along with the total charge distribution, were generated (Fig. 8.3).

The most favourable DFT result for the Gr_CAT_C system corresponds to the lowest binding energy of 0.62 eV. The electron density distributions of the HOMO and LUMO orbitals, shown in Fig. 8.3, closely resemble those of pristine graphene. This similarity suggests that the interaction is predominantly governed by physisorption, consistent with the calculated binding energies. This conclusion is further supported by the total charge density plot, which clearly shows charge localization on both the graphene and catechin molecules, without significant overlap or electronic hybridization. A similar behaviour was observed for the GO_CAT_A system.

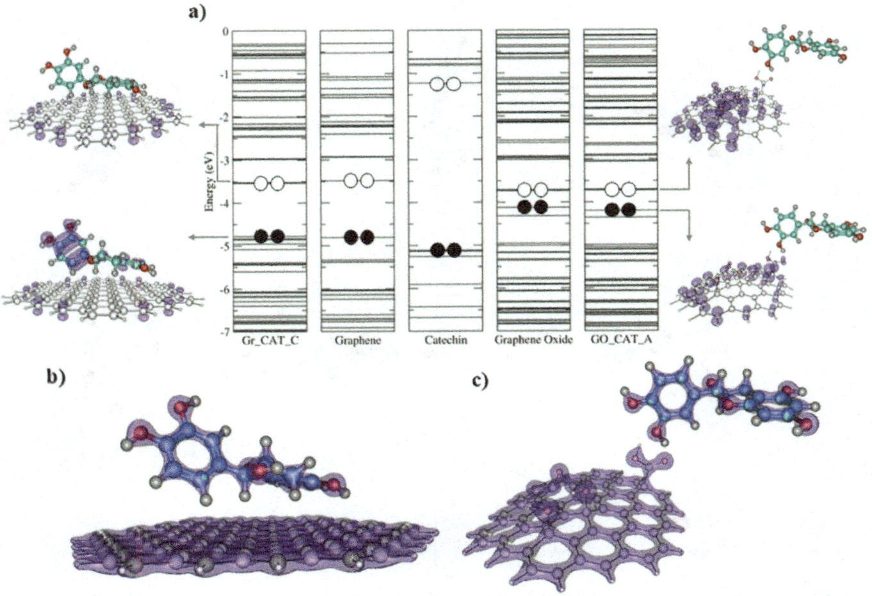


**Fig. 8.3** **a** Electronic energy levels of isolated CAT, graphene (Gr), and graphene oxide (GO), along with their most stable adsorption configurations (Gr_CAT_C and GO_CAT_A); **b** charge distribution in the HOMO and LUMO orbitals; **c** total charge density maps for the respective configurations

These results are consistent with those reported in one of our previous studies, which investigated the interactions between GO and biomolecules. The study demonstrated that GO interacts with apigenin and orientin primarily through π-π stacking. Molecular docking simulations further suggested that both molecules exhibit potential inhibitory activity against the spike proteins of the SARS-CoV-2 Omicron and Delta variants [37].

We analyse the interaction between the CAT and the systems formed by the biomolecule, with graphene and GO as ligands to the pyrin domain (PYD) of NLRP3. The chosen systems were Gr_CAT_C and GO_CAT_A, which were related to the most stable systems for each configuration between the molecule and nanostructure. The results were analysed in comparison to those obtained from simulating MCC950 as a ligand since it is a known NLRP3 inhibitor. The MCC950 is a small synthetic molecule capable of blocking NLRP3 activation at nanomolar concentrations [42]. We attempted to understand whether Gr would improve or worsen the inhibition. The simulations were performed using the AutoDock Vina software [43] through the graphical tool AMDock [44]. The chosen three-dimensional structure of the pyrin (PYD) domain of NLRP3 was obtained from the Protein Data Bank, identified as 2NAQ [45]. The simulation was performed at a basic pH of 7.4, as specified in the software, and the summarized results are presented in Table 8.2. AutoDock Vina is one of the most widely used programs for molecular docking, which calculates a score for several proposed ligand poses with the molecule. One of the basic metrics to analyze is the Root Mean Square Deviation (RMSD), which was calculated using PyMol software between the best docking poses for each ligand and the results of all performed simulations, which presented values lower than 2 Å. This analysis is essential for consistency analysis within the methodology [46].

The 2D maps of interaction were visualized using Discovery Studio software, which enabled an understanding of the resulting interactions between the protein and ligand. The MCC950 exhibited a binding affinity of −7.8 kcal/mol and a ligand efficiency of −0.28, with only one hydrogen bond (HB) between the protein and ligand, which aligns with our goal as an inhibitor. The results for the biomolecule, graphene, and GO, isolated, are presented in Fig. 8.4. The molecular docking simulations revealed that the proposed ligands yielded similar results, aligning with the findings for MCC95. The ligand efficiency of the biomolecule as a ligand is −0.33 with only one hydrogen bond. At the same time, catechin had an overbinding affinity of −6.9 kcal/mol.

It is essential to highlight the 2D map for MCC950 in Fig. 8.5b, as MCC950 is a known inhibitor of NLRP3. The 2D maps indicate the interactions between the target (PYD NLRP3) and the ligand, detailing which amino acid residues are involved in the interaction. In our results, the amino acid residues present in the interactions are Phenylalanine (PHE), Isoleucine (ILE), Leucine (LEU), Proline (PRO), Aspartic Acid (ASP), Glutamine (GLN), Lysine (LYS), Glutamic Acid (GLU), and Serine (SER). The interaction is presented by its name, followed by the protein chain, where the amino acid is located, and its position in the protein chain sequence. The isolated catechin exhibits a π-π stacked interaction with PHE (A: 59); therefore, catechin and NLRP3 interact through phenylalanine located at position 59 in the protein chain A. About the CNs, graphene only exhibits an alkyl interaction with ILE (A: 37), whereas GO displays a π-π stacking interaction with PHE (A: 59), as observed in the referenced MCC950. On the other hand, the simulation of systems formed by graphene and the biomolecule revealed that the carbon nanomaterial is responsible

**Table 8.2** Binding affinity, ligand efficiency, and number of hydrogen bonds and π–π stacked interactions for the molecular docking of catechin, MCC950, and the systems Gr_CAT_C and GO_CAT_A with the PYD NLRP3

| Ligand | Binding affinity (kcal/mol) | Ligand efficiency | Number of hydrogen bonds | Number of π–π stacked interactions |
|---|---|---|---|---|
| CAT | −6.9 | −0.33 | 1 | 0 |
| Gr | −13.6 | −0.14 | 0 | 0 |
| GO | −10 | −0.16 | 1 | 1 |
| Gr_CAT_C | −12.7 | −0.11 | 0 | 8 |
| GO_CAT_A | −8.8 | −0.11 | 2 | 0 |
| MCC950 | −7.8 | −0.28 | 1 | 1 |

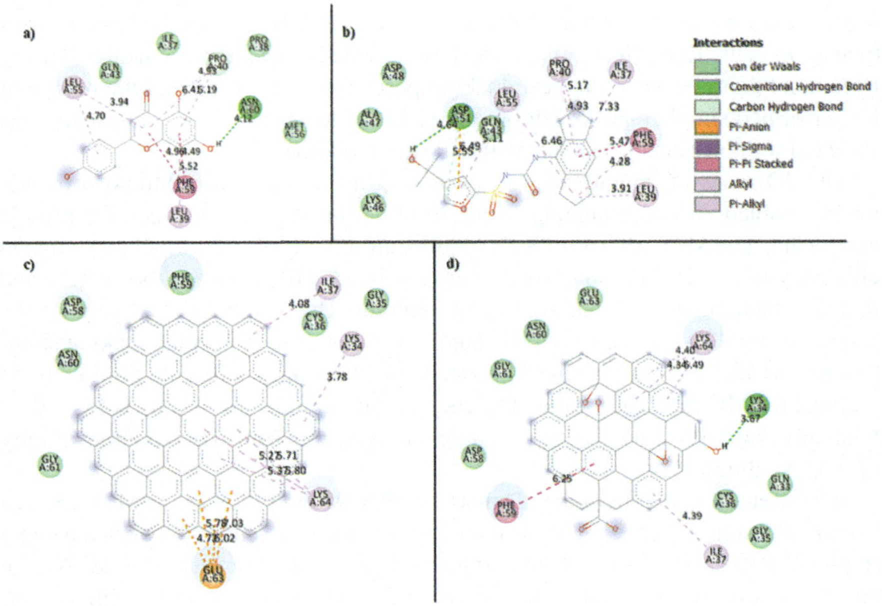


**Fig. 8.4** 2D maps of interaction for the systems: **a** CAT, **b** MCC950, **c** Graphene, and **d** GO with types of interaction with the PYD NLRP3

for the worst results in NLRP3 inhibition, as shown in the 2D maps presented in Fig. 8.5.

The graphene system interacts with the protein through several π-π stacked interactions with PHE (A: 59), while MCC950 exhibits only one with the same amino acid residue; however, the hydrogen bond was broken, which could explain the lower values for both binding affinity and ligand efficiency. The GO systems did not exhibit any common interactions with NLRP3. The GO system has values of binding affinity and ligand efficiency closer to the ones for MCC950, and the 3D map illustrates two

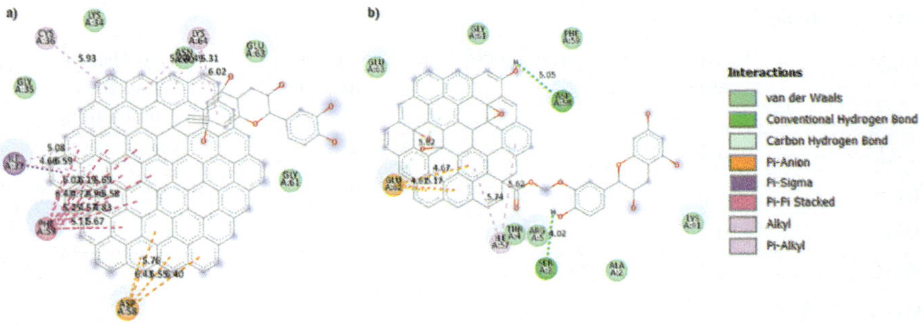


**Fig. 8.5** 2D maps of interaction for the systems: **a** Gr_CAT_C and **b** GO_CAT_A with types of interaction highlighted with the PYD NLRP3

different hydrogen bonds with ASP (A: 58) and SER (A: 3), but as long-distance interactions with values of 5.05 Å and 4.02 Å, respectively.

In this work, since NLRP3 inhibition was the goal, the biomolecule-graphene systems were not suitable ligands. However, the results still help us understand the chemistry of the interactions. We observed that the best binding affinity was achieved for the isolated biomolecule. In addition to graphene and GO, which presented several $\pi$-$\pi$ stacked interactions, the most effective ligands were those with higher ligand efficiency, similar to MCC950. We highlight that the best results among the carbon nanostructures refer to isolated GO, which is capable of interacting with NLRP3 through hydrogen bonding and shares a $\pi$-$\pi$ stacking interaction with PHE (A:59) similar to that of MCC950. At the same time, the proposed biomolecule show similar aminoacid residues are presented in the interactions map, such as GLN (A:43), LEU (A: 39, 55), ILE (A: 37), PHE (A:59) and PRO (A: 40), which indicate that the local of interaction is very similar for MCC950 and catechin.

In general, the DFT simulations were crucial for understanding the stability of the proposed ligands, where we observed that graphene and GO interact with the biomolecule primarily through $\pi$-$\pi$ stacking, resulting in the most stable configurations. In the same direction, molecular docking is essential for determining whether our systems are good or bad inhibitors for a biological application, and these results provide a basis for avoiding high-cost, practical research.

## 3 Overview of Graphene Use in Biosensors

Graphene, a single layer of carbon atoms arranged in a two-dimensional honeycomb lattice, has emerged as a revolutionary material in biosensor development due to its exceptional physical, chemical, and electronic properties [47]. These characteristics make graphene highly advantageous for the sensitive and selective detection of biomarkers in various clinical applications. These are a high surface area, providing a sizeable active surface for biomolecule immobilization, enhancing sensor sensitivity; excellent conductivity, facilitating efficient electron transfer, crucial for electrochemical biosensors; high mechanical strength and flexibility, enabling durable and adaptable sensor designs for various platforms; and biocompatibility, supporting the integration of biological recognition elements without compromising their activity. This unique set of characteristics has been utilized in recent years to enhance the performance of various biosensors [47, 48]. In Electrochemical Biosensors, graphene enhances the detection of low-concentration biomarkers, such as glucose, DNA, and proteins, by improving signal transduction and reducing detection limits [49]. In Optical Biosensors, functionalized graphene with fluorescent labels, nanoparticles, or biomolecules is used in fluorescence, surface plasmon resonance (SPR) [50], and Raman-based sensors to improve optical signal stability and specificity [51]. In Field-Effect Transistor (FET)-based Biosensors, graphene's conductive properties enable real-time, label-free detection of biomolecular interactions, making it ideal for detecting DNA, RNA, and proteins [46, 52]. Integrating one or more types of sensors,

Point-of-Care devices are now utilizing graphene-based materials in portable and wearable biosensors for real-time monitoring of health parameters [53]. Graphene-based biosensors offer enhanced sensitivity due to graphene's unique structure, which enables ultra-low limits of detection, even at the single-molecule level [54]. They also provide versatility because graphene can be easily functionalized with biomolecules, such as enzymes, antibodies, and nucleic acids, to detect diverse targets relatively cheaply, making it suitable for scalable and affordable diagnostic solutions.

In summary, graphene's remarkable properties make it a transformative material for biosensor applications, paving the way for more accurate, reliable, and accessible diagnostic tools in healthcare. However, challenges such as standardizing functionalization techniques, achieving production scalability, and integrating them into commercial devices need to be addressed [55]. Ongoing research is focused on overcoming these limitations and unlocking graphene's potential in biosensing technologies.

## *3.1 Electrolyte-Gate Graphene Field-Effect Transistor Biosensors*

In recent years, the need for highly sensitive, fast, label-free, portable, and low-cost biosensing platforms has motivated extensive research and different transduction strategies [6, 48, 54–56]. Among them, the most promising type of FET-based graphene biosensor is the electrolyte-gate FET (EGFET). The device features a standard FET, where the graphene channel and the recessed gate electrode are separated by a liquid electrolyte that covers the graphene surface and serves as the gate dielectric. Graphene is polarized through interfacial capacitances forming at the solid–liquid interfaces when a voltage is applied to the gate ($V_{GS}$). In an undoped graphene channel, the transfer curve forms a "V" with a Dirac point and an associated Dirac voltage ($V_{Dirac}$) at the point of maximal resistance (the "V" apex) at $V_{GS} = 0$. When a positive $V_{GS}$ is applied, negative charges (electrons) are induced in the graphene. Therefore, the Fermi energy ($E_F$), which accounts for the electron filling of the graphene density of states (DOS), shifts above the Dirac point, resulting in n-doping of the graphene. Conversely, positive charges are induced in the graphene when a negative $V_{GS}$ is applied – p-type doping and $E_F$ will shift below the Dirac point. Similarly, biomolecules with a negative charge (e.g., DNA) that approach the graphene channel, directly contacting it or attaching to a specific molecular probe previously immobilized on the graphene (functionalization), will p-dope graphene, while positively charged molecules n-dope graphene. This proximity effect is known as local gating, the core transduction mechanism in graphene field-effect transistor biosensing.

To describe the EGFET transducing behaviour quantitatively, we start by noting that the gate voltage couples electrostatically to the channel through a capacitance consisting of a series association of the electrical double-layer and the graphene quantum capacitance:

$$V_{GS} - V_{Dirac} = ne\frac{\lambda_D}{\varepsilon} + \frac{hv_F}{2e\sqrt{\pi}}\sqrt{n} \quad (8.2)$$

where $e$ is the elementary charge, $n$ is the graphene carrier concentration, $h$ is Planck's constant, $v_F$, is the Fermi velocity, $\varepsilon = \varepsilon_R\varepsilon_0$, the electrical permittivity of the electrolyte and $\lambda_D$ is the Debye length, which gives the electrical double-layer thickness of the interfacial capacitance. The first term in the Eq. 8.2 is the interfacial capacitance, and the second is the graphene quantum capacitance. Whenever charged or polar molecules attach to the graphene surface, local gating occurs, changing $V_{Dirac}$ to higher or lower values, $V^*_{Dirac}$ depending on the sign of the charge. The measure of $\Delta V = V_{Dirac} - V^*_{Dirac}$ is the biosensor's signal (see Fig. 8.6).

Graphene EGFETs require a biorecognition element, like any other biosensor, to make them specific for the target molecule. This is accomplished by immobilizing a single DNA strand, an antibody, an aptamer, or any other active biomolecule against a biomarker on the graphene surface, a complicated process due to graphene's inherent chemical inertness. Functionalization typically requires a linker that binds to graphene at one end and a molecular probe at the opposite end. Such hetero-bi-linkers can bind covalently or through van der Waals stacking to graphene.

Non-covalent functionalization [57] has been proposed as a superior alternative for constructing graphene EGFET-based biosensors, as it is argued that covalent chemical functionalization compromises the aromatic lattice of the graphene network, thereby affecting its electrical performance. However, some studies indicate that

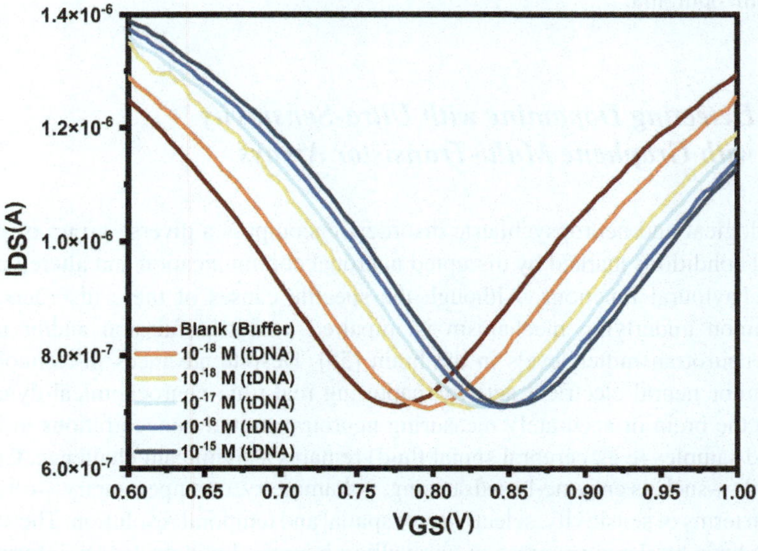


**Fig. 8.6** A set of graphene transfer curves where the movement of the Dirac point (see text) is evident as the target DNA (tDNA) concentration increases over four orders of magnitude

covalent functionalization can be an alternative, provided it is performed appropriately, offering improved stability, reliability, and specificity [58]. Standard non-covalent binders are pyrene, acridine, and fluorenyl methyl derivatives. The most commonly used is 1-pyrene-butyric acid succinimidyl ester (PBSE), but fluorenyl methyl succinimidyl carbonate (FSC) and Acridine Orange (AO) succinimidyl ester, as well as others, are sometimes used. PBSE provides higher surface coverage and fast immobilization kinetics. A recent study [21, 57] shows that single-stranded DNA probes (ssDNA) immobilized on graphene via PBSE linker provide higher DNA probe surface density (up to $1.31 \times 10^{13}$ molecules per $cm^{-2}$) than using FSC or AO ($1.03 \times 10^{13}$ and $1.18 \times 10^{13}$ molecules $cm^{-2}$, respectively) and hybridization efficiencies of at least 70% with 20 to 30% of ssDNA directly adsorbed non-specifically on the graphene surface. This result also shows that, after probe immobilization, it is crucial to passivate the remaining non-reacted linker molecules to avoid false positives. This can be achieved using a small molecule, such as ethanolamine, with an N-Hydroxysuccinimide (NHS) blocking reaction that exhibits a binding efficiency of 80%.

Once the functionalization is accomplished, the solution under study is brought into contact with the graphene EGFET channel. This can be achieved by dropping a few microliters of solution onto the multi-transistor biosensing chip or by coupling a microfluidic device to the graphene chip. The chip is designed in such a way that its entire surface is passivated against liquids, ions, and molecules, leaving only the loci of the graphene channels and the gate contact surface exposed to the solution.

The following describes an example of a real-world application, specifically a highly sensitive and reliable sensing platform for the detection of dopamine, a key neurotransmitter implicated in diverse brain disorders, such as Parkinson's Disease and schizophrenia.

### 3.2 Detecting Dopamine with Ultra-Sensitivity with Graphene Multi-Transistor Arrays

Neurological and neuropsychiatric disorders encompass a diverse group of pathological conditions marked by disrupted neuronal communication and altered circuit and behavioural functions. Although the specific causes of these disorders vary, a common underlying mechanism is impaired neurotransmission and/or unbalanced neurotransmitter levels in the brain [59]. Despite advances in technologies to monitor neural electrical activity, capturing real-time neurochemical dynamics within the brain or accurately measuring neurotransmitter concentrations in brain-derived samples (e.g., cerebral spinal fluid) remains a significant challenge. Current methods—such as enzyme-based sensing, voltammetry, or amperometry—often fall short in terms of sensitivity, selectivity, or spatial and temporal resolution. These limitations have hindered progress in unravelling how the brain processes information and develops pathology. Emerging technologies that can accurately track the brain's molecular signals could accelerate scientific discovery, power the development of next-generation brain-machine interfaces, and drive progress in neuromorphic computing.

However, detecting neurotransmitters is particularly demanding. These are tiny molecules that appear only transiently in the brain, at low concentrations, and within a milieu crowded with structurally similar compounds. Any biosensor designed to measure them must therefore achieve high levels of sensitivity and selectivity. A range of experimental approaches currently exists, including cyclic voltammetry and amperometry for electroactive neurotransmitters, positron emission tomography, single-photon emission computed tomography, optical fiber sensors, fluorescent indicators, surface-enhanced Raman spectroscopy, and high-performance liquid chromatography of microdialyzed cerebrospinal fluid [60]. Nevertheless, none of these techniques delivers the combination of selectivity, sensitivity, and resolution required to capture fast neurochemical events at physiologically relevant scales. While genetically encoded fluorescent indicators offer some promise, their reliance on viral vectors or protein dyes introduces neuroethical concerns and limits their translational potential [61]. Furthermore, most of these methods require bulky, expensive instrumentation, cannot be integrated into miniaturized devices, and often involve materials incompatible with chronic implantation in the body.

To address these limitations, we developed a highly sensitive and reliable sensing platform based on EGFETs for the detection of dopamine [62], a key neurotransmitter involved in numerous neurological conditions. This platform is built upon EGFET multitransistor arrays (gMTAs) functionalized with a dopamine-specific DNA aptamer through a non-covalent linker (PBSE). Using scalable, wafer-level fabrication, we integrated multiple graphene transistors with a co-planar gate onto compact $4.5 \times 4.5$ mm$^2$ chips. The array configuration allows for multiple, independent readings from the same sample, thereby improving the robustness of the data and reducing measurement variability. This design enables ultra-sensitive and reproducible detection of dopamine from small-volume samples (2–20 $\mu$L)—even in high-ionic-strength biological media—at concentrations as low as one attomolar ($10^{-18}$ M). The device operates across a broad dynamic range spanning ten orders of magnitude (up to 100 $\mu$M) and demonstrates a peak sensitivity of 22 mV/decade in artificial cerebrospinal fluid (aCSF) (Fig. 8.7).

Notably, the platform maintained its attomolar sensitivity in highly complex biological matrices, including brain homogenate samples – arguably one of the most challenging biological matrices due to the presence of many interfering and competing molecules (ions, DNA, proteins, hormones, etc.) in the solution. In dopamine-depleted brain homogenates, spiked with known concentrations of dopamine, the sensor array maintained an attomolar limit of detection. It showed a wide dynamic range up to ~ 100 $\mu$M, alongside peak sensitivity values consistent with those observed in aCSF. Importantly, control experiments confirmed that the sensor's response was specific to dopamine and not affected by other biomolecules or incubation time. This marks a significant advance over prior ion-sensitive sensors, which typically suffer from sensitivity and selectivity reduction in such environments.

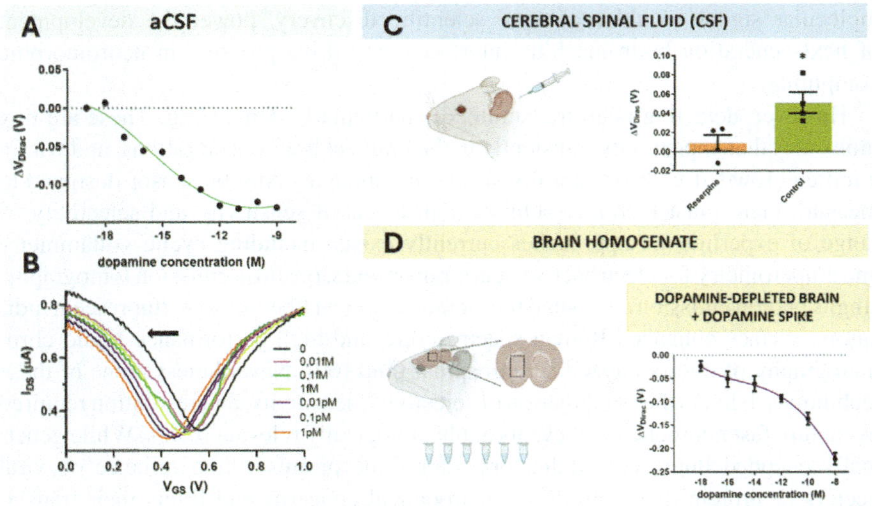


**Fig. 8.7** **A** Dopamine detection in vitro with gMTAs. Calibration curve for dopamine detection in undiluted artificial cerebrospinal fluid (1 × aCSF) (data is mean ± sem, with fourth-order polynomial line fit). **B** Representative transfer curve shifts as a function of increasing dopamine concentrations for the linear detection range of one EG-gFET in 1 × aCSF, with VDirac moving towards less positive gate voltages (VGS) (black arrow indicates the direction of VDirac shifts). Dopamine detection with gMTAs in biological samples from a mouse model of Parkinson's Disease (PD). **C** Response of gMTAs to biological cerebral spinal fluid (CSF) samples obtained from PD (reserpine) and control mice (data is mean ± sem; paired t-test, $p < 0.05$). **D** gMTAs response to dopamine-depleted brain homogenate samples obtained from PD mice spiked with increasing dopamine concentrations (data is mean ± sem, with 4th order polynomial line fit). (Adapted from [62] under Creative Commons Attribution (CC BY 4.0) license)

The sensor also allowed the identification of Parkinson's Disease in a pharmacological mouse model of this neurological disease, by detecting a significant reduction of dopamine levels in unprocessed 2 μL cerebrospinal fluid (CSF) samples, when compared with healthy mice (Fig. 8.7C, D). These results underscore the sensor's precision and reliability even in real-world biological scenarios, which is crucial for diagnostic applications and disease monitoring.

Together, these findings establish our gMTA-based biosensor as a promising tool for neuroscience research, pharmaceutical development, and potentially even clinical diagnostics, offering unmatched sensitivity and selectivity in neurotransmitter detection across a spectrum of biological environments.

## 4 Final Remarks

This chapter of the book examines the significant role of carbon nanomaterials, specifically graphene and GO, in biomedical applications, with a focus on their use in controlled release systems and biosensor development (Fig. 8.8). It highlights how graphene's unique physicochemical properties, including its two-dimensional structure and high electron mobility, make it a key material for various nanotechnology applications. The details of how both pristine and functionalized forms of graphene interact with biomolecules are presented, highlighting the importance of chemical functionalization in enhancing the selectivity, stability, and efficiency of adsorption processes. Through computational simulations like DFT and molecular docking, the research investigates the adsorption of molecules, such as the adsorption of natural compounds, such as catechin, a flavonoid known as an anti-inflammatory biomolecule.

Graphene has become a transformative platform for biosensor development. Its high surface area and excellent conductivity enhance sensitivity and facilitate efficient electron transfer, making it suitable for different types of sensors, including electrochemical, optical, and FET-based biosensors. Notably, EGFETs utilize graphene's capability to modulate charge carriers to detect biomolecular interactions in real time, without the need for labelling. Functionalizing graphene with biorecognition elements, such as DNA probes or antibodies, enables high specificity and selectivity. However, achieving stable and efficient functionalization remains a challenge, necessitating the careful selection of linkers to preserve graphene's electronic properties. Among non-covalent linkers, PBSE has demonstrated superior probe density and hybridization efficiency.

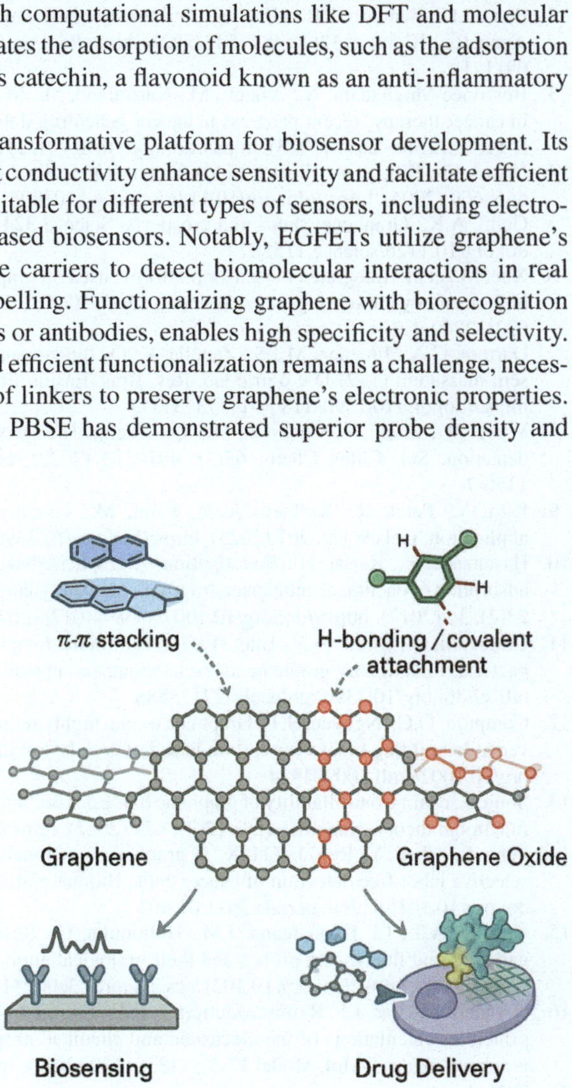


**Fig. 8.8** Graphical summary of the main mechanisms underlying molecular interactions between graphene-based nanomaterials and biologically relevant molecules

In summary, this chapter emphasizes the potential of graphene-based materials as versatile platforms for advancing diagnostics, monitoring, and therapeutic strategies. It highlights the importance of understanding molecular interactions for the development of innovative nanotechnology approaches in healthcare.

## References


1. Mahor, A., et al.: Carbon-based nanomaterials for delivery of biologicals and therapeutics: a cutting-edge technology. C **7**(1), 19 (2021). https://doi.org/10.3390/c7010019
2. Ban, D.K., Bandaru, P.R.: Graphene and two-dimensional materials for biomolecule sensing. Annu. Rev. Biophys. **52**(1), 487–507 (2023). https://doi.org/10.1146/annurev-biophys-111622-091121
3. Borzooee Moghadam, N., Avatefi, M., Karimiand, M., Mahmoudifard, M.: Graphene family in cancer therapy: recent progress in cancer gene/drug delivery applications. J. Mater. Chem. B **11**(12), 2568–2613 (2023). https://doi.org/10.1039/D2TB01858F
4. Novoselov, K.S., et al.: Electric field effect in atomically thin carbon films. Science **306**(5696), 666–669 (2004). https://doi.org/10.1126/science.1102896
5. Geim, A.K.: Graphene: status and prospects. Science **324**(5934), 1530–1534 (2009). https://doi.org/10.1126/science.1158877
6. Xue, M., et al.: Integrated biosensor platform based on graphene transistor arrays for real-time high-accuracy ion sensing. Nat. Commun. **13**(1), 5064 (2022). https://doi.org/10.1038/s41467-022-32749-4
7. Dartora, C.A., Jimenez, M.J.S., Zanella, F.: Os fundamentos da física dos férmions de Dirac sem massa em (1+2)-D e o grafeno. Rev. Bras. Ensino Fís. **37**(3), 3301–1–3301–13 (2015). https://doi.org/10.1590/S1806-11173731713
8. Wang, W., Zhai, W., Chen, Y., He, Q., Zhang, H.: Two-dimensional material-based virus detection. Sci. China Chem. **65**(3), 497–513 (2022). https://doi.org/10.1007/s11426-021-1150-7
9. Kim, Y., Patel, R., Kulkarni, C.V., Patel, M.: Graphene-based aerogels for biomedical application. Gels **9**(12), 967 (2023). https://doi.org/10.3390/gels9120967
10. Hasanzade, Z., Raissi, H.: Investigation of graphene-based nanomaterial as nanocarrier for adsorption of paclitaxel anticancer drug: a molecular dynamics simulation study. J. Mol. Model. **23**(2), 36 (2017). https://doi.org/10.1007/s00894-017-3207-1
11. Gollavelli, G., Ghule, A.V., Ling, Y.-C.: Multimodal imaging and phototherapy of cancer and bacterial infection by graphene and related nanocomposites. Molecules **27**(17), 5588 (2022). https://doi.org/10.3390/molecules27175588
12. Compton, O.C., Nguyen, S.T.: Graphene oxide, highly reduced graphene oxide, and graphene: versatile building blocks for carbon-based materials. Small **6**(6), 711–723 (2010). https://doi.org/10.1002/smll.200901934
13. Song, J., et al.: Controllability of graphene oxide doxorubicin loading capacity based on density functional theory. Nanomaterials **12**(3), 479 (2022). https://doi.org/10.3390/nano12030479
14. Feng, L., Chen, Y., Ren, J., Qu, X.: A graphene functionalized electrochemical aptasensor for selective label-free detection of cancer cells. Biomaterials **32**(11), 2930–2937 (2011). https://doi.org/10.1016/j.biomaterials.2011.01.002
15. Krasley, A.T., Li, E., Galeana, J.M., Bulumulla, C., Beyene, A.G., Demirer, G.S.: Carbon nanomaterial fluorescent probes and their biological applications. Chem. Rev. **124**(6), 3085–3185 (2024). https://doi.org/10.1021/acs.chemrev.3c00581
16. Hernández Rosas, J.J., Ramírez Gutiérrez, R.E., Escobedo-Morales, A., Chigo Anota, E.: First principles calculations of the electronic and chemical properties of graphene, graphane, and graphene oxide. J. Mol. Model **17**(5), 1133–1139 (2011). https://doi.org/10.1007/s00894-010-0818-1

17. Zaboli, A., Raissi, H., Hashemzadeh, H., Farzad, F.: Graphene oxide hosting a pH-sensitive prodrug: an in silico investigation of graphene oxide-based nanovehicle toward cancer therapy. ACS Appl. Bio Mater. **6**(7), 2826–2836 (2023). https://doi.org/10.1021/acsabm3c00276
18. Khakpour, E., Salehi, S., Naghib, S.M., Ghorbanzadeh, S., Zhang, W.: Graphene-based nanomaterials for stimuli-sensitive controlled delivery of therapeutic molecules. Front. Bioeng. Biotechnol. **11**, 1129768 (2023). https://doi.org/10.3389/fbioe2023.1129768
19. Srivastava, A., Azad, U.P.: Nanobioengineered surface comprising carbon based materials for advanced biosensing and biomedical application. Int. J. Biol. Macromol. **253**, 126802 (2023). https://doi.org/10.1016/j.ijbiomac.2023.126802
20. Al Ragib, A., et al.: Multifunctional carbon dots in nanomaterial surface modification: a descriptive review. Carbon Res. **2**(1), 37 (2023). https://doi.org/10.1007/s44246-023-00069-x
21. Domingues, T., et al.: Tailoring DNA surface interactions on single-layer graphene: comparative analysis of pyrene, acridine, and fluorenyl methyl linkers. Langmuir **41**(1), 263–273 (2025). https://doi.org/10.1021/acs.langmuir.4c03420
22. Xu, Z., et al.: Graphene-assisted electrochemical sensor for detection of pancreatic cancer markers. Front. Chem. **9**, 733371 (2021). https://doi.org/10.3389/fchem2021.733371
23. Cui, G., et al.: Graphene-based nanomaterials for breast cancer treatment: promising therapeutic strategies. J. Nanobiotechnol. **19**(1), 211 (2021). https://doi.org/10.1186/s12951-021-00902-8
24. Chen, Y.W., Su, Y.L., Hu, S.H., Chen, S.Y.: Functionalized graphene nanocomposites for enhancing photothermal therapy in tumor treatment. Adv. Drug Deliv. Rev. **105**, 190–204 (2016). https://doi.org/10.1016/j.addr.2016.05.022
25. Kaymaz, S.V., Nobar, H.M., Sarıgül, H., Soylukan, C., Akyüz, L., Yüce, M.: Nanomaterial surface modification toolkit: principles, components, recipes, and applications. Adv. Coll. Interface. Sci. **322**, 103035 (2023). https://doi.org/10.1016/j.cis.2023.103035
26. Machado, M., et al.: Graphene biosensors—a molecular approach. Nanomaterials **12**(10), 1624 (2022). https://doi.org/10.3390/nano12101624
27. Zhang, X., et al.: Ultrasensitive field-effect biosensors enabled by the unique electronic properties of graphene. Small **16**(15), 1902820 (2020). https://doi.org/10.1002/smll.201902820
28. Tonel, M.Z., Zanella, I., Fagan, S.B.: Theoretical study of small aromatic molecules adsorbed in pristine and functionalised graphene. J. Mol. Model. **27**(6), 193 (2021). https://doi.org/10.1007/s00894-021-04806-0
29. Tonel, M.Z., González-Durruthy, M., Zanella, I., Fagan, S.B.: Interactions of graphene derivatives with glutamate-neurotransmitter: a parallel first principles—docking investigation. J. Mol. Graph. Model. **88**, 121–127 (2019). https://doi.org/10.1016/j.jmgm.2019.01.007
30. Tonel, M.Z., Martins, M.O., Zanella, I., Pontes, R.B., Fagan, S.B.: A first-principles study of the interaction of doxorubicin with graphene. Comput. Theor. Chem. **1115**, 270–275 (2017). https://doi.org/10.1016/j.comptc.2017.07.004
31. Trusek, A., Kijak, E., Granicka, L.: Graphene oxide as a potential drug carrier—chemical carrier activation, drug attachment and its enzymatic controlled release. Mater. Sci. Eng. C **116**(6), 111240 (2020). https://doi.org/10.1016/j.msec.2020.111240
32. COVID-19 deaths | WHO COVID-19 dashboard. datadot. Acesso em: 9 de janeiro de 2025. Disponível em: https://data.who.int/dashboards/covid19/cases
33. Unal, M.A., et al.: Graphene oxide nanosheets interact and interfere with SARS-CoV-2 surface proteins and cell receptors to inhibit infectivity. Small **17**(25), 2101483 (2021). https://doi.org/10.1002/smll.202101483
34. Srivastava, D., Atluri, S.N.: Computational nanotechnology: a current perspective. CMES Comput. Model. Eng. Sci. **3**(5), 531–538 (2002)
35. Dos Santos, A.F., et al.: Evaluation interaction of graphene oxide with heparin for antiviral blockade: a study of ab initio simulations, molecular docking, and experimental analysis. J. Mol. Model. **29**(8), 235 (2023). https://doi.org/10.1007/s00894-023-05645-x
36. Jobim, M.L., et al.: Açai (Euterpe oleracea, Mart.), an Amazonian fruit has antitumor effects on prostate cancer cells. Arch. Biosc. Health **1**(1), 61–76 (2019). https://doi.org/10.18593/abh.16966

37. Schultz, J.V., Tonel, M.Z., Martins, M.O., Fagan, S.B.: Graphene oxide and flavonoids as potential inhibitors of the spike protein of SARS-CoV-2 variants and interaction between ligands: a parallel study of molecular docking and DFT. Struct. Chem. (2023). https://doi.org/10.1007/s11224-023-02135-x
38. Perdew, J.P., Zunger, A.: Self-interaction correction to density-functional approximations for many-electron systems. Phys. Rev. B **23**(10), 5048–5079 (1981). https://doi.org/10.1103/PhysRevB.23.5048
39. Boys, S.F., Bernardi, F.: The calculation of small molecular interactions by the differences of separate total energies. Some procedures with reduced errors. Mol. Phys. **19**(4), 553–566 (1970). https://doi.org/10.1080/00268977000101561
40. De Matos, C.F., Leão, M.B., Vendrame, L.F.O., Jauris, I.M., Zanella, I., Fagan, S.B.: Unlocking the paracetamol adsorption mechanism in graphene tridimensional-based materials: an experimental-theoretical approach. Front. Carbon **3**, 1305183 (2024). https://doi.org/10.3389/frcrb.2024.1305183
41. Davidson, C.B., et al.: Euterpe oleracea Mart. Bioactive molecules: promising agents to modulate the NLRP3 inflammasome. Biology **13**(9), 729 (2024). https://doi.org/10.3390/biology13090729
42. Coll, R.C., et al.: A small-molecule inhibitor of the NLRP3 inflammasome for the treatment of inflammatory diseases. Nat. Med. **21**(3), 248–255 (2015). https://doi.org/10.1038/nm.3806
43. Trott, O., Olson, A.J.: AutoDock Vina: improving the speed and accuracy of docking with a new scoring function, efficient optimization, and multithreading. J. Comput. Chem. **31**(2), NA–NA (2009). https://doi.org/10.1002/jcc21334
44. Valdés-Tresanco, M.S., Valdés-Tresanco, M.E., Valiente, P.A., Moreno, E.: AMDock: a versatile graphical tool for assisting molecular docking with Autodock Vina and Autodock4. Biol. Direct **15**(1), 12 (2020). https://doi.org/10.1186/s13062-020-00267-2
45. Berman, H.M., et al.: The protein data bank www.rcsb.org. Nucleic Acids Res. **28**(1), 235–242 (2000). https://doi.org/10.1093/nar/28.1.235
46. Martins, M.O., Silva, I.Z.D., Fagan, S.B., Santos, A.F.D.: Docking fundamentals for simulation in nanoscience. DS-CNT **22**(3), 67–76 (2021). https://doi.org/10.37779/nt.v22i3.4106
47. Ohno, Y., Maehashi, K., Yamashiro, Y., Matsumoto, K.: Electrolyte-gated graphene field-effect transistors for detecting pH and protein adsorption. Nano Lett. **9**(9), 3318–3322 (2009). https://doi.org/10.1021/nl901596m
48. Wu, Y., Zhou, Y., Leng, Y., Lai, W., Huang, X., Xiong, Y.: Emerging design strategies for constructing multiplex lateral flow test strip sensors. Biosens. Bioelectron. **157**, 112168 (2020). https://doi.org/10.1016/j.bios.2020.112168
49. Krishnan, S.K., Nataraj, N., Meyyappan, M., Pal, U.: Graphene-based field-effect transistors in biosensing and neural interfacing applications: recent advances and prospects. Anal. Chem. **95**(5), 2590–2622 (2023). https://doi.org/10.1021/acs.analchem.2c03399
50. Khodaie, A., Heidarzadeh, H.: Ultra-sensitive surface plasmon resonance sensor integrating MXene (Ti3C2TX) and graphene for advanced carcinoembryonic antigen detection. Sci. Rep. **15**(1), 13571 (2025). https://doi.org/10.1038/s41598-025-97853-z
51. Cucuiet, V., Maniu, D., Astilean, S., Lamy De La Chapelle, M., Focsan, M.: Graphene-mediated surface enhanced Raman spectroscopy for DNA detection & hybridization: breakthroughs and challenges. Biosens. Bioelectron. **286**, 117610 (2025). https://doi.org/10.1016/j.bios.2025.117610
52. Seo, G., et al.: Rapid detection of COVID-19 causative virus (SARS-CoV-2) in human nasopharyngeal swab specimens using field-effect transistor-based biosensor. ACS Nano **14**, 5142 (2020). https://doi.org/10.1021/acsnano.0c02823
53. Prattis, I., Hui, E., Gubeljak, P., Kaminski Schierle, G.S., Lombardo, A., Occhipinti, L.G.: Graphene for biosensing applications in point-of-care testing. Trends Biotechnol. **39**(10), 1065–1077 (2021). https://doi.org/10.1016/j.tibtech.2021.01.005
54. Campos, R., et al.: Attomolar label-free detection of DNA hybridization with electrolyte-gated graphene field-effect transistors. ACS Sens. **4**(2), 286–293 (2019). https://doi.org/10.1021/acssensors.8b00344

55. Islam, M.A., et al.: Recent achievement of graphene in biomedicine: advancements by integrated microfluidics system and conventional techniques. Sens. Int. **5**, 100293 (2024). https://doi.org/10.1016/j.sintl.2024.100293
56. Balderston, S., et al.: Discrimination of single-point mutations in unamplified genomic DNA via Cas9 immobilized on a graphene field-effect transistor. Nat Biomed Eng **5**(7), 713–725 (2021). https://doi.org/10.1038/s41551-021-00706-z
57. Saltzgaber, G., et al.: Scalable graphene field-effect sensors for specific protein detection. Nanotechnology **24**(35), 355502 (2013). https://doi.org/10.1088/0957-4484/24/35/355502
58. Palacio, I., et al.: Attomolar detection of hepatitis C virus core protein powered by molecular antenna-like effect in a graphene field-effect aptasensor. Biosens. Bioelectron. **222**, 115006 (2023). https://doi.org/10.1016/j.bios.2022.115006
59. Nimgampalle, M., Chakravarthy, H., Sharma, S., Shree, S., Bhat, A.R., Pradeepkiran, J.A., Devanathan, V.: Neurotransmitter systems in the etiology of major neurological disorders: emerging insights and therapeutic implications. Age. Res. Rev. **89**, 101994 (2023). https://doi.org/10.1016/j.arr.2023.101994
60. Yi, S., Bian, S., Sawan, M.: Real-time in vivo detection techniques for neurotransmitters: a review. Analyst **145**(19), 6193–6210 (2020). https://doi.org/10.1039/D0AN01175D
61. Dunham, K.E., Jill Venton, B.: Electrochemical and biosensor techniques to monitor neurotransmitter changes with depression. Anal. Bioanal. Chem. **416**, 2301–2318 (2024). https://doi.org/10.1007/s00216-024-05136-9
62. Shalini Devi, K.S., Tsujimura, S. (eds): Future approaches to electrochemical sensing of neurotransmitters, vol. 30. https://doi.org/10.1039/9781837675593
63. Nasri, B., et al.: Hybrid CMOS-graphene sensor array for subsecond dopamine detection. IEEE Trans. Biomed. Circuits Syst. **11**(6), 1192–1203 (2017). https://doi.org/10.1109/TBCAS.2017.2778048
64. Cisneros-Fernandez, J., et al.: A 1024-Channel 110-Bit 36-μW/ch CMOS ROIC for multiplexed GFET-only sensor arrays in brain mapping. IEEE Trans. Biomed. Circuits Syst. **15**(5), 860–876 (2021). https://doi.org/10.1109/TBCAS.2021.3113556
65. Domingues, T., et al.: Graphene surface functionalization with pyrene-, acridine-, and fluorenyl-linkers for DNA hybridization studies using electrolyte-gate graphene field effect transistors. Langmuir, p. acs.langmuir.5c01900 (2025). https://doi.org/10.1021/acs.langmuir.5c01900
66. Thierheimer, M., Cioffi, G., Waite, K.A., Kruchko, C., Ostrom, Q.T., Barnholtz-Sloan, J.S.: Mortality trends in primary malignant brain and central nervous system tumors vary by histopathology, age, race, and sex. J. Neurooncol **162**(1), 167–177 (2023). https://doi.org/10.1007/s11060-023-04279-6
67. Louis, D.N., et al.: The 2016 world health organization classification of tumors of the central nervous system: a summary. Acta Neuropathol. **131**(6), 803–820 (2016). https://doi.org/10.1007/s00401-016-1545-1
68. Huang, H.-M., Shih, Y.-Y.: Pushing CT and MR imaging to the molecular level for studying the 'Omics': current challenges and advancements. Biomed. Res. Int. **2014**, 1–17 (2014). https://doi.org/10.1155/2014/365812
69. Bex, A., Mathon, B.: Advances, technological innovations, and future prospects in stereotactic brain biopsies. Neurosurg. Rev. **46**(1), 5 (2022). https://doi.org/10.1007/s10143-022-01918-w
70. Haque, A., Banik, N.L., Ray, S.K.: Molecular alterations in glioblastoma, em Progress in Molecular Biology and Translational Science, vol. 98, pp. 187–234. Elsevier (2011). https://doi.org/10.1016/B978-0-12-385506-0.00005-3
71. Dong, C., Zheng, Y., Long-Iyer, K., Wright, E.C., Li, Y., Tian, L.: Fluorescence imaging of neural activity, neurochemical dynamics, and drug-specific receptor conformation with genetically encoded sensors. Annu. Rev. Neurosci. **45**, 273–294 (2022). https://doi.org/10.1146/annurev-neuro-110520-031137
72. Abrantes, M., Rodrigues, D., Domingues, T., Nemala, S.S., Monteiro, P., Borme, J., Alpuim, P., Jacinto, L.: Ultrasensitive dopamine detection with graphene aptasensor multitransistor arrays. J. Nanobiotechnol. **20**(1), art. no. 495 (2022). https://doi.org/10.1186/s12951-022-01695-0

# Utilizing Artificial Intelligence Models to Optimize Adsorptive Processes

**Ana Carolina Ferreira Piazzi Fuhr, Guilherme Luiz Dotto, and Fernando Machado Machado**

**Abstract** The growing concern about environmental pollution, notably water pollution, has driven the development of more efficient and sustainable wastewater treatment technologies. Among them, adsorption stands out for its effectiveness, low cost, and the possibility of regenerating adsorbent materials. However, optimizing this process using traditional methods demands time, cost, and a high number of experiments. In this context, artificial intelligence (AI) emerges as a promising alternative for modeling and predicting the behavior of adsorptive systems with high precision. This chapter presents the main AI models applied to adsorption, including artificial neural networks, support vector machines, and models based on decision trees, discussing their advantages, such as greater efficiency, resource savings, and the ability to identify relevant variables, compared to conventional experimental methods. The limitations of using AI are also addressed, such as the need for robust databases and the risk of overfitting. Computational tools, such as Python, R, and MATLAB, are discussed as means of practical implementation. Finally, examples of AI applications in adsorption processes are presented, highlighting its potential in predicting adsorptive capacity and optimizing operating conditions.

**Keywords** Wastewater treatment · Adsorption · Computational tools · Machine learning

A. C. F. P. Fuhr (✉) · F. Machado Machado
Technology Development Center, Federal University of Pelotas (UFPel), Pelotas, RS, Brazil
e-mail: carolpiazzi@yahoo.com.br

F. Machado Machado
e-mail: fernando.machado@ufpel.edu.br

G. L. Dotto
Research Group on Adsorptive and Catalytic Process Engineering (ENGEPAC), Federal University of Santa Maria, Santa Maria, RS, Brazil
e-mail: guilherme_dotto@yahoo.com.br

© The Author(s), under exclusive license to Springer Nature Switzerland AG 2026

C. P. Bergmann and F. Machado Machado (eds.), *Carbon Nanomaterials as Adsorbents for Environmental and Biological Applications*, Carbon Nanostructures,
https://doi.org/10.1007/978-3-032-08669-3_9

## Abbreviations

| | |
|---|---|
| AB | AdaBoosting |
| AI | Artificial intelligence |
| ANFIS | Adaptive neuro-fuzzy inference systems |
| ANN | Artificial neural networks |
| DL | Deep learning |
| DT | Decision tree |
| GB | Gradient boosting |
| KNN | K-nearest neighbors |
| ML | Machine learning |
| RF | Random forest |
| SHAP | Shapley Additive Explanation |
| SVM | Support vector machine |

## 1 Introduction

The growing concern about water pollution highlights the urgency of improving wastewater treatment techniques. Several methodologies, such as flocculation/coagulation, membrane filtration, biological processes, advanced oxidation, and adsorption, have been investigated to minimize environmental impacts [1–3]. However, operational complexity, high costs, and high energy consumption still limit their large-scale application [2, 4].

Thus, adsorption has proven to be a promising alternative, recognized for its high efficiency, reliability, and low cost [2, 4]. Furthermore, it allows the regeneration and reuse of the adsorbent, expanding its economic and environmental appeal. Several materials, including activated carbon, zeolites, graphene, silica gel, and chitosan, can be used [5, 6], and their physical and chemical properties, combined with operational variables (initial adsorbate concentration, dosage, temperature, pH, and contact time), directly influence the performance and adsorption mechanisms [5, 7].

However, the traditional adsorption optimization methodology, based on batch experiments, has significant limitations: it requires a considerable investment of time and resources, high reagent consumption, and an unpredictable number of trials to determine the ideal conditions [8–10]. Given these obstacles, it is crucial to explore approaches that expedite the process and minimize costs.

In recent years, artificial intelligence (AI) computing tools have been developed to learn directly from experimental data [11, 12]. AI stands out for overcoming the limitations of traditional methods, as it can identify correlations and patterns in the data and allows the extraction of the relevance of the variables involved in the process [13], which makes it possible to reduce the number of experiments and the time, research costs, and effort for researchers [14, 15]. In particular, machine learning (ML), an AI sub-area, has proven effective in predicting relationships between input and output parameters of adsorptive processes, providing deeper and more accurate analyses [10, 12, 16].

In recent years, the AI and ML application for optimization in various processes has grown significantly, emphasizing its use in the investigation of adsorptive processes. As illustrated in Fig. 9.1, there is a notable increase in studies that employ AI and ML in different areas (Fig. 9.1A), as well as a specific growth in the number of research studies aimed at using this technology in adsorption processes (Fig. 9.1B).

## 2 Artificial Intelligence and Machine Learning—Definition

Artificial intelligence (AI) there is still no specific, widely accepted definition, so in this way, it can be defined in several ways [17]. AI can generally be understood as the ability to interpret data through learning, aiming to use it to achieve objectives and perform specific tasks [18, 19]. Problem-solving by AI occurs through simulations, which allow systems to learn from experiences and adapt to environmental changes. Furthermore, AI can simulate reasoning and decision-making processes [20]. Furthermore, AI reduces the operational costs of a process since; by learning from the available data of a process, AI can predict target variables whose direct measurement would be unfeasible or economically impractical using traditional equipment. In this context, AI acts as a virtual sensor, taking advantage of the available information to make accurate predictions of variables that are difficult to measure [21].

AI encompasses two main subfields: machine learning (ML) and deep learning (DL). However, literature predominantly highlights ML application techniques in adsorption processes [9]. ML allows machines to learn directly from data without relying on models based on predetermined equations [10, 12, 16], and encompasses four primary modalities: supervised, semi-supervised, unsupervised learning and reinforcement learning [9, 22, 23]. In supervised learning, algorithms learn patterns within a labeled dataset composed of input (independent) and output (dependent) variables. Learning occurs in two stages: the training stage, in which the algorithm identifies patterns and transforms them into a mathematical model, and the inference stage, in which the obtained model is used to analyze and predict data that has never been observed. In contrast to supervised learning, unsupervised learning operates with unlabeled data, using only input variables. In this way, algorithms use input information to identify patterns, trends, and underlying structures. Their goal is not to predict outputs but rather to detect and extract patterns in the data [24–26]. Complementing these approaches, semi-supervised learning combines a large volume of available input data and a limited number of output data [25]. Finally, reinforcement learning involves the interaction of an agent with the environment, adjusting its parameters based on rewards and penalties to model adaptive decisions and optimize behaviors in dynamic scenarios [24–26].

ML algorithms are easily adaptable and progressively adapted to improve predictions and minimize errors according to the data available for the training process. The

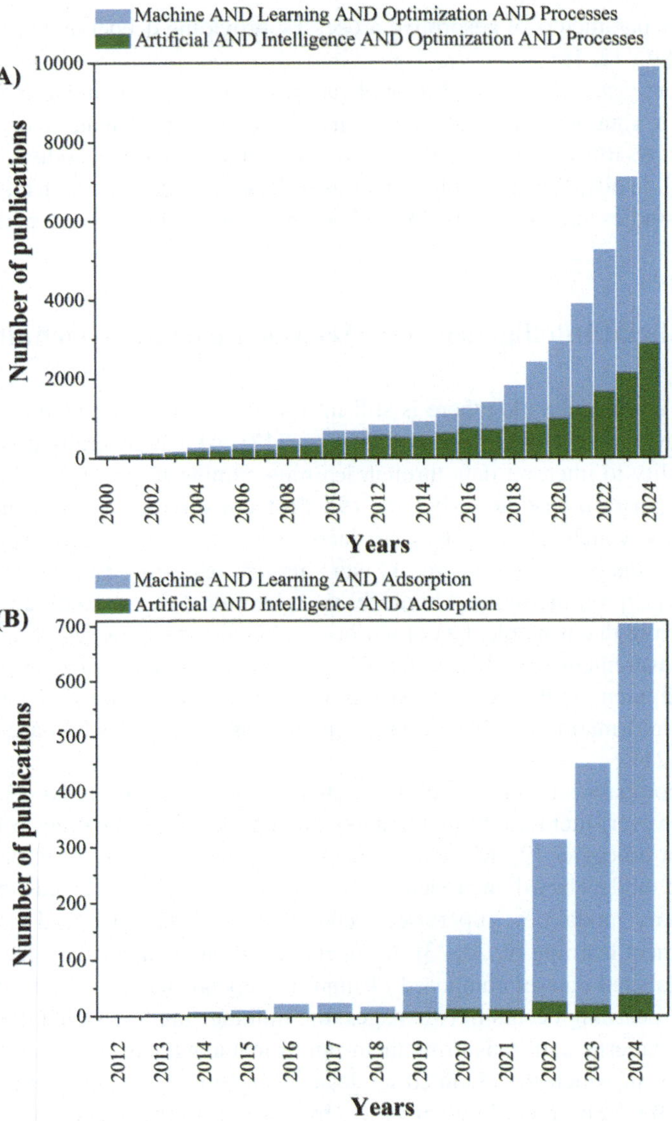


**Fig. 9.1** Prospecting the use of AI and ML over the last 25 years (between 2000 and 2024) in the SCOPUS database. **A** Increase in the number of research projects involving AI and ML in process optimization. **B** Increase in the number of research projects involving AI and ML in adsorption

effectiveness of ML models depends directly on the quality and quantity of the data used in their construction [12]. Through iterations, algorithms refine their architecture, learn from errors, and reproduce the desired results and generalize the model to deal with new and unprecedented data [10, 27]. Furthermore, ML is a methodology that enables decision-making [12].

## 3 Artificial Intelligence and Machine Learning Steps

The AI and ML application involves some fundamental steps, as illustrated in Fig. 9.2. These steps include data collection, exploratory data analysis, feature engineering, algorithm selection, model validation and evaluation, and variable analysis [28, 29].

1. **Data collection**: the quantity and quality of data are essential to ensure the effectiveness of algorithms. The type of data used, whether numerical or symbolic, can increase the complexity of the model, making it necessary to choose an appropriate and representative data set. Furthermore, if the data set is insufficient in quantity and representativeness of the process, the reliability and predictive power of the models are compromised. Therefore, it is clearly essential to ensure that the volume and quality of the selected data are adequate to the needs of the analysis [10, 30, 31].
2. **Exploratory data analysis**: this stage involves data cleaning, which means that the data will be analyzed to identify incorrect, missing, noisy, duplicated, or irrelevant data in order to correct or remove them from the data set [31]. After cleaning the data, it is essential to perform a correlation analysis between the variables selected for the data set. The presence of highly correlated variables in the data set causes the model to misinterpret the data, causing multicollinearity problems leading to overfitting [28, 32].
3. **Feature engineering**: Feature engineering typically occurs after data collection and cleaning and involves both feature selection (removal of redundant or irrelevant variables, resulting in a subset of the original features, performed before or during model training) and feature extraction (reduction of dimensionality by creating new features, such as linear combinations in PCA); these dimension reduction techniques are often applied or explored at this stage [33, 34].
4. **Algorithm selection**: Several AI algorithms have been developed to meet a wide range of applications, each with its particular characteristics, strengths, and limitations. These algorithms are designed to address various types of learning problems, ranging in complexity, data nature, and modeling objectives. Given this diversity, it is essential to conduct systematic comparisons between the different methods to identify the one that presents the best performance for the specific problem under analysis. Selecting the best algorithm is crucial [22, 25, 35].

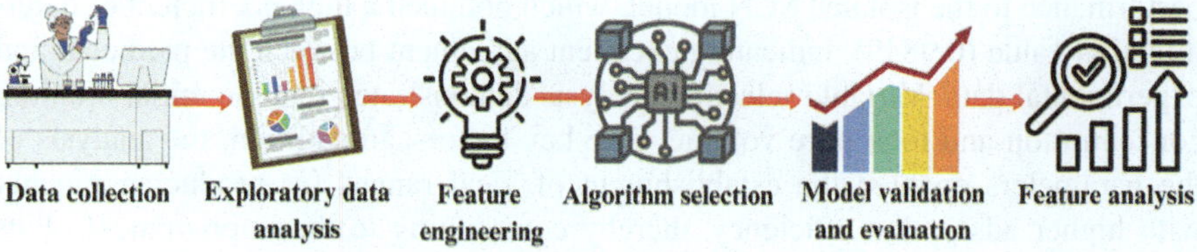


**Fig. 9.2** General flowchart of the main steps for applying AI

5. **Model validation and evaluation**: The appropriate choice of model evaluation techniques and algorithms is crucial. The estimated performance of a model should ideally reflect its ability to generalize to unseen data since correctly predicting future data is often the central goal of AI. To select the most effective model, it is necessary to estimate its performance accurately. This process requires the judicious use of evaluation techniques, such as cross-validation. Additionally, statistical tests for comparing algorithms and strategies, as well as methods for correcting for multiple comparisons, are warranted [36].
6. **Feature analysis**: AI models often operate as "black boxes" without clearly explaining how they make decisions. While some, such as decision trees, are inherently interpretable, most do not reveal their decision-making criteria. Feature analysis and additive explanation techniques are employed to enhance transparency, facilitating a deeper understanding of the factors that influence outcomes and promoting greater trust in AI systems [37].

# 4 Main Machine Learning Algorithms

## *4.1 AdaBoosting*

AdaBoosting (AB) is a supervised learning technique that uses a combination of low-prediction models to generate a model with greater predictive power [22, 38–40]. AB creates a set of weak models, adds weights to the training data, and adjusts the weight values after each iteration of the weak model in an adaptive manner. The weights of the training samples that reduce the predictive power of the current model will be increased, and the weights of the samples that improve the predictive power will be decreased [22, 39, 40].

AB has several weak models, called weak learners, which, when combined in large numbers, can form a robust prediction model. The process of using a new model to optimize the results of the predecessor model is repeated until a robust model is built [22, 38, 39]. In this way, AB causes weak learners to be adjusted later to favor samples that were erroneously predicted [22, 41], as shown in Fig. 9.3a.

Qu et al. [42] used the Adaboost–SCN (stochastic configuration network) model to predict uranium adsorption by biochar. The Adaboost–SCN model showed superior performance to the isolated SCN model, which obtained a high coefficient of determination value (0.9849), indicating excellent agreement between the predicted and experimental data. Variable selection analysis demonstrated that the initial uranium concentration and total pore volume were key factors. In addition, the analysis of the parameters enabled the establishment of ideal ranges for producing biochar with higher adsorption efficiency, thereby contributing to the improvement of its manufacturing process [42].

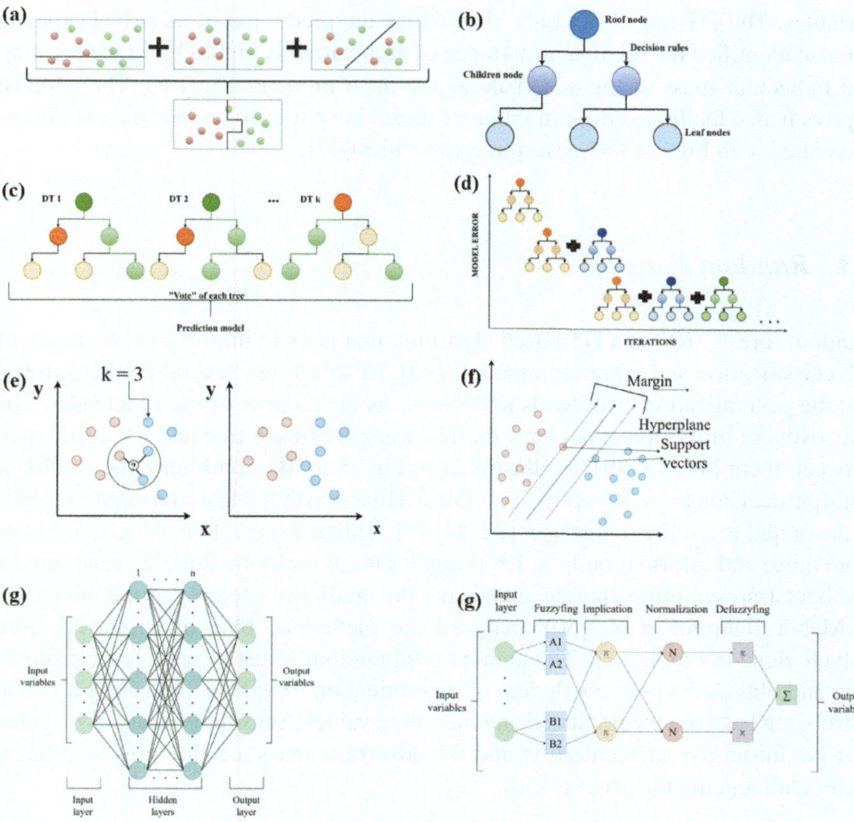


**Fig. 9.3** **a** AB. **b** DT. **c** RF. **d** GB. **e** KNN. **f** SVM. **g** ANN. **h** ANFIS

## *4.2 Decision Tree*

Decision trees (DT) are an ML technique used to build predictive models from experimental data. This method generates a simple mathematical model graphically represented as a decision tree [43–45]. The structure of the tree resembles a flow diagram that reflects the logic of human reasoning, making regression trees intuitive and easy to interpret. Just like in a flowchart, the tree has different components: the root node (located at the top) is responsible for initiating the division of data based on an attribute; the internal nodes (or children) represent attributes or variables; the branches indicate the decision rules; and the leaf nodes correspond to the final results [22, 46], as illustrated in Fig. 9.3b.

Bezerra and Altino [47] analyzed the adsorption behavior of light gases by microporous materials, such as zeolites, metal–organic frameworks, and activated carbons, by analyzing data extracted from 22 articles published between 1974 and 2022. Using a DT algorithm, they constructed a database with 3297 data points related to 8 input variables. The DT model not only allowed for the prediction of adsorbed capacity but also identified the relative importance of each variable, highlighting the pressure and molecular mass of the adsorbate as the most influential factors. The adopted approach also facilitated the extraction of useful heuristics to understand conditions associated with high or low adsorption capacities [47].

## *4.3 Random Forest*

Random forests (RF) is a DT-based algorithm that aims to improve the accuracy of DT classification and regression models [48]. RF combines several predictor trees, and the generalization error tends to decrease as the number of trees increases. The effectiveness of the forest depends on the strength of each tree and the correlation between them [44, 48, 49], as illustrated in Fig. 9.3c. By combining the results of multiple decision trees, RF seeks to establish a function that relates the input variables to the output in a more robust way [22, 44, 48]. Unlike a single tree, which can suffer from noise and information loss, RF mitigates these issues by utilizing randomness and bootstrap sampling, thereby enhancing the predictive capacity of the model.

Mehdi Bahrami et al. [50] analyzed the methylene blue dye removal from polyethylene microplastics. The authors used random forest to predict the removal data and obtained a high coefficient of determination (97.55%), indicating a strong correlation between the predicted and measured values. Sensitivity analysis revealed that the initial dye concentration and the adsorbent mass are the most significant factors influencing the process [50].

## *4.4 Gradient Boosting*

The Gradient Boosting (GB) algorithm is a DT-based ML technique. It is a method that combines multiple DTs to form a robust model with high predictive performance [22, 51, 52]. Unlike methods such as Random Forest, which operate in parallel, GB adopts a sequential approach, in which each subsequent tree is trained to correct the errors made by previous trees, as illustrated in Fig. 9.3d. This strategy leads to a progressive reduction in the model's error, thereby enhancing its predictive capacity [51].

GB is especially effective in heterogeneous data sets with correlated and categorical variables, which reinforces its applicability in various analytical contexts. Additionally, the algorithm can be implemented in multiple ways, which enhances its flexibility and adaptability to diverse computational and statistical demands [53].

Zhao et al. [54] investigated four AI models to predict the maximum adsorption capacity of hydrochar and identify the key factors influencing the process. Among the models analyzed, GB showed excellent predictive capacity, with a coefficient of determination of 0.93. Additionally, the model enabled the analysis of process variables to identify the primary factors influencing the process. The results indicate that material properties, such as carbon, oxygen, nitrogen, and hydrogen contents, play an essential role in controlling the adsorption capacity. This study demonstrated that the ML model is effective in both predicting the adsorption of different heavy metals on hydrochar and optimizing the material preparation conditions [54].

## *4.5 K-Nearest Neighbors*

The K-nearest neighbors (KNN) model is a simple, non-parametric, instance-based ML algorithm widely used in classification and regression tasks. Its performance depends on the measure of distance or similarity between the test data and the training data [55]. KNN makes no assumptions about the distribution of the data [56] and operates based on the proximity between samples, assuming that similar data belong to the same class [57, 58], as shown in Fig. 9.3e. To make predictions, the algorithm identifies the k closest neighbors of the target point in the training set [58]. In classification problems, the most frequent class among the neighbors determines the prediction (majority vote); in regression, the output is estimated by the average of the values of these neighbors [57].

Zamouche et al. [59] used KNN to predict the adsorption capacity of phenol in activated carbon. They used batch experimental data, and the process variables analyzed were adsorbent dosage, temperature, initial phenol concentration, stirring speed, and pH. The KNN model showed high predictive power to predict the adsorption capacity when compared to traditional techniques (adsorption isotherms), providing more accurate values and very high statistical coefficients [59].

## *4.6 Support Vector Machine*

The Support Vector Machine (SVM) algorithm is widely used due to its accuracy and simplicity and is effective in classification, regression, and clustering tasks. It operates by searching for the hyperplane that best separates the data with the largest margin between the support vectors, as shown in Fig. 9.3f. In problems with non-linear separable data, the SVM uses a soft margin and kernel functions to transform the data into a space where linear separation is possible. The choice of the kernel function and the penalty parameter is crucial for the model's performance [22, 25].

Guediri et al. [60] applied an SVM model to accurately predict the adsorption behavior of methylene blue on sodium alginate-encapsulated beads synthesized from orange peel. The SVM model proved to be effective in modeling the process, allowing the optimization of operating conditions for maximum dye removal efficiency [60].

## 4.7 Artificial Neural Networks

Artificial Neural Networks (ANNs) are computational models inspired by biological neural processes developed to represent the structure and functions of the brain mathematically. They learn and adapt from experience, extracting knowledge from it. ANNs reproduce the interaction between brain neurons through a simple and robust system of information transmission and processing [22, 61, 62]. Composed of elements called neurons, these models can describe nonlinear behaviors of dynamic systems, depending on the complexity of the connections between neurons. Layers form their structure: an input layer (with process variables), one or more hidden layers (processing layers), and an output layer (with model results). Neurons act as fundamental information processing units [61, 62], as shown in Fig. 9.3g.

Deivayanai et al. [63] evaluated the effectiveness of carbonated pineapple peel residues integrated with magnetic nanoparticles in the removal of synthetic dyes Congo red and methylene blue from wastewater and applied ANN for predictive modeling. After the physicochemical characterization of the materials and analysis of the experimental data, ANN was used to predict the adsorption capacity of the dyes with high accuracy. The models exhibited high correlation coefficients (0.9948 for methylene blue and 0.9939 for Congo red), validating the reliability of the experimental data and demonstrating the ability of ANNs to represent the dynamics of the adsorption process accurately [63].

## 4.8 Adaptive Neuro-Fuzzy Inference Systems

Adaptive neuro-fuzzy inference systems (ANFIS) combine the adaptive control of fuzzy systems with the learning capacity of ANNs, combining the advantages of both minimizing the error between input data and model predictions. Its typical architecture consists of five layers: fuzzy, product, normalization, defuzzification, and general output, as shown in Fig. 9.3h. The system automatically adjusts the premise and consequence parameters during training, which are crucial for model performance. ANFIS is effective in representing complex relationships between variables, with high modeling and generalization capacity [9, 22, 64].

Fuhr et al. [28] used data related to the removal of 2,4-dichlorophenoxyacetic acid and atrazine by pyrolyzed soybean hulls. The authors evaluated different AI models, including GB, ANN, and ANFIS, in terms of their predictive performance and generalization capacity in predicting pollutant removal capacity. The results obtained indicated ANFIS as the model with the best predictive capacity. The analysis of the variables' importance indicated that time and initial concentration are the most relevant factors [28].

## 5 Main Techniques for Evaluating the Performance of AI Models

Evaluating the performance of AI models is essential to verify their predictive capacity, as well as their accuracy and reliability. To this end, it is common to use statistical metrics that help validate the quality of predictions [65], in addition to contributing to the detection of problems such as overfitting and underfitting.

Several statistical techniques are recommended to measure the performance of models, varying according to the type of task, regression or classification [66]. In classification problems, metrics such as accuracy, precision, recall, F-score, and confusion matrix are frequently used. In regression models, the most common metrics include the coefficient of determination, correlation coefficient, mean absolute error, mean squared error, and root mean squared error [65]. The choice of the most appropriate metrics depends directly on the nature of the problem and the specific objectives of the analysis.

## 6 Main Feature Analysis

Most AI models operate as "black boxes," even though some models, such as decision trees, can be interpreted through their design. AI models do not provide the clarity with which the model made decisions. Thus, the use of feature analysis methodologies is a way to make the model more interpretable, thereby obtaining more information about the model's decisions [37].

One of the variable interpretation methodologies that has gained prominence in the analysis of artificial intelligence models is the Shapley Additive Explanation (SHAP). This approach, based on cooperative game theory, enables the quantification of the individual contribution of each input variable to the model's prediction, both at a global and local level [67, 68], as shown in Fig. 9.4.

The SHAP methodology is highly flexible and can be applied to various AI algorithms, regardless of whether they focus on regression or classification tasks. Its main difference lies in the assignment of SHAP values to each input variable, which represents, in a consistent and interpretable way, the impact of each variable on the output predicted by the model [69]

In addition, SHAP supports both local explainability, which analyzes each prediction individually, and global explainability, which evaluates the overall behavior of the model concerning the variables. This capability makes the methodology a robust tool for understanding and justifying the decisions made by complex machine-learning models [68, 69].

The application of ML techniques, combined with variable selection, has a high potential for investigating the interrelationships between factors that influence the adsorption process. This approach significantly contributes to process optimization, as it enables the efficient identification of the most relevant variables, thereby

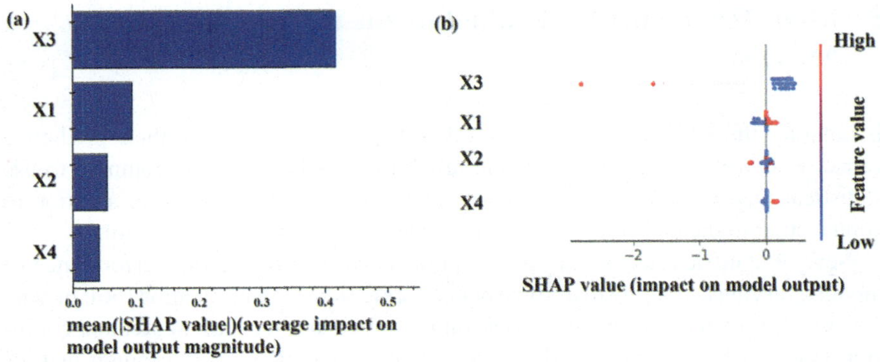


**Fig. 9.4** **a** Generic representation of the graphical representation of the global SHAP value. **b** Generic representation of the graphical representation of the local SHAP value

reducing model complexity and enhancing its predictive capacity [9, 70]. In addition, ML techniques stand out for their low computational cost, high robustness, and high predictive performance, which makes them particularly advantageous in industrial and environmental applications related to adsorption [70].

Shayanmehr et al. [71] used the SHAP technique in conjunction with the ML model to interpret the influence of variables on the adsorption capacity of thiophene compounds by metal–organic frameworks materials. The SHAP analysis identified that the initial sulfur concentration and the contact time were the main factors that impacted the efficiency of the desulfurization process. The approach also allowed for the classification of variables into three categories—process conditions, adsorbent characteristics, and adsorbate characteristics, with process conditions being the most relevant group [71].

## 7 Computational Tools Available

To enable the development and implementation of various AI models, several computational tools have been widely used and continually improved over time, including Python, MATLAB, R, and graphical interface-based environments. The appropriate choice of these tools can directly influence the efficiency, scalability, and applicability of AI solutions, thereby becoming a critical factor in their application.

Among the various programming languages applied to the development of AI-based solutions, Python stands out as the most widely adopted language today. This predominance is mainly due to the wide availability of specialized libraries that meet different levels of complexity. It is an open-source language, completely free, which expands its access and use in academic and industrial environments [22, 72].

Python is a high-level scripting language that integrates explanatory, interpreted, interactive, and object-oriented features, favoring readability, modularity, and code maintenance. With constant updates and the progressive incorporation of new features, it has been increasingly used in the development of independent applications, as well as in large-scale and highly complex projects [72]. The main Python libraries for developing AI models are *Scikit-learn, TensorFlow, Keras* [22, 72], *PyTorch, Pandas, Spark MLlib, Theano, NumPy, SciPy*, and *Matplotlib* [22].

R is an open-source language and environment for statistical analysis and ML derived from the S language. Its functionality is extended through thousands of packages available on CRAN (Comprehensive R Archive Network), which serve different purposes, including popular AI libraries such as *RandomForest, XGBoost, e1071*, and *TensorFlow* [72, 73].

MATLAB is an engineering-oriented software that enables the integration of AI into the design, development, and operation of systems. Engineers and scientists widely use it to create datasets, develop domain-specific AI models, and test them in integrated systems contexts [74]. However, it is proprietary software with a high cost. In addition, its standard functionalities are already largely replicated in open-source languages, such as Python [72].

Choosing the right computational tools becomes a strategic element for the success of their implementation. Understanding the potential and limitations of each tool is crucial in guiding technical decisions and optimizing the performance of AI-based solutions.

## 8 Advantages and Disadvantages of Using AI Versus Traditional Methods

AI models have several advantages over traditional adsorption models, particularly due to their ability to identify complex patterns and correlations in experimental data [10, 75]. Unlike conventional models, which rely on mathematical equations based on idealized assumptions, AI models learn directly from data without the need for predefined analytical formulations [20]. This characteristic allows AI to represent systems with greater fidelity to experimental reality, even in highly nonlinear and multidimensional contexts.

In addition, machine learning techniques offer high predictive performance when trained with adequate data and enable the analysis of variable importance (feature importance), contributing to the interpretation of models and the optimization of adsorption processes [70]. Another significant advantage is the ability to predict system behavior under previously untested conditions, thereby reducing experimental costs and accelerating the development of new strategies [14, 15].

However, the use of AI also has limitations. The effectiveness of models depends heavily on the availability of large and representative datasets. In scenarios with scarce data, there is a greater susceptibility to overfitting, compromising the model's generalization capacity [10, 30, 31]. In addition, the development and application of AI models require specialized technical knowledge, computational resources, and additional steps, such as cross-validation and hyperparameter optimization. Without these precautions, the model's performance may be compromised, limiting its practical applicability.

## 9 Application Examples

Several studies in the literature demonstrate the effectiveness of these tools in describing nonlinear phenomena and in reducing experimental time and costs. Below, representative examples of the application of AI models in adsorption systems are presented, highlighting their contribution to the advancement of contaminated water treatment and environmental remediation technologies.

Rajput et al. [76] used ML-based predictive models to estimate the adsorptive capacity of biochar in the removal of methylene blue dye from aqueous solutions. Eight regression algorithms were evaluated: Linear Regression, Support Vector Machine, Decision Tree, Random Forest, Ridge Regression, AdaBoost, Histogram Gradient Boosting, and Light Gradient Boosting. The dataset used consisted of eleven independent variables related to the properties of biochar and the conditions of the adsorption process, namely: total carbon content, molar ratio of hydrogen to carbon, molar ratio of oxygen and nitrogen to carbon, molar ratio of oxygen to carbon, specific surface area, pore diameter, pore volume, and pH at the point of zero charge, solution pH, temperature, and initial dye concentration concerning the biochar dosage. Among the models tested, Random Forest presented the best performance, with a coefficient of determination of 0.9415, mean squared error of 0.0536, and root mean squared error of 0.2315. The analysis of the variables' importance indicated that the initial dye concentration concerning the biochar dosage is the most influential factor in determining the adsorption capacity. The results demonstrate the potential of ML models as efficient tools to support water treatment plants in formulating optimized strategies for MB removal, thereby contributing to cost reduction and environmental impact mitigation [76].

Fuhr et al. [28] investigated the application of AI models in predicting the adsorptive capacity of two pesticides, atrazine and 2,4-dichlorophenoxyacetic acid. The study emphasized the importance of correlation analysis between the variables involved in the process, highlighting that the identification and exclusion of highly correlated variables are fundamental steps to improve the predictive performance of the models. For this, a correlation analysis was previously performed to eliminate redundancies and select the most representative variables. Then, three machine learning algorithms were evaluated: GB, ANN, and ANFIS. All models presented satisfactory statistical performance; however, ANFIS stood out for its superior predictive capacity compared to the others. The analysis of the variables' importance was performed using the SHAP methodology and revealed that contact time and the initial concentration of the pesticide are the most determining factors for the removal process. The results demonstrate that the quality and relevance of input variables can significantly influence the accuracy of models, often exceeding the complexity of the algorithm used. Thus, the study reinforces the need for careful selection of parameters to ensure that models can present robust, reliable, and reproducible results [28].

Liu et al. [77] analyzed the use of RF, GB, and SVM to predict the adsorption efficiency of 11 heavy metals on 7 iron oxides/oxyhydroxides. Based on a comprehensive database, the models showed high predictive performance, with RF being the best predictive model. An analysis of the variables' importance was performed, indicating that the main factors influencing adsorption were solution pH, mineral surface area, metal concentration, and mineral concentration. The study presents a data-driven approach to understanding adsorption mechanisms at various scales and supports the environmental management of heavy metals [77].

## 10 Conclusions

The application of artificial intelligence, particularly machine learning, in adsorption processes represents a significant advancement in the area of wastewater treatment. By overcoming the limitations of traditional methods, such as high cost, long time, and the need for numerous experimental tests, AI offers an efficient, fast, and accurate approach for modeling, predicting, and optimizing adsorptive systems. AI models have demonstrated great capacity to identify complex relationships between operational variables and properties of adsorbent materials, contributing to a better understanding and control of the process.

Despite the advances, relevant challenges persist. The quality and representativeness of experimental data directly influence the performance of predictive models. The absence of standardized and robust data sets compromises the accuracy of models. In this sense, it is essential to invest in the construction of comprehensive and well-structured databases.

Among the prospects, the integration of AI with complementary methods, such as molecular simulations, computational dynamics, and multi-objective optimization strategies, is particularly notable. This combination has the potential to expand predictive capacity and accelerate the development of new adsorbents with optimized properties. Another critical aspect is the appropriate selection of variables that represent the phenomena involved in a non-redundant manner. The use of advanced feature engineering techniques, including additive explanation methods (feature importance/analysis), can not only improve model performance but also provide clearer interpretations of the factors that govern adsorption mechanisms.

Finally, it is worth noting that traditional methods, such as batch tests and isotherm modeling, have limitations in terms of time and scalability. The incorporation of ML into these classic models offers a promising alternative, with the potential to accelerate experimental testing, reduce costs, and reveal patterns, especially in complex systems.

Thus, the convergence between artificial intelligence and adsorptive processes represents an innovative and sustainable strategy with great potential to transform the scenario of pollutant treatment and contribute to more effective solutions to current environmental challenges.

**Acknowledgements** The authors gratefully acknowledge financial support from the Brazilian funding agencies: Conselho Nacional de Desenvolvimento Científico e Tecnológico (CNPq), and Fundação de Amparo à Pesquisa do Estado do Rio Grande do Sul (FAPERGS). Prof. Fernando M. Machado thanks CNPq for research grant support (Processes 308203/2022-4, 408413/2022-0, 443031/2023-1, and 403522/2025-0) and FAPERGS (Process 23/2551-0001863-2). Dr. Ana Carolina F. P. Fuhr thanks FAPEGRS for research grant support (Process 23/2551-0001863-2).

## References

1. Rashid, R., Shafiq, I., Akhter, P., et al.: A state-of-the-art review on wastewater treatment techniques: the effectiveness of adsorption method. Environ. Sci. Pollut. Res. **28**, 9050–9066 (2021). https://doi.org/10.1007/s11356-021-12395-x
2. Chai, W.S., Cheun, J.Y., Kumar, P.S., et al.: A review on conventional and novel materials towards heavy metal adsorption in wastewater treatment application. J. Cleaner Prod. **296**, 126589 (2021). https://doi.org/10.1016/j.jclepro.2021.126589
3. Sathya, K., Nagarajan, K., Carlin Geor Malar, G., et al.: A comprehensive review on comparison among effluent treatment methods and modern methods of treatment of industrial wastewater effluent from different sources. Appl Water Sci **12**, 70 (2022). https://doi.org/10.1007/s13201-022-01594-7
4. Rajendran, S., Priya, A.K., Senthil Kumar, P., et al.: A critical and recent developments on adsorption technique for removal of heavy metals from wastewater-a review. Chemosphere **303**, 135146 (2022). https://doi.org/10.1016/j.chemosphere.2022.135146
5. Rathi, B.S., Kumar, P.S.: Application of adsorption process for effective removal of emerging contaminants from water and wastewater. Environ. Pollut. **280**, 116995 (2021). https://doi.org/10.1016/j.envpol.2021.116995
6. Shahid, M.K., Kashif, A., Fuwad, A., et al.: Current advances in treatment technologies for removal of emerging contaminants from water—a critical review. Coord. Chem. Rev. **442**, 213993 (2021). https://doi.org/10.1016/j.ccr.2021.213993
7. Cheng, N., Wang, B., Wu, P., et al.: Adsorption of emerging contaminants from water and wastewater by modified biochar: a review. Environ. Pollut. **273**, 116448 (2021). https://doi.org/10.1016/j.envpol.2021.116448
8. Adegoke, K.A., Akinnawo, S.O., Ajala, O.A., et al.: Progress and challenges in batch and optimization studies on the adsorptive removal of heavy metals using modified biomass-based adsorbents. Bioresour. Technol. Rep. **19**, 101115 (2022). https://doi.org/10.1016/j.biteb.2022.101115
9. Zhang, W., Huang, W., Tan, J., et al.: Modeling, optimization and understanding of adsorption process for pollutant removal via machine learning: recent progress and future perspectives. Chemosphere **311**, 137044 (2023). https://doi.org/10.1016/j.chemosphere.2022.137044
10. Taoufik, N., Boumya, W., Achak, M., et al.: The state of art on the prediction of efficiency and modeling of the processes of pollutants removal based on machine learning. Sci. Total. Environ. **807**, 150554 (2022). https://doi.org/10.1016/j.scitotenv.2021.150554
11. Cateni, S., Colla, V.: Variable selection for efficient design of machine learning-based models: efficient approaches for industrial applications. In: Jayne, C., Iliadis, L. (eds.) Engineering Applications of Neural Networks, pp. 352–366. Springer International Publishing, Cham (2016)
12. Mekonnen, Y., Namuduri, S., Burton, L., et al.: Review—machine learning techniques in wireless sensor network based precision agriculture. J. Electrochem. Soc. **167**, 037522 (2020). https://doi.org/10.1149/2.0222003JES

13. Zhong, S., Zhang, K., Bagheri, M. et al.: Machine learning: new ideas and tools in environmental science and engineering. Environ. Sci. Technol. acs.est.1c01339 (2021). https://doi.org/10.1021/acs.est.1c01339
14. Sadek, A.H., Fahmy, O.M., Nasr, M., Mostafa, M.K.: Predicting Cu(II) adsorption from aqueous solutions onto nano zero-valent aluminum (nZVAl) by machine learning and artificial intelligence techniques. Sustainability **15**, 2081 (2023). https://doi.org/10.3390/su15032081
15. Anjum, M., Khan, K., Ahmad, W., et al.: New SHapley Additive ExPlanations (SHAP) approach to evaluate the raw materials interactions of steel-fiber-reinforced concrete. Materials **15**, 6261 (2022). https://doi.org/10.3390/ma15186261
16. Jakhar, D., Kaur, I.: Artificial intelligence, machine learning and deep learning: definitions and differences. Clin. Exp. Dermatol. **45**, 131–132 (2020). https://doi.org/10.1111/ced.14029
17. Wang, P.: On defining artificial intelligence. J. Artif. Gen. Intell. **10**, 1–37 (2019). https://doi.org/10.2478/jagi-2019-0002
18. Kaplan, A., Haenlein, M.: Siri, Siri, in my hand: who's the fairest in the land? On the interpretations, illustrations, and implications of artificial intelligence. Bus. Horiz. **62**, 15–25 (2019). https://doi.org/10.1016/j.bushor.2018.08.004
19. Visvikis, D., Cheze, L.R.C., Jaouen, V., et al.: Artificial intelligence, machine (deep) learning and radio(geno)mics: definitions and nuclear medicine imaging applications. Eur. J. Nucl. Med. Mol. Imaging **46**, 2630–2637 (2019). https://doi.org/10.1007/s00259-019-04373-w
20. Muthukrishnan, N., Maleki, F., Ovens, K., et al.: Brief history of artificial intelligence. Neuroimaging Clin. N. Am. **30**, 393–399 (2020). https://doi.org/10.1016/j.nic.2020.07.004
21. Paepae, T., Bokoro, P., Kyamakya: A virtual sensing concept for nitrogen and phosphorus monitoring using machine learning techniques. Sensors **22**, 7338 (2022). https://doi.org/10.3390/s22197338
22. Thai, H.-T.: Machine learning for structural engineering: a state-of-the-art review. Structures **38**, 448–491 (2022). https://doi.org/10.1016/j.istruc.2022.02.003
23. Barbierato, E., Gatti, A.: The challenges of machine learning: a critical review. Electronics **13**, 416 (2024). https://doi.org/10.3390/electronics13020416
24. Mobarak, M.H., Mimona, M.A., Islam, Md.A., et al.: Scope of machine learning in materials research—a review. Appl. Surf. Sci. Adv. **18**, 100523 (2023). https://doi.org/10.1016/j.apsadv.2023.100523
25. Stergiou, K., Ntakolia, C., Varytis, P., et al.: Enhancing property prediction and process optimization in building materials through machine learning: a review. Comput. Mater. Sci. **220**, 112031 (2023). https://doi.org/10.1016/j.commatsci.2023.112031
26. Bertolini, M., Mezzogori, D., Neroni, M., et al.: Machine learning for industrial applications: a comprehensive literature review. Expert Syst. Appl. **175**, 114820 (2021). https://doi.org/10.1016/j.eswa.2021.114820
27. El Naqa, I., Murphy, M.J.: What is machine learning? In: El Naqa, I., Li, R., Murphy, M.J. (eds.) Machine Learning in Radiation Oncology, pp. 3–11. Springer International Publishing, Cham (2015)
28. Fuhr, A.C.F.P., Netto, M.S., Vieira, Y., et al.: Enhancing AI model robustness in organic pollutant adsorption forecasting: insights from feature analysis. Separ. Purif. Technol. **359**, 130497 (2025). https://doi.org/10.1016/j.seppur.2024.130497
29. Xia, W., Jiang, Y., Chen, X., Zhao, R.: Application of machine learning algorithms in municipal solid waste management: a mini review. Waste Manag. Res. **40**, 609–624 (2022). https://doi.org/10.1177/0734242X211033716
30. Xu, X., Aggarwal, D., Shankar, K.: Instantaneous property prediction and inverse design of plasmonic nanostructures using machine learning: current applications and future directions. Nanomaterials **12**, 633 (2022). https://doi.org/10.3390/nano12040633
31. Maharana, K., Mondal, S., Nemade, B.: A review: data pre-processing and data augmentation techniques. Global Transit. Proc. **3**, 91–99 (2022). https://doi.org/10.1016/j.gltp.2022.04.020
32. Wu, S.: Multi-collinearity in regression. In: Medium. https://towardsdatascience.com/multi-collinearity-in-regression-fe7a2c1467ea(2021). Accessed 5 Sep 2023

33. Verdonck, T., Baesens, B., Óskarsdóttir, M., et al.: Special issue on feature engineering editorial. Mach. Learn. **113**, 3917–3928 (2024). https://doi.org/10.1007/s10994-021-06042-2
34. Mumuni, A., Mumuni, F.: Automated data processing and feature engineering for deep learning and big data applications: a survey. J. Inf. Intell. **3**, 113–153 (2025). https://doi.org/10.1016/j.jiixd.2024.01.002
35. Zhu, M., Wang, J., Yang, X., et al.: A review of the application of machine learning in water quality evaluation. Eco-Environ. Health **1**, 107–116 (2022). https://doi.org/10.1016/j.eehl.2022.06.001
36. Raschka, S.: Model evaluation, model selection, and algorithm selection in ma-chine learning (2018). https://doi.org/10.48550/ARXIV.1811.12808
37. Confalonieri, R., Coba, L., Wagner, B., et al.: A historical perspective of explainable artificial intelligence. WIREs Data Mining Knowl. Discov. **11**, e1391 (2021). https://doi.org/10.1002/widm.1391
38. Schapire, R.E.: Explaining AdaBoost. In: Schölkopf, B., Luo, Z., Vovk, V. (eds.) Em-piricalk Inference. Springer, Berlin (2013)
39. Wang, R.: AdaBoost for feature selection, classification and its relation with SVM, a review. Phys. Procedia **25**, 800–807 (2012). https://doi.org/10.1016/j.phpro.2012.03.160
40. Xu, Z., Sun, S.: An algorithm on multi-view Adaboost. In: Wong, K.W., Mendis, B.S.U., Bouzerdoum, A. (eds.) Neural Information Processing. Theory and Algorithms, pp 355–362. Springer Berlin Heidelberg, Berlin, Heidelberg (2010)
41. Chang, V., Li, T., Zeng, Z.: Towards an improved Adaboost algorithmic method for computational financial analysis. J. Parallel Distrib. Comput. **134**, 219–232 (2019). https://doi.org/10.1016/j.jpdc.2019.07.014
42. Qu, Z., Wang, W., He, Y.: Prediction of biochar adsorption of uranium in wastewater and inversion of key influencing parameters based on ensemble learning. Toxics **12**, 698 (2024). https://doi.org/10.3390/toxics12100698
43. Loh, W.: Classification and regression trees. WIREs Data Mining Knowl. Discov. **1**, 14–23 (2011). https://doi.org/10.1002/widm.8
44. Rodriguez-Galiano, V., Sanchez-Castillo, M., Chica-Olmo, M., et al.: Machine learning predictive models for mineral prospectivity: an evaluation of neural networks, random forest, regression trees and support vector machines. Ore Geol. Rev. **71**, 804–818 (2015). https://doi.org/10.1016/j.oregeorev.2015.01.001
45. Somvanshi, M., Chavan, P., Tambade, S., et al.: A review of machine learning techniques using decision tree and support vector machine. In: 2016 International Conference on Computing Communication Control and automation (ICCUBEA), pp. 1–7. IEEE, Pune (2016)
46. Navlani, A.: Decision tree classification in python tutorial. In: Datacamp. https://www.datacamp.com/tutorial/decision-tree-classification-python (2018)
47. Bezerra, T.D.R., Altino, S.A.: Investigation of light gas adsorption by microporous materials using data analysis and decision tree machine learning algorithm. OLEL 22:e3668 (2024). https://doi.org/10.55905/oelv22n3-066
48. Breiman, L.: Random forests. Mach. Learn. **45**, 5–32 (2001). https://doi.org/10.1023/A:1010933404324
49. Cutler, A., Cutler, D.R., Stevens, J.R.: Random forests. In: Zhang, C., Ma, Y. (eds.) Ensemble Machine Learning, pp. 157–175. Springer, US, Boston, MA (2012)
50. Bahrami, M., Amiri, M.J., Rajabi, S., et al.: The removal of methylene blue from aqueous solutions by polyethylene microplastics: modeling batch adsorption using random forest regression. Alexandria Eng. J. **95**, 101–113 (2024). https://doi.org/10.1016/j.aej.2024.03.100
51. Ayyadevara, V.K.: Gradient boosting machine. In: Pro Machine Learning Algorithms, pp. 117–134. Apress, Berkeley, CA (2018)
52. Friedman, J.H.: Stochastic gradient boosting. Comput. Stat. Data Anal. **38**, 367–378 (2002). https://doi.org/10.1016/S0167-9473(01)00065-2
53. Lu, H., Karimireddy, S.P., Ponomareva, N. et al.: Accelerating gradient boosting machines. In: Proceedings of Machine Learning Research, pp. 516–526. Palermo (2020)

54. Zhao, F., Tang, L., Jiang, H., et al.: Prediction of heavy metals adsorption by hydrochars and identification of critical factors using machine learning algorithms. Bioresour. Technol. **383**, 129223 (2023). https://doi.org/10.1016/j.biortech.2023.129223
55. Abu Alfeilat, H.A., Hassanat, A.B.A., Lasassmeh, O., et al.: Effects of distance measure choice on k-nearest neighbor classifier performance: a review. Big Data **7**, 221–248 (2019). https://doi.org/10.1089/big.2018.0175
56. Taunk, K., De, S., Verma, S., et al.: A brief review of nearest neighbor algorithm for learning and classification. In: 2019 International Conference on Intelligent Computing and Control Systems (ICCS), pp. 1255–1260. IEEE, Madurai, India (2019)
57. IBM K-nearest neighbors algorithm. In: What is the k-nearest neighbors algorithm? https://www.ibm.com/topics/knn#:~:text=The%20k%2Dnearest%20neighbors%20algorithm%2C%20also%20known%20as%20KNN%20or,of%20an%20individual%20data%20point. Accessed 16 Dec 2022
58. Zhang, S., Cheng, D., Deng, Z., et al.: A novel k NN algorithm with data-driven k parameter computation. Pattern Recognit. Lett. **109**, 44–54 (2018). https://doi.org/10.1016/j.patrec.2017.09.036
59. Zamouche, M., Chermat, M., Kermiche, Z., et al.: Predictive model based on k-nearest neighbor coupled with the gray wolf optimizer algorithm (KNN_GWO) for estimating the amount of phenol adsorption on powdered activated carbon. Water **15**, 493 (2023). https://doi.org/10.3390/w15030493
60. Guediri, A., Bouguettoucha, A., Tahraoui, H., et al.: Thermodynamic study and the development of a support vector machine model for predicting adsorption behavior of orange peel-derived beads in wastewater treatment. J. Mol. Liq. **403**, 124860 (2024). https://doi.org/10.1016/j.molliq.2024.124860
61. González-Figueredo, C., Alejandro Flores-Estrella, R., Rojas-Rejón O, A.: Fermen-tation: Metabolism, Kinetic Models, and Bioprocessing. In: Shiomi, N. (ed.) Current Topics in Biochemical Engineering. IntechOpen (2019)
62. Sewsynker-Sukai, Y., Faloye, F., Kana, E.B.G.: Artificial neural networks: an efficient tool for modelling and optimization of biofuel production (a mini review). Biotechnol. Biotechnol. Equip. **31**, 221–235 (2017). https://doi.org/10.1080/13102818.2016.1269616
63. Deivayanai, V.C., Karishma, S., Thamarai, P., et al.: Artificial neural network modeling for adsorption of congo red and methylene blue dye removal using pineapple waste-mediated magnetic nanoparticles. Appl. Nanosci. **15**, 9 (2025). https://doi.org/10.1007/s13204-025-03084-z
64. Pimenov, D.Y., Bustillo, A., Wojciechowski, S., et al.: Artificial intelligence systems for tool condition monitoring in machining: analysis and critical review. J. Intell. Manuf. **34**, 2079–2121 (2023). https://doi.org/10.1007/s10845-022-01923-2
65. Abdualgalil, B., Abraham, S.: Applications of machine learning algorithms and performance comparison: a review. In: 2020 International Conference on Emerging Trends in Information Technology and Engineering (ic-ETITE), pp. 1–6. IEEE, Vellore, India (2020)
66. Wang, W.-C., Chau, K.-W., Cheng, C.-T., et al.: A comparison of performance of several artificial intelligence methods for forecasting monthly discharge time series. J. Hydrol. **374**, 294–306 (2009). https://doi.org/10.1016/j.jhydrol.2009.06.019
67. Lundberg, S., Lee, S.-I.: A unified approach to interpreting model predictions (2017). https://doi.org/10.48550/ARXIV.1705.07874
68. Shapley, L.S.: 17. A value for n-person games. In: Kuhn, H.W., Tucker, A.W. (eds.) Contributions to the theory of games (AM-28), vol. II, pp 307–318. Princeton University Press (1953)
69. Lu, Y., Fan, X., Zhang, Y., et al.: Machine learning models using SHapley additive exPlanation for fire risk assessment mode and effects analysis of stadiums. Sensors **23**, 2151 (2023). https://doi.org/10.3390/s23042151
70. Yang, H., Huang, K., Zhang, K., et al.: Predicting heavy metal adsorption on soil with machine learning and mapping global distribution of soil adsorption capacities. Environ. Sci. Technol. **55**, 14316–14328 (2021). https://doi.org/10.1021/acs.est.1c02479

71. Shayanmehr, M., Aarabi, S., Ghaemi, A., et al.: A data driven machine learning approach for predicting and optimizing sulfur compound adsorption on metal organic frameworks. Sci. Rep. **15**, 3138 (2025). https://doi.org/10.1038/s41598-025-86689-2
72. Jiao, Z., Hu, P., Xu, H., et al.: Machine learning and deep learning in chemical health and safety: a systematic review of techniques and applications. ACS Chem. Health Saf. **27**, 316–334 (2020). https://doi.org/10.1021/acs.chas.0c00075
73. Giorgi, F.M., Ceraolo, C., Mercatelli, D.: The R language: an engine for bioinformatics and data science. Life **12**, 648 (2022). https://doi.org/10.3390/life12050648
74. What is MATLAB? https://www.mathworks.com/discovery/what-is-matlab.html. Accessed 25 Jun 2025
75. Alam, G., Ihsanullah, I., Mu, N., et al.: Applications of artificial intelligence in water treatment for optimization and automation of adsorption processes: recent advances and prospects. Chem. Eng. J. **427**, 130011 (2022). https://doi.org/10.1016/j.cej.2021.130011
76. Rajput, P., Yadav, S., Liu, C., et al.: Predicting biochar adsorption capacity for methylene blue removal using machine learning. J. Water Process Eng. **69**, 106749 (2025). https://doi.org/10.1016/j.jwpe.2024.106749
77. Liu, J., Zhao, J., Du, J., et al.: Machine learning predicts heavy metal adsorption on iron (oxyhydr)oxides: a combined insight into the adsorption efficiency and binding configuration. Sci. Total. Environ. **950**, 175370 (2024). https://doi.org/10.1016/j.scitotenv.2024.175370

The manufacturer's authorised [illegible]
Nature Customer Service Center GmbH [illegible]
[illegible]
[illegible]

[illegible]

The manufacturer's authorised representative in the EU is Springer Nature Customer Service Centre GmbH, Europaplatz 3, 69115 Heidelberg, Germany. If you have any concerns regarding our products, please contact ProductSafety@springernature.com

Printed and bound by CPI Group (UK) Ltd, Croydon, CR0 4YY
05/10/2026
02228057-0002